Universitext

To Sofia
　　　　H.S.M.
To my family
　　　　C.M.R.

Henning S. Mortveit
Christian M. Reidys

An Introduction to Sequential Dynamical Systems

Henning S. Mortveit
Department of Mathematics and
Virginia Bioinformatics Institute 0477
Virginia Polytechnic Institute and State
    University
1880 Pratt Drive
Blacksburg, Virginia 24061
henning.mortveit@gmail.com

Christian M. Reidys
Center for Combinatorics, LPMC
Nankai University
Tianjin 300071
P.R. China
reidys@nankai.edu.cn

ISBN-13: 978-0-387-30654-4   e-ISBN-13: 978-0-387-49879-9

Library of Congress Control Number: 2007928150

Mathematics Subject Classification (2000): Primary: 37B99; Secondary: 05Cxx, 05Axx

Printed on acid-free paper.

9 8 7 6 5 4 3 2 1

springer.com

Preface

The purpose of this book is to give a comprehensive introduction to sequential dynamical systems (SDS). This is a class of dynamical systems defined over graphs where the dynamics arise through functional composition of local dynamics. As such, we believe that the concept and framework of SDS are important for modeling and simulation of systems where causal dependencies are intrinsic.

The book is written for mathematicians, but should be readily accessible to readers with a background in, e.g., computer science or engineering that are interested in analysis, modeling, and simulation of network dynamics. We assume the reader to be familiar with basic mathematical concepts at an undergraduate level, and we develop the additional mathematics needed.

In contrast to classical dynamical systems, the theory and analysis of SDS are based on an interplay of techniques from algebra, combinatorics, and discrete mathematics in general. To illustrate this let us take a closer look at SDS and their structure. An SDS is a triple that consists of a finite graph Y where each vertex has a state taken from a finite set K, a vertex-indexed sequence of Y-local maps $(F_{v,Y})_v$ of the form $F_v \colon K^n \longrightarrow K^n$, and a word $w = (w_1, \ldots, w_k)$ over the vertex set of Y. The associated dynamical system is the SDS-map, and it is given by the composition of the local maps $F_{v,Y}$ in the order specified by w.

SDS generalize the concept of, for example, cellular automata (CA). Major distinctions from CA include (1) SDS are considered over arbitrary graphs, (2) for SDS the local maps can be applied multiple times while with CA the rules are applied exactly once, and (3) the local maps of an SDS are applied sequentially while for CA the rules are typically applied in parallel.

Much of the classical theory of dynamical systems over, e.g., $\mathbb{R}^n$ is based on continuity and derivatives of functions. There are notions of derivatives for the discrete case as well, but they do not play the same central role for SDS or other finite dynamical systems. On a conceptual level the theory of SDS is much more shaped by algebra and combinatorics than by the classical dynamical systems theory. This is quite natural since the main research questions for SDS involve properties of the base graph, the local maps, and the ordering on the one hand, and the structure of the discrete phase space on the other hand. As an example, we will use Sylow's theorems to prove the existence of SDS-maps with specific phase-space properties.

To give an illustration of how SDS connects to algebra and combinatorics we consider SDS over words. For this class of SDS we have the dependency graph $G(w, Y)$ induced by the graph Y and the word w. It turns out that there is a purely combinatorial equivalence relation $\sim_Y$ on words where equivalent words induce equivalent SDS. The equivalence classes of $\sim_Y$ correspond uniquely to certain equivalence classes of acyclic orientations of $G(w, Y)$ (induced by a natural group action). In other words, there exists a bijection $W_k / \sim_Y \longrightarrow \bigcup_{\varphi \in \Phi} \left[\mathsf{Acyc}(G(\varphi, Y)) / \sim_{\mathsf{Fix}(\varphi)} \right]$, where W_k is the set of words of length k and Φ is a set of representatives with respect to the permutation action of S_k on words W_k.

The book's first two chapters are optional as far as the development of the mathematical framework is concerned. However, the reader interested in applications and modeling may find them useful as they outline and detail why SDS are oftentimes a natural modeling choice and how SDS relate to existing concepts.

In the book's first chapter we focus on presenting the main conceptual ideas for SDS. Some background material on systems that motivated and shaped SDS theory is included along with a discussion of the main ideas of the SDS framework and the questions they were originally designed to help answer.

In the second chapter we put the SDS framework into context and present other classes of discrete dynamical systems. Specifically, we discuss cellular automata, finite-state machines, and random Boolean networks.

In Chapter 3 we provide the mathematical background concepts required for the theory of SDS presented in this book. In order to keep the book self-contained, we have chosen to include some proofs. Also provided is a list of references that can be used for further studies on these topics.

In the next chapter we present the theory of SDS over permutations. That is, we restrict ourselves to the case where the words w are permutations of the vertex set of Y. In this setting the dependency graph $G(w, Y)$ is isomorphic to the base graph Y, and this simplifies many aspects significantly. We study invertible SDS, fixed points, equivalence, and SDS morphisms.

Chapter 5 contains a collection of results on SDS phase-space properties as well as results for specific classes of SDS. This includes fixed-point characterization and enumeration for SDS and CA over circulant graphs based on a deBruijn graph construction, properties of threshold-SDS, and the structure of SDS induced by the Boolean nor function.

In Chapter 6 we consider w-independent SDS. These are SDS where the associated SDS-maps have periodic points that are independent of the choice of word w. We will show that this class of SDS induces a group and that this group encodes properties of the phase-space structures that can be generated by varying the update order w.

Chapter 7 analyzes SDS over words. Equivalence classes of acyclic orientations of the dependency graph now replace acyclic orientations of the base graph, and new symmetries in the update order w arise. We give several combinatorial results that provide an interpretation of equivalence of words and the corresponding induced SDS.

We conclude with Chapter 8, which is an outline of current and possible research directions and application areas for SDS ranging from packet routing protocols to gene-regulatory networks. In our opinion we have only started to uncover the mathematical gems of this area, and this final chapter may provide some starting points for further study.

A Guide for the Reader: The first two chapters are intended as background and motivation. A reader wishing to proceed directly to the mathematical treatment of SDS may omit these. Chapter 3 is included for reference to make the book self-contained. It can be omitted and referred to as needed in later chapters. The fourth chapter presents the core structure and results for SDS and is fundamental to all of the chapters that follow. Chapter 6 relies on results from Chapter 5, but Chapter 7 can be read directly after Chapter 4.

Each chapter comes with exercises, many of which include full solutions. The anticipated difficulty level for each problem is indicated in bold at the end of the problem text. We have ranked the problems from 1 (easy, routine) through 5 (hard, unsolved). Some of the problems are computational in the sense that some programming and use of computers may be helpful. These are marked by the additional letter 'C'.

We thank Nils A. Baas, Chris L. Barrett, William Y. C. Chen, Anders Å. Hansson, Qing H. Hou, Reinhard Laubenbacher, Matthew Macauley, Madhav M. Marathe, and Bodo Pareigis for discussions and valuable suggestions. Special thanks to the researchers of the Center of Combinatorics at Nankai University. We also thank the students at Virginia Tech University who took the course *4984 Mathematics of Computer Simulations* — their feedback and comments plus the lecture preparations helped shape this book. Finally, we thank Vaishali Damle, Julie Park, and Springer for all their help in preparing this book.

Blacksburg, Virginia, January 2007   *Henning S. Mortveit*
Tianjin, China, January 2007   *Christian M. Reidys*

Contents

1

What is a Sequential Dynamical System?

The purpose of this chapter is to give an idea of what sequential dynamical systems (SDS)[1] are and discuss the intuition and rationale behind their structure without going into too many technical details. The reader wishing to skip this chapter may proceed directly to Chapter 4 and refer to background terminology and concepts from Chapter 3 as needed.

The structure of SDS is influenced by features that are characteristic of computer simulation systems and general dynamical processes over graphs. To make this more clear we have included short descriptions of some of the systems that motivated the structure of SDS. Specifically, we will discuss aspects of the TRANSIMS urban traffic simulation system, transport computations over irregular grids, and optimal scheduling on parallel computing architectures. Each of these areas is a large topic in itself, so we have necessarily taken a few shortcuts and made some simplifications. We have chosen to focus on the aspects of these systems that apply to SDS.

Enjoy the ride!

1.1 Sequential Dynamical Systems: A First Look

To illustrate what we mean by an SDS, we consider the following example. First, let Y be the circle graph on the four vertices $0, 1, 2,$ and 3. We denote this graph as Circ_4—it is shown in Figure 1.1. To each vertex i of the graph we assign a state x_i from the state set $K = \{0, 1\}$, and we write $x = (x_0, x_1, x_2, x_3)$ for the *system state*. We also assign each vertex the symmetric, Boolean function $\mathrm{nor}_3 \colon K^3 \longrightarrow K$ defined by

$$\mathrm{nor}_3(x, y, z) = (1 + x)(1 + y)(1 + z) \, ,$$

---

[1] We will write SDS in singular as well as plural form. The plural abbreviation "SDSs" does not seem right from an aesthetic point of view. Note that the abbreviation SDS is valid in English, French, German, and Norwegian!

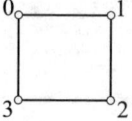

Fig. 1.1. The circle graph on four vertices, Circ_4.

where addition and multiplications are modulo 2. You may recognize nor_3 as the standard logical nor function that returns 1 if all its arguments are zero and that returns zero otherwise. We next define functions $\mathrm{Nor}_i \colon K^4 \longrightarrow K^4$ for $0 \le i \le 3$ by

$$\begin{aligned}
\mathrm{Nor}_0(x_0, x_1, x_2, x_3) &= (\mathrm{nor}_3(x_3, x_0, x_1), x_1, x_2, x_3), \\
\mathrm{Nor}_1(x_0, x_1, x_2, x_3) &= (x_0, \mathrm{nor}_3(x_0, x_1, x_2), x_2, x_3), \\
\mathrm{Nor}_2(x_0, x_1, x_2, x_3) &= (x_0, x_1, \mathrm{nor}_3(x_1, x_2, x_3), x_3), \\
\mathrm{Nor}_3(x_0, x_1, x_2, x_3) &= (x_0, x_1, x_2, \mathrm{nor}_3(x_2, x_3, x_0)) \,.
\end{aligned}$$

We see that the function Nor_i may only change the state of vertex i, and it does so based on the state of vertex i and the states of the neighbors of i in the graph Circ_4. Finally, we prescribe an ordering $\pi = (0, 1, 2, 3)$ of the vertices of Circ_4. All the quantities are shown in Figure 1.2. This is how the

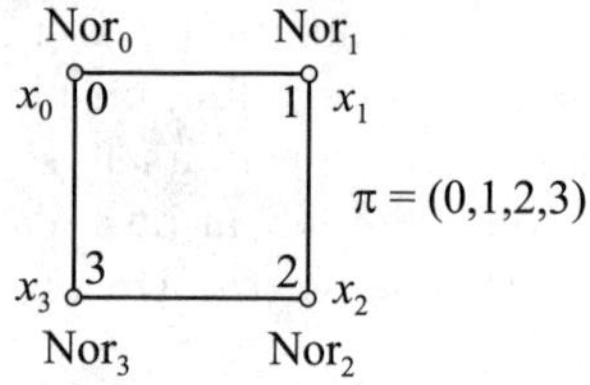

Fig. 1.2. Core constituents of an SDS: a graph (Circ_4), vertex states (x_0 through x_3), functions (Nor_0 through Nor_3), and an update order ($\pi = (0, 1, 2, 3)$).

dynamics arise: By applying the four maps Nor_i to, for example, the state $x = (x_0, x_1, x_2, x_3) = (1, 1, 0, 0)$ in the order given by π, we get (as you should verify)

$$(1,1,0,0) \overset{\mathrm{Nor}_0}{\longmapsto} (0,1,0,0) \overset{\mathrm{Nor}_1}{\longmapsto} (0,0,0,0) \overset{\mathrm{Nor}_2}{\longmapsto} (0,0,1,0) \overset{\mathrm{Nor}_3}{\longmapsto} (0,0,1,0) \,.$$

In contrast to what would be the case for a synchronous or parallel update scheme, note that the output from Nor_0 is the input to Nor_1, the output from Nor_1 is the input to Nor_2, and so on. Effectively we have applied the composed map

$$\mathrm{Nor}_3 \circ \mathrm{Nor}_2 \circ \mathrm{Nor}_1 \circ \mathrm{Nor}_0 \tag{1.1}$$

to the given state $(1, 1, 0, 0)$. This composed function is the SDS-map of the SDS over the graph Circ_4 induced by nor functions with update order $(0, 1, 2, 3)$.

We will usually write $[(\mathrm{Nor}_{i, \mathsf{Circ}_4})_i, (0, 1, 2, 3)]$ or $[\mathbf{Nor}_{\mathsf{Circ}_4}, (0, 1, 2, 3)]$ for the SDS-map. In other words, we have

$$[\mathbf{Nor}_{\mathsf{Circ}_4}, (0, 1, 2, 3)](1, 1, 0, 0) = (0, 0, 1, 0) .$$

If we apply $[\mathbf{Nor}_{\mathsf{Circ}_4}, (0, 1, 2, 3)]$ repeatedly, we get the sequence of points $(1, 1, 0, 0)$, $(0, 0, 1, 0)$, $(1, 0, 0, 0)$, $(0, 1, 0, 1)$, $(0, 0, 0, 0)$, $(1, 0, 1, 0)$, $(0, 0, 0, 1)$, $(0, 1, 0, 0)$, and $(0, 0, 1, 0)$, which then repeats. This is an example of an *orbit*. You can see this particular orbit in Figure 1.3. Readers with a background in classical dynamical systems should be on familiar grounds now and can probably foresee many of the questions we will address in later chapters.

Although it may be obvious, we want to point out that the new vertex states x_i were calculated in a sequential order. You may want to verify that $(1, 1, 0, 0)$ maps to $(0, 0, 0, 0)$ if the new vertex states are computed synchronously or "in parallel." The sequential update order is a unique feature of SDS. Sequential and synchronous update schemes generally may produce very different dynamical behavior.

The above example is, of course, a very specific and simple instance of an SDS, but exhibits all core features:

- a finite *graph Y*,
- a *state* for each vertex v,
- a *function* F_v for each vertex v,
- an *update order* of the vertices.

In general, an SDS is constructed from a graph Y of order n, say, with vertex states in a finite set or field K, a vertex-indexed family of functions $(F_v)_v$, and a word update order $w = (w_1, \ldots, w_k)$ where $w_i \in \mathrm{v}[Y]$. The SDS is the triple $(Y, (F_v)_v, w)$, and we write the resulting SDS-map as $[\mathbf{F}_Y, w] : K^n \longrightarrow K^n$. It is given by

$$[\mathbf{F}_Y, w] = F_{w_k, Y} \circ \cdots \circ F_{w_1, Y} , \tag{1.2}$$

and it is a time- and space-discrete dynamical system. Here is some terminology we will use in the following: The application of the map F_v is the *update* of the state x_v, and the application of $[\mathbf{F}_Y, w]$ to $x = (x_v)_v$ is a *system update*. The *phase space* of the map $[\mathbf{F}_Y, w]$ is the directed graph Γ defined by

$$\mathrm{v}[\Gamma] = \{x \in K^n\},$$
$$\mathrm{e}[\Gamma] = \{(x, [\mathbf{F}_Y, w](x)) \mid x \in \mathrm{v}[\Gamma]\} ,$$

where $\mathrm{v}[Y]$ and $\mathrm{e}[Y]$ denote the vertex set of Y and the edge set of Y, respectively. Since the number of states is finite, it is clear that the graph Γ is a finite union of finite, unicyclic, directed graphs. You may want to verify that

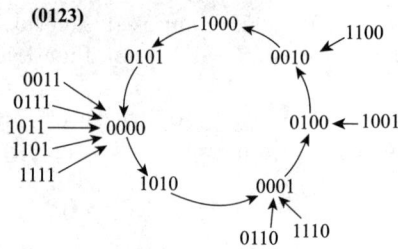

Fig. 1.3. The phase space of the SDS-map $[\mathbf{Nor}_{\mathrm{Circ}_4}, (0, 1, 2, 3)]$.

the directed graph in Figure 1.3 is indeed the phase space of the SDS-map in the above example. Further instances of SDS phase spaces are displayed in Figure 1.12. The phase space of an SDS encodes all of its dynamics. The goal in the study of SDS is to derive as much information about the structure of the phase space Γ as possible based on the properties of the graph Y, the functions $(F_v)_v$, and the update order w. Since the global dynamics is generated by composition of local dynamics, the analysis often has a local-to-global character.

1.2 Motivation

In this section we provide some motivation for studying sequential dynamical systems. The reader anxious to start exploring the theory may omit the remainder of this section.

Let us start with the *graph* Y of an SDS. The graph structure is a natural way to represent interacting entities, agents, brokers, biological cells, molecules, and so on. A vertex v represents an entity, and an edge $\{v, v'\}$ encodes the fact that the entities corresponding to v and v' can interact in some way. An example of such a graph is an electrical power network. Physical components in this network typically include power *generators*, distribution stations or *buses*, *loads* (consumers), and *lines*. The meanings of these terms are self-explanatory. In such networks generators, buses, and loads are represented as vertices. Lines connect the other three types of components and naturally represent edges. Only components connected by an edge can affect each other directly. A particular (small) power grid is given in Figure 1.4.

Another example of an SDS graph is the social contact network for the people living in some city or geographical region. In this network the individuals of the population are the vertices. There are various ways to connect people by edges. One way that is relevant for epidemiology is to connect any pair of individuals that were in contact or were at the same location for a minimal duration on some given day. Clearly, this is a natural structure to consider for the disease dynamics.

A third example arises in the context of traffic. We will study this in detail in the next section. Here we just note that one way to represent traffic

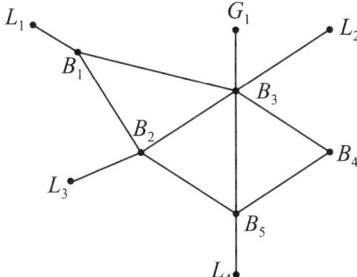

Fig. 1.4. An example of a small electrical power network. Generators are labeled G, buses are labeled B, and loads are labeled L. Edges represent physical lines.

by a graph is to consider vehicles as vertices and consider any two that are sufficiently close on the road to be adjacent. In this particular case the graph typically varies with time.

The *function* f_v of a vertex v in an SDS abstracts the behavioral characteristics of the corresponding entity. The input to this function is the state of the entity itself and the state of its neighbors in the graph. In the electrical power network the vertex state would typically include current and voltage. A vertex function f uses voltage differences to its neighbors and the respective currents to compute its new voltage or current level so that Kirchhoff's laws are satisfied locally at that vertex.

If we are studying disease propagation across a social contact network, then the function f_v could compute the total exposure to the contagious disease throughout a day and use that to determine if an uninfected individual will become infected. Since the infection process inherently has some random elements, one could think of making f_v a random variable and thus obtain a *stochastic system*.

For the traffic system the position and velocity are natural quantities to include in the vertex state. Based on the open space in the lane ahead and in the neighbor lanes, the function f_v may determine if a vehicle will increase its speed, slow down, change lanes, or move forward.

The *update order* of an SDS specifies the sequence in which the entities have their states updated. In this book and for SDS in general, we oftentimes consider update orders that are permutations or finite words over the vertex set of the graph Y. Other choices of update schemes include, for example, parallel update and infinite words. An infinite word corresponds closely to the structure of *event-driven simulations* [1, 2]. There are several reasons behind our choice for the update order of SDS. Having a fixed and finite update order gives us a dynamical system in a straightforward way: The composition of the functions F_v as specified by the permutation or word is a map $F \colon X \longrightarrow X$ and this map can be applied iteratively to states. However, if the update order is given by some infinite word $(w_i)_0^\infty$, then it is not so easy to identify such a map F, and it is not obvious what the phase space should be.

With a sequential or asynchronous update scheme we can naturally include causal order. Related events do typically not happen simultaneously—one event triggers another event, which in turn may trigger more events. With a parallel or synchronous update scheme all events happen simultaneously. This may be justified when modeling systems such as an ideal gas, but it is easy to think of systems where the update order is an essential part that cannot easily be ignored. Note also that the "sequential" in sequential dynamical system does not imply a complete lack of parallelism. We will return to this in more detail in Section 1.3.2. For now simply note that if we use the update order $\pi = (0, 2, 1, 3)$ for the SDS over Circ_4 in the introductory example, then we may perform the update of vertices 0 and 2 in parallel followed by a parallel update of vertices 1 and 3. Informally speaking, the SDS update is typically somewhere between strictly parallel and strictly sequential.

We are not advocating the use of sequential update orders: It is obvious that it is crucial to determine what gives the best description of the system one is trying to describe. Further aspects that potentially influence the particular choice of model are to encompass efficient analysis and prediction. Simply ignoring the modeling aspect and using a parallel update order because that may map more easily to current high-performance computing hardware can easily lead to models where validity becomes more than just a concern.

Note also that any system that is updated in parallel can be implemented as a sequential system. This is not a very deep observation and can be thought of as implementing one-step memory. The principle of "doubling" the graph as shown in Figure 1.5 can easily be used to achieve this. The process should be clear from the figure.

Fig. 1.5. Simulating a parallel system with a sequential system through "graph-doubling."

Returning to our traffic example, we see that the choice of scheduling makes a difference for both modeling and dynamics. Consider a highway with three parallel lanes with traffic going in the same direction. The situation where two vehicles from the outer lanes simultaneously merge to the same position in the middle lane requires special implementation care in a parallel update scheme. With simultaneous lane changes to the left and right it is easy to get collisions. Unless one has intentionally planned to incorporate collisions, this typically leads to states that are overwritten in memory and cars "disappear." For a sequential update scheme this problem is simply non-existent. There may, of course, be other situations that favor a parallel update order. However, this just shows one more time that modeling is a nontrivial process.

Readers familiar with transport computations and sweep scheduling on irregular grids [3], a topic that we return to in Section 1.3.2, will know how important scheduling can be for convergence rates. As we will see, choosing a good permutation order leads to computation convergence rates order of magnitudes better than poorly chosen update orders. As it turns out, a parallel update scheme would in fact give the slowest convergence rate for this particular class of problems.

1.3 Application Paradigms

In this section we describe two application and simulation frameworks that motivated SDS and where SDS-based models are used. The first application we will look at is TRANSIMS, which is a simulation system used for analyzing traffic in large urban areas. The second application is from transport computations. This example will show the significance of sequential update schedules, and it naturally leads to a general, SDS-based study of optimal scheduling on parallel computing architectures.

1.3.1 TRANSIMS

TRANSIMS [4–8], an acronym for *TR*ansportation *AN*alysis *SIM*ulation *S*ystem, is a large-scale computer simulation system that was developed at Los Alamos National Laboratory. This system has been used to simulate and analyze traffic at a resolution level of individual travelers in large U.S. metropolitan areas. Examples of such urban areas include Houston, Chicago, and Dallas/Ft. Worth. TRANSIMS was one of the systems that motivated the design of SDS. In this section we will give a fairly detailed description of this simulation system with an emphasis on the car dynamics and the driving rules. Hopefully, this may also serve to demonstrate some of the strengths of discrete modeling.

TRANSIMS Overview

To perform a TRANSIMS analysis of an urban area, one needs (1) a population, (2) a location-based activity plan for each person for the duration of the simulation, and (3) a network description of all transportation pathways of the area that is being analyzed. We will not go into details about how these data are gathered and prepared. It suffices to say that the data in (1) and (2) are generated based on extensive surveys and other information sources so as to be statistically indistinguishable from the available data. The network representation is essentially a complete description of the real transportation network of the given urban area, and it includes roadways, walkways, public transportation systems, and so on.

The TRANSIMS simulation system is composed of two main modules: the TRANSIMS *router* and the cellular automaton-based *micro-simulator*. The router translates each activity plan for each individual into a detailed travel route that can include several modes of travel and transportation. The travel routes are then passed to the micro-simulator, which is responsible for executing the travel routes and takes each individual through the transportation network so that its activity plan is carried out. This is typically done on a 1-second time scale and in such a way that all constraints imposed on individuals from traffic driving rules, road signaling, fellow travelers, and public transportation schedules are respected.

For the first iteration this typically leads to travel times that are too high compared to real travel times as measured by survey data. This is because too many routes involve common road segments such as highways, which leads to congested traffic. In the second pass of the simulation a certain fraction of the individuals that had too high travel times are rerouted. Their new routes are handed to the micro-simulator, which is then run again. This iterative feedback loop is repeated until one has realistic and acceptable travel times. Note that the fraction of individuals that is rerouted decreases with each iteration pass.

The TRANSIMS Micro-Simulator

The micro-simulator is constructed as a large, finite dynamical system. In this section we will show some of the details behind this module. Admittedly, this is a complex model, and we will make some simplifications. For instance, TRANSIMS can handle many modes of transportation such as car travel, public transportation, and walking. We will only consider car dynamics. The vehicles in TRANSIMS can also have different lengths, but for simplicity we will only consider "standard" vehicles.

We first need to explain the *road network representation*. The initial description of the network is in terms of *links* and *nodes*. Intersections are typical examples of nodes, but there are also nodes where there are changes in road structure such as at a lane merging point. A link is a road segment between two nodes. A link has a certain length, a certain number of lanes in each traffic direction, and possibly one or more lanes for merging and turning. For each node there are a description of lane-link connectivity across nodes and also a description of traffic signals if there are any. All other components found in realistic road networks such as reversible lanes and synchronized signals are, of course, handled too, but again, going into these details is beyond the point of this overview.

This network description is turned into a cell-network description as follows. Each lane of every link is discretized into cells. A cell corresponds to a 7.5-meter lane segment, and a cell can have up to four neighbor cells (front,

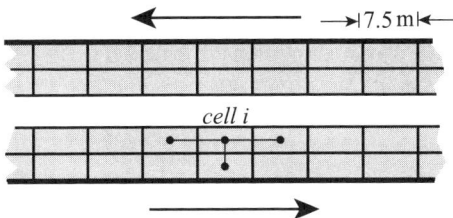

Fig. 1.6. A TRANSIMS cell-network description. The figure shows a link with two lanes in both directions. Cell i and its immediate neighbor cells are depicted in the lower link.

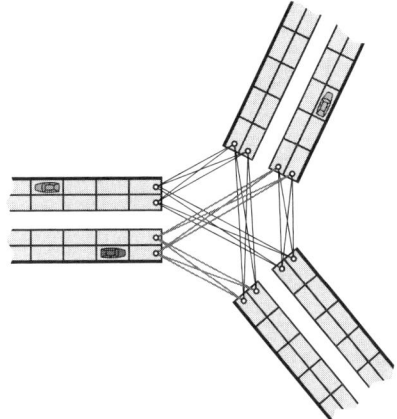

Fig. 1.7. Link and lane connectivity across a TRANSIMS node.

left, back, and right) as shown in Figure 1.6. A cell can hold at most one vehicle. Link connectivity is specified across nodes as in Figure 1.7.

The *vehicle dynamics* is specified as follows. First, vehicles travel with discrete velocities that are either 0, 1, 2, 3, 4, or 5 measured in cells per update time step. Each update time step brings the simulation 1 second forward in time, and thus the maximal speed of $v_{\max} = 5$ corresponds to an actual speed of $5 \times 7.5\,\mathrm{m/s} = 37.5\,\mathrm{m/s} = 135\,\mathrm{kmh}$, or approximately 83.9 mph.

The micro-simulator executes three functions for each vehicle in every update: (1) lane changing, (2) acceleration, and (3) movement. In the description here we have ignored intersections and we only consider straight road segments such as highways. This can be implemented through four *cellular automata* (see Chapter 2):

- Φ_1 — lane change decision,
- Φ_2 — lane change execution,
- Φ_3 — acceleration/deceleration,
- Φ_4 — movement.

These four cellular automata maps are applied in the order they are listed. The maps Φ_1 and Φ_2 that take care of lane changing are, of course, only applied when there is more than one lane in a given direction. For this reason we start with the acceleration/deceleration pass, which is always performed.

Velocity Update/Acceleration

A vehicle has limited positive acceleration and can increase its speed by at most 1 cell per second per second. However, if the road ahead is blocked, the vehicle can come to a complete stop in 1 second. The map that is applied to each cell i that has a car can be specified as the following two-step sequence:

1. $v := \min(v+1, v_{\max}, \Delta(i))$ (acceleration),
2. if [UniformRandom() $< p_{\text{break}}$] and [$v > 0$], then $v := v - 1$ (stochastic deceleration).

Here $\Delta(i)$ is free space in front of cell i measured in cells, and p_{break} is a parameter. The reason for including stochastic deceleration is that this gives driving behavior that matches real traffic patterns significantly better than what is the case if this element is ignored. All the cells in the network are updated synchronously in this pass.

Position Update/Movement

The update pass that handles vehicle movement takes place after the acceleration pass. It is executed as follows:

- If cell i has a car with velocity $v > 0$, then the state of cell i is set to zero.
- If cell i is empty and if there is a car $\delta(i)$ cells behind cell i with velocity $\delta(i) + 1$, then this car and its state are assigned to the state of cell i.
- In all other cases the cell states are updated using the identity update.

Here $\delta(i)$ denotes the free space measured in cells behind cell i. The nature of the velocity update pass guarantees that there will be no collisions. Again, all the cells are updated synchronously.

Lane Changing

With multilane traffic, vehicles can change lanes. This is more complex and requires that we specify rules for passing. Here we will make it simple and assume that vehicles can pass other vehicles on both the left and the right sides. The lane changes are done in parallel, and this requires some care. We want to avoid having two vehicles change lanes with a common target cell. The way this is handled in TRANSIMS is to only allow lane changes to the left (right) on odd (even) time steps.

In order to describe the lane change in terms of SDS or cellular automata, we need two stages: the *lane-changing decision* and the *lane-changing execution*. This is because an SDS-map or a cellular automaton rule is only allowed to change the state of the cell that it is applied to. Of course, in an implementation these two stages can easily be combined with no change in semantics.

Lane Change Decision. The case where the simulation time t is an odd integer is handled as follows: If cell i has a car and a left lane change to cell j is *desirable* ($\Delta(i) < v_{max}$ and $\Delta(j) > \Delta(i)$, and thus the car can go faster in the target lane) and *permissible* ($\delta(j) > v_{max}$ so that there is sufficient space for a safe lane change), set this cell's lane change state to 1 and to 0 otherwise. In all other circumstances the cell's lane change state is set to 0. The situation for even-numbered time steps is handled analogously with left and right interchanged.

Lane Change: The case where the simulation time t is an odd integer is handled as follows: If there is a car in cell i and this cell's lane change state is 1, then set the state of cell i to zero. Otherwise, if there is no car in cell i, and if the right neighbor cell j of cell i has its lane change state set to 1, then set the state of cell i to the state of cell j. In all other circumstances the cell state is updated using the identity map. The case of even time steps t is handled in the obvious manner with left and right interchanged.

Some of the update rules are illustrated in Figure 1.8 for the cell occupied by the darker vehicle. Here $\delta = \Delta = 1$ and we have $\Delta(l) = 2$ and $\Delta(r) = 4$, while $\delta(l)$ and $\delta(r)$ are at least 5. Here l and r refer to the left and right cell of the given cell containing the darker vehicle, respectively.

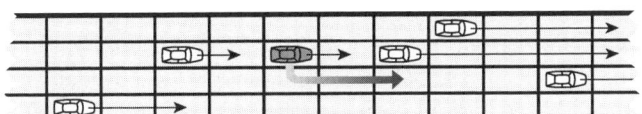

Fig. 1.8. Lane changing in the TRANSIMS cell network.

The overall computation performed by the micro-simulator update pass is the composition of the four cellular automata maps given above and is given by

$$\Phi_4 \circ \Phi_3 \circ \Phi_2 \circ \Phi_1 . \tag{1.3}$$

Notes

The basic structure of sequential dynamical systems is clearly present in the TRANSIMS micro-simulator. There is a graph where vertices correspond to cells. Two vertices v and v' are connected if their lane numbers differ by at most one and if their position along the road differs by at most v_{max} cells.

Each cell has a state that includes a vehicle ID, velocity, and a lane-changing state. There is a collection of four different functions for each vertex that are used for the four different update passes.

Although the four update passes are executed sequentially, we note that there is no sequential update order within each update pass — they are all done synchronously. So how does this relate to *sequential* dynamical systems? To explain this, consider the road configuration shown in Figure 1.9. In Figure 1.9 a line of vehicles is waiting for the light to turn from red to green at a traffic light. Once the light turns green, we expect the first row of vehicles to start, followed by a short delay, then the next row of vehicles starts, and so on. If we use a front-to-back (as seen from the traffic light) sequential update order, we see that all the vehicles start moving in the first update pass. This perfect predictive behavior is not realistic. If we use a back-to-front sequential update order, we see that this more resembles what is observed in realistic traffic. Here is the key observation: For this configuration the parallel update scheme gives dynamics that coincides precisely with the back-to-front sequential update order dynamics. Thus, even though the implementation of the model employs a synchronous update scheme, it has the semantics of a sequential model. This also serves to point out that modeling and implementation are separate issues.

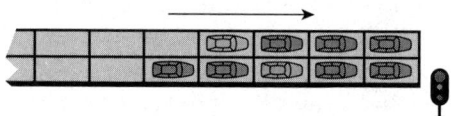

Fig. 1.9. A line of vehicles waiting for a green light at a traffic light.

Finally, we remark that this is a *cell-based description* or model of the traffic system. It is also possible to formulate this as a *vehicle-based model*. However, the cell-based formulation has the large advantage that the neighborhood structure of each cell is fixed. This is clearly not the case in a vehicle-based description where vertices would encode vehicles. In this case the graph Y would be dynamic.

Discrete Modeling

As we have just seen, TRANSIMS is built around a discrete mathematical model. In applied mathematics, and in science in general, continuous models are much more common. What follows is a short overview of why TRANSIMS uses discrete models and what some application features are that favor such models.

It is an understatement to say that the PDE- (partial differential equations) based approach to mathematical modeling has proved itself as an efficient method for both qualitative and quantitative analysis. Using, for example, conservation laws, one can quickly pass from a system description to a mathematical description based on PDEs or integral equations. For the resulting systems there are efficient and well-established mathematical results and techniques that allow one to analyze the systems both analytically and numerically. This works very well for describing a wide range of phenomena such as diffusion processes, fluid flows, or anywhere where the scales or dimensions warrant the use of such a macroscopic approach.

Conservation laws and PDEs have been used to study models of traffic configurations [9]. These models can capture and predict, for example, the movement of traffic jams as shocks in hyperbolic PDEs. However, for describing realistic road systems such as those encountered in urban traffic at the level of detail found in TRANSIMS, the PDE approach is not that useful or applicable. In principle, even if one could derive the set of all coupled PDEs describing the traffic dynamics of a reasonably sized urban area, there is, for example, no immediate way to track the movement of specific individuals.

The interaction between vehicles is more naturally specified in terms of entity functions as they occur in SDS and cellular automata. As pointed out in [10], we note that SDS- or cellular automata-based models can be implemented more or less directly in a computational model or computer program. This is in contrast to the PDE approach, which typically starts by deriving a PDE or integral formulation of the phenomenon based on various hypotheses. This is followed by a space and time discretization (i.e., model approximation) and implementation using various numerical algorithms and error bounds to compute the final "answer." This final implementation actually has much in common with an SDS or cellular automaton model: There is a graph (the discretization grid), there are states at vertices, and there is a local function at each vertex.

Some other advantages of discrete models are that they readily map to software and hardware, they typically scale very well, and they can be implemented on specialized and highly efficient hardware such as in [11].

This discussion on modeling is not meant to imply that discrete models are "better" than continuous models. The purpose is simply to point out that there are many phenomena or systems that can be described more naturally and more efficiently through discrete models than through continuous models. In the next section we describe a class of systems that naturally incorporate the notion of update order.

1.3.2 Task Scheduling and Transport Computations

A large class of computational problems has the following structure. The overall task has a collection $\mathcal{T}$ of N subtasks τ_i that are to be executed. The subtasks are ordered as vertices in a directed acyclic graph G, and a task τ_i cannot be executed unless all tasks that precede it in G have been executed. The subtasks are executed on a parallel computing architecture with M processors where each processor can execute zero or one subtask per processor cycle. Each subtask is assigned to a processor,[2] and the goal is to minimize the overall number of processor cycles required to complete the whole task by ordering the subtasks "appropriately" on their respective processors.

To illustrate the problem, consider the directed acyclic graph in Figure 1.10. The overall task has four subtasks, and there are two processors. We have assigned τ_1 and τ_2 to processor 1 and τ_3 and τ_4 to processor 2. With

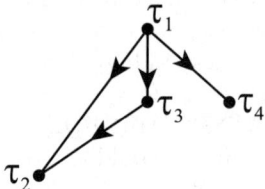

Fig. 1.10. Four tasks to be executed on two processors constrained by the directed acyclic graph shown.

our assignment it is easy to see that tasks τ_3 and τ_4 can be ordered any way we like on processor 2 since these tasks are independent. But it is also clear that executing τ_3 prior to τ_4 allows us to cut the total number of processor cycles needed by one since processor 1 can be put to better use in this case.

	τ_1,τ_2	τ_4,τ_3
Pass 1	1	—
Pass 2	—	4
Pass 3	—	3
Pass 4	2	—

	τ_1,τ_2	τ_3,τ_4
Pass 1	1	—
Pass 2	—	3
Pass 3	2	4
Pass 4	—	—

Admittedly, this is a trivial example. However, as the number of tasks grows and the directed acyclic graph becomes more complex, it is no longer obvious how to order the tasks. In the next section we will see how this problem comes up in transport computations on irregular grids.

Transport Computations

Here we show an example of how the scheduling problem arises in *transport computations*. We will also show how the entire algorithm used in the transport computation can be cast as an SDS. Our description is based on [3]. Without going into too many details we can describe the transport problem to be solved as follows. We are given some three-dimensional volume or region of space that consists of a given material. Some form of transport (e.g., photons

---

[2] Here we assume that the processor assignment is given. We have also ignored interprocess communication costs.

or radioactive radiation) is passing through the volume and is being partially absorbed. The goal could be to find the steady-state levels throughout the volume.

In one numerical algorithm that is used to solve this problem the region is first partitioned into a set of tetrahedra $\{T_1, \ldots, T_r\}$. Since the geometry of the volume can be arbitrary, there is generally no regularity in the tetrahedral partition or *mesh*. The numerical method used in [3] to solve the problem uses a set of three-dimensional vectors $D = \{D_0, \ldots, D_k\}$ where each D_i is a unit vector in $\mathbb{R}^3$. These vectors are the *sweep directions*. Each sweep direction D_i induces a directed acyclic graph G_i over the tetrahedra as shown in Figure 1.11.[3] Two tetrahedra T_a and T_b that have a common face will be connected by a directed edge in G. If T_a occur "before" T_b as seen from the direction D_i, then the edge is (T_a, T_b). Otherwise the edge is (T_b, T_a). Each

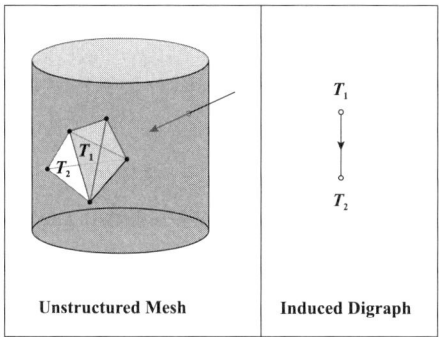

Fig. 1.11. Induced directed acyclic graphs in transport computations.

iteration of the numerical algorithm makes a pass over all the tetrahedra for all directions at each execution step. The function f that is evaluated for a tetrahedron and a direction is basically computing fluxes over the boundaries and absorption amounts. The algorithm stops when consecutive iterations give system states that are close enough as measured by some suitable metric.

For each direction D_i the tetrahedra are updated in an order consistent with the directed acyclic graph G_i induced by the given sweep direction D_i. This is intuitively what one would do in order to have, e.g., radiation pass through the volume efficiently in the numerical algorithm. If we were to update the tetrahedron states in parallel, we would expect slower convergence rates. (Why?) If we now distribute the tetrahedra on a set of M processors, we see that we are back at the situation we described initially on scheduling.

---

[3] This is almost true. Some degenerate situations can, in fact, give rise to cycles. These cycles will have to be broken so that we can get an acyclic directed graph.

It should be clear that one pass of the numerical algorithm for a given direction D_i corresponds precisely to the application of an SDS-map $[F_Y, \pi]$ where Y is the graph obtained from G_i by making G_i undirected, and π is a linear order or permutation compatible with the directed acyclic graph G_i induced by D_i. In general, there are several permutations π compatible with D_i. As we saw in the previous section, different linear orders may lead to different execution times. We thus have an optimization problem for the computation time of the algorithm where the optimization is over all linear orders compatible with G_i. In Chapters 3 and 4 we will introduce the notion of update graph. The component structure of this graph, which is also central to the theory and study of SDS, is precisely what we need to understand for this optimization problem. We note that the optimization problem can be approached in the framework of evolutionary optimization; see Section 8.3.

1.1. How does the numerical Gauss–Seidel algorithm relate to SDS and the transport computation we just described? If you are unfamiliar with this numerical algorithm you may want to look it up in [12] or in a numerical analysis text such as [13].   [2-]

1.4 SDS: Characteristics and Research Questions

Having constructed SDS from a graph, a sequence of vertex functions, and a word, it is natural to ask how these three quantities are reflected in the SDS-map and its phase space. Of course, it is also natural to ask what motivated the SDS axiomatization itself, but we leave that question for the next section.

1.4.1 Update Order Dependencies

A unique aspect of SDS is the notion of update order, and one of the first questions we addressed in the study of SDS was when is $[\mathbf{F}_Y, w] = [\mathbf{F}_Y, w']$? In other words, if we keep the graph and the functions fixed, when do two different update orders yield the same composed map? In general, the answer to this question depends on the graph, the functions, and the update order. As an example of how the update order may affect the SDS-map, consider the phase spaces of the four SDS-maps $[\mathbf{Nor}_{\mathrm{Circ}_4}, (0, 1, 2, 3)]$, $[\mathbf{Nor}_{\mathrm{Circ}_4}, (3, 2, 1, 0)]$, $[\mathbf{Nor}_{\mathrm{Circ}_4}, (0, 1, 3, 2)]$, and $[\mathbf{Nor}_{\mathrm{Circ}_4}, (0, 2, 1, 3)]$, which are displayed in Figure 1.12. It is clear from Figure 1.12 that the phase space of $[\mathbf{Nor}_{\mathrm{Circ}_4}, (0, 1, 2, 3)]$ is different from all the other phase spaces. In fact, no two phase spaces are identical. However, it is not hard to see that the phase spaces of $[\mathbf{Nor}_{\mathrm{Circ}_4}, (0, 1, 2, 3)]$ and $[\mathbf{Nor}_{\mathrm{Circ}_4}, (3, 2, 1, 0)]$ are the same if we ignore the states or labels. In this case we say that the two SDS-maps are dynamically equivalent.

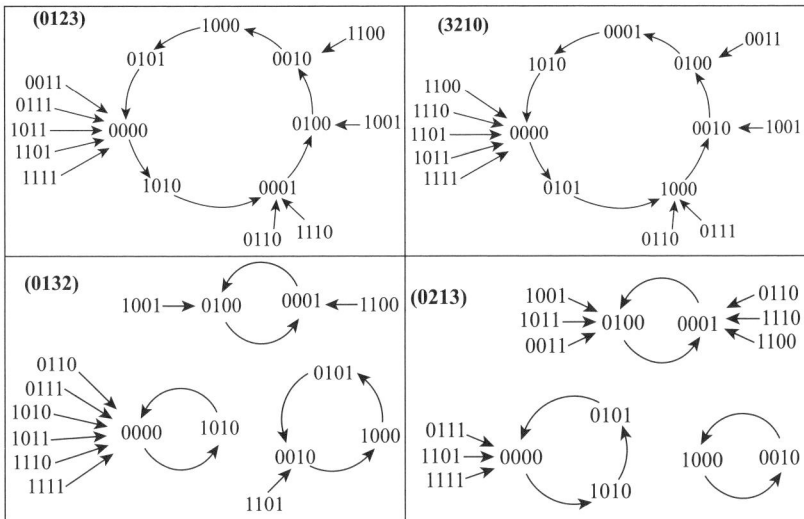

Fig. 1.12. Phase spaces for SDS-maps over the graph Circ_4 where all functions are given by nor_3. The update orders are $(0, 1, 2, 3)$ (upper left), $(3, 2, 1, 0)$ (upper right), $(0, 1, 3, 2)$ (lower left), and $(0, 2, 1, 3)$ (lower right).

In Chapter 4 we will show that if the update order is a permutation of the vertices of Circ_4, then we can create at most 14 different SDS-maps of the form $[\mathbf{Nor}_{\mathrm{Circ}_4}, \pi]$ by varying the update order π. Moreover, we will show that of these 14 different SDS-maps, there are only 3 non-isomorphic phase space structures, all of which are represented in Figure 1.12. We leave the verification of all these statements as Problem 1.2.

1.2. (*a*) Give a simple argument for the fact that $[\mathbf{Nor}_{\mathrm{Circ}_4}, (0, 1, 3, 2)]$ and $[\mathbf{Nor}_{\mathrm{Circ}_4}, (0, 3, 1, 2)]$ are identical as functions. Does your argument depend on the particular choice of nor as vertex function? (*b*) Prove that the phase spaces of the SDS-maps $[\mathbf{Nor}_{\mathrm{Circ}_4}, (0, 1, 2, 3)]$ and $[\mathbf{Nor}_{\mathrm{Circ}_4}, (3, 2, 1, 0)]$ are identical as unlabeled, directed graphs. [**1+**]

1.4.2 Phase-Space Structure

A question of a different character that often occurs is the following: What are the states $x = (x_v)_v$ such that

$$[\mathbf{F}_Y, w](x) = x \ ?$$

Such a state is called a *fixed state* or a *fixed point*. Once a system reaches a fixed point, it clearly will remain there. A fixed point is an example of an *attractor* or *invariant set* of the system. More generally, we may ask for states x such that

$$[\mathbf{F}_Y, w]^k(x) = x \ , \tag{1.4}$$

where $[\mathbf{F}_Y, w]^k(x)$ denotes the k-fold composition of the SDS-map $[\mathbf{F}_Y, w]$ applied to x. Writing $\phi = [\mathbf{F}_Y, w]$, the k-fold composition applied to x is defined recursively by $\phi^1(x) = \phi(x)$ and $\phi^k(x) = \phi(\phi^{k-1}(x))$. The points x that satisfy (1.4) are the *periodic points* of $[\mathbf{F}_Y, w]$. Fixed points and periodic points are of interest since they represent long-term behavior of the dynamical system. As a particular example, consider SDS over the graph Circ_6 where each vertex function f_v is the majority function $\mathrm{majority}_3 : \{0,1\}^3 \longrightarrow \{0,1\}$. This function is given by the *function table* below. Note that the indices are computed modulo 6.

$(x_{i-1}x_ix_{i+1})$	111	110	101	100	011	010	001	000
$\mathrm{majority}_3$	1	1	1	0	1	0	0	0

It is easy to see that the majority function is symmetric. We now ask for a characterization of all the fixed points of such a permutation SDS. As we will show later, the fixed points of this class of SDS do not depend on the update order. It turns out that the labeled graph in Figure 1.13 fully describes the fixed points. As we will see in Chapter 5, the vertex labels of this graph

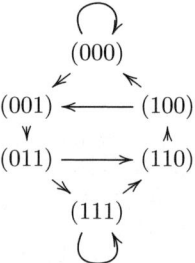

Fig. 1.13. Fixed points of majority-SDS-map over the graph Circ_6.

correspond to all possible *local fixed points*, and a closed cycle of length n corresponds to a unique global fixed point of the SDS-map $[\mathbf{Majority}_{\mathsf{Circ}_n}, \pi]$.

1.5 Computational and Algorithmic Aspects

Although the focus of this book is on the mathematical properties of SDS, we want to point out that there is also a computational theory for SDS and finite dynamical systems. To do this topic justice would require a book on its own, and we will not pretend to attempt that here. Nevertheless, we would like to give a quick view of some of the problems and questions that are studied in this area.

One of the central questions is the *reachability problem* [14]. In its basic form it can be cast as follows: We are given system states x and y and an SDS-map $\phi = [\mathbf{F}_Y, \pi]$. Starting from the system state x, can we reach the system state y? In other words, does there exist an integer $r > 0$ such that

$\phi^r(x) = y$? Of course, one way to find out is to compute the orbit of x and check if it includes y, but even in the simplest case where we have states in $\{0,1\}$ the running time of this (brute-force) algorithm is exponential in the number of graph vertices $n = |\mathrm{v}[Y]|$. The worst-case scenario for this is when all system states are on one orbit and y is mapped to x. For this situation and with binary states we would need to compute $2^n - 1$ iterates of ϕ before we would get to y. A related problem is the *fixed-point reachability problem*, in which case we are given a system state x and the question is if there exists an integer $r > 0$ such that $\phi^{r+1}(x) = \phi^r(x)$.

We would, of course, like to devise algorithms that allow us to answer these questions more efficiently than by the brute-force approach above. So are there more efficient algorithms? Yes and no. The reachability problem is computationally intractable[4] even in the special case of SDS with Boolean symmetric vertex functions. So in the general case we are left with the brute-force approach. However, more efficient algorithms can be constructed if we, for example, restrict the classes of graphs and functions that are considered. For instance, for SDS induced by nor vertex functions [see Eq. (4.9)] it is known that the reachability problem can be solved by an algorithm with polynomial running time [14]. The same holds for the fixed-point reachability problem in the case of linear vertex functions over a finite field or a semi-ring with unity. We have also indicated efficient ways to determine and count fixed points in Section 1.4.2 when we have restrictions on the classes of graphs that we consider.

Other computational problems for SDS include the *permutation-existence problem* [16]. In this situation we are given states x and y, a graph Y, and vertex functions $(f_v)_v$. Does there exist a permutation (i.e., update order) π such that $[\mathbf{F}_Y, \pi]$ maps x to y in one step? That is, does there exist an SDS update order π such that $[\mathbf{F}_Y, \pi](x) = y$? Naturally, we would also like to construct efficient algorithms to answer this if possible. The answer to this problem is similar to the answer for the reachability problem. For SDS with Boolean threshold vertex functions (see Definition 5.11), the problem is NP-complete, but for nor vertex functions it can be answered efficiently. Note that the reachability problem can be posed for many other types of dynamical systems than SDS, but the permutation existence problem is unique to SDS.

The last computational problem we mention is the *predecessor-existence problem* [16]: Given a system state x and an SDS-map $\phi = [\mathbf{F}_Y, \pi]$, does there exist a system state z such that $\phi(z) = x$? Closely related to this is the *#predecessor problem*, which asks for the number of predecessors of a system state x. This problem has also been studied in the context of *cellular automata* (see Section 2.1) in, for example, [17]. Exactly as for the previous problems the predecessor existence problem is NP-complete in the general case, but can be solved efficiently for restricted classes of vertex functions and/or graphs. Examples include SDS where the vertex functions are given by logical And

---

[4] The problem is PSPACE-complete; see, for example, [15].

functions and SDS where the graphs have *bounded tree-width* [16]. Locating the combined function/graph complexity boundary for when such a problem goes from being polynomially solvable to NP-complete is an interesting research question.

For more results along the same lines and for results that pertain to computational universality, we refer the interested reader to, for example, [14, 16, 18–20].

1.6 Summary

The notion of geographically or computationally distributed systems of interacting entities calls for models based on dynamical systems over graphs. The fact that real applications typically have events or decisions that trigger other events and decisions makes the use of an update sequence a natural choice. The update order or scheduling component is an aspect that distinguishes SDS from most other models, some of which are the topic of the next chapter.

A Note on the Problems

You will find exercises throughout the book. Many of them come with full solutions, and some include comments about how they relate to open problems or to possible research directions. Inspired by [21] we have chosen to grade the difficulty level of each problem from 1 through 5. A problem at level 1 should be fairly easy, whereas the solution to a problem marked 5 could probably form the basis for a research article.

Some of the exercises are also marked by a "C." This is meant to indicate that some programming can be helpful when solving these problems. Computers are particularly useful in this field since in most cases the state values are taken from some small set of integers and we do not have to worry about round-off problems. The use of computers allows one to explore a lot more of the dynamics, and it can be a good source for discovering general properties that can be turned into proofs. Naturally, it can also be an effective method for discovering counterexamples. In our work we have used everything from C++ to Maple, Mathematica, and Matlab. Although we do not have any particular recommendation for what tools to use, we do encourage you to try the computational problems.

Problems

1.3. *Coupled map lattices (CML)* [22, 23] are examples of "classical" discrete dynamical systems that have been used to study spatio-temporal chaos. In this setting we have n lattice sites (vertices) labeled 0 through $n-1$, and each site i has a state $x_i \in \mathbb{R}$. Moreover, we have a map $f : \mathbb{R} \longrightarrow \mathbb{R}$. In, e.g., [22] the state of each site is updated as

$$x_i(t+1) = (1 - \epsilon)f(x_i(t)) + (\epsilon/2)\big[f(x_{i+1}(t)) + f(x_{i-1}(t))\big] , \qquad (1.5)$$

where $\epsilon \geq 0$ is a *coupling parameter*, and where site labels i and $i+n$ are identified. This can easily be interpreted as a discrete dynamical system defined over a graph Y. What is this graph?   [1+]

Answers to Problems

1.1. In their basic forms both the Gauss–Seidel and the Gauss–Jacobi algorithms attempt to solve the matrix equation $Ax = b$ by iteration. For simplicity let us assume that A is a real $n \times n$ matrix, that $x = (x_1, \ldots, x_n) \in \mathbb{R}^n$, and that $(x_1^0, \ldots, x_n^0)$ is the initial value in the iteration. Whereas the Gauss–Jacobi scheme successively computes

$$x_i^k = \left(b_i - \sum_{j \neq i} a_{ij} x_j^{k-1} \right) / a_{ii} \, ,$$

the Gauss–Seidel scheme computes

$$x_i^k = \left(b_i - \sum_{j < i} a_{ij} x_j^k - \sum_{j > i} a_{ij} x_j^{k-1} \right) / a_{ii} \, .$$

In other words, as one pass of the Gauss–Seidel algorithm progresses, the new values for x_i^k are immediately used in the later stages of the pass. For the Gauss–Jacobi scheme only the old values x_i^{k-1} are used. The Gauss–Seidel algorithm may therefore be viewed as a real-valued SDS-map over the complete graph with update order $(1, 2, \ldots, n)$.

1.2. (a) The two update orders differ precisely by a transposition of the two consecutive vertices 1 and 3. Since $\{1, 3\}$ is not an edge in Circ_4, there is no way that the new value of x_1 can influence the update of the state x_3, or vice versa. It is not specific to the particular choice of vertex function. (b) The map $\gamma \colon \{0, 1\}^4 \longrightarrow \{0, 1\}^4$ given by $\gamma(s, t, u, v) = (v, u, t, s)$ is a bijection that maps the phase space of $[\mathbf{Nor}_{\mathsf{Circ}_4}, (0, 1, 2, 3)]$ onto the phase space of $[\mathbf{Nor}_{\mathsf{Circ}_4}, (3, 2, 1, 0)]$. This means that the two phase spaces look the same up to relabeling. We will return to this question in Chapter 4.

1.3. The new value of a site is computed based on its own current value and the current value of its two neighbors. Since site labels are identified modulo n, the graph Y is the circle graph on n vertices (Circ_n).

In later work as in, for example, [23] the coupling scheme is more liberal and the states are updated as

$$x_i(t+1) = (1 - \epsilon) f(x_i(t)) + \frac{\epsilon}{N} \sum_{k=1}^{N} f(x_k(t)),$$

where k is understood to run over the set of neighbors of site i. As you can see, this corresponds more closely to a real-valued discrete dynamical system where the coupling is defined by a graph on n vertices. In [24] real-valued discrete dynamical systems over arbitrary finite directed graphs are studied. We will discuss real-valued SDS in Section 8.5.

2

A Comparative Study

As we pointed out in the previous chapter, several frameworks and constructions relate to SDS, and in the following we present a short overview. This chapter is not intended to be a complete survey — the list of frameworks that we present is not exhaustive, and for the concepts that we discuss we only provide enough of an introduction to allow for a comparison to SDS. Specifically, we discuss *cellular automata*, *random Boolean networks*, and *finite-state machines*. Other frameworks related to SDS that are not discussed here include *interacting particle systems* [25] and *Petri nets* [26].

2.1 Cellular Automata

2.1.1 Background

Cellular automata, or CA[1] for short, were introduced by von Neumann and Ulam around 1950 [27]. The motivation for CA was to obtain a better formal understanding of biological systems that are composed of many identical components and where each component is relatively simple, at least as compared to the full system. The design and structure of the first computers were another motivation for the introduction of CA.

The global dynamics or pattern evolution of a cellular automaton is the result of interactions of its components or cells. Questions such as to which patterns can occur for a given CA (computational universality) and which CA that, in an appropriate sense, can be used to construct descriptions of other CA (universal construction) were central in the early phases [27, 28]. Cellular automata have been studied from a dynamical systems perspective (see, for example, [29–33]), from a logic, automata, and language theoretic perspective (e.g., [28, 34, 35]), and through ergodic theory and in probabilistic settings

---

[1] Just as for SDS we use the abbreviation CA for both the singular and plural forms. It will be clear from the context which form is meant.

(e.g., [36–39]). Applications of cellular automata can be found, for example, in the study of biological systems (see [40]), in hydrodynamics in the form of lattice gases (see, for example, [41–43]), in information theory, and in the construction of codes [44], and in many other areas. For further details and overviews of the history and theory of CA, we refer to, e.g., [18, 45–47].

2.1.2 Structure of Cellular Automata

Cellular automata have many features in common with SDS. There is an underlying cell or lattice structure where each lattice point or cell v has a state state x_v taken from some finite set. Each lattice point has a function defined over a collection of states associated to nearby lattice points. As a dynamical system, a cellular automaton evolves in discrete time steps by the synchronous application of the cell functions.

Notice that the lattice structure is generally not the same as the base graph of SDS. As we will explain below, the notion of what constitutes adjacent vertices is determined by the lattice structure *and* the functions. Note that in contrast to SDS it is not uncommon to consider cellular automata over infinite lattices.

One of the central ideas in the development of CA was uniform structure, and in particular this includes translation invariance. As a consequence of this, the lattice is typically regular such as, for example, $\mathbb{Z}^k$ for $k \geq 1$. Moreover, translation invariance also implies that the functions f_v and the state spaces S_v are the same for all lattice points v. Thus, there are a common function f and a common set S such that $f_v = f$ and $S_v = S$ for all v. Additionally, the set S usually has some designated zero element or *quiescent state* s_0. Note that in the study of CA dynamics over infinite structures like $\mathbb{Z}^k$, one considers the system states[2] $x = (x_v)_v$ where only a finite number of the cell states x_v are different from s_0. Typically, $S = \{0, 1\}$ and $s_0 = 0$.

Each vertex v in Y has a *neighborhood* $n[v]$, which is some sequence of lattice points. Again for uniformity reasons all the neighborhoods $n[v]$ exhibit the same structure. In the case of $\mathbb{Z}^k$ the neighborhood is constructed from a sequence $N = (d_1, \ldots, d_m)$ where $d_i \in \mathbb{Z}^k$, and each neighborhood is given as $n[v] = v + N = (v + d_1, \ldots, v + d_m)$. A *global CA state*, system state, or CA configuration is an element $x \in S^{\mathbb{Z}^k}$. For convenience we write $x[v] = (x_{v+d_1}, \ldots, x_{v+d_m})$ for the subconfiguration associated with the neighborhood $n[v]$.

Definition 2.1 (Cellular automata over $\mathbb{Z}^k$). Let S, N, and f be as above. The cellular automaton with states in S, neighborhood N, and function f is the map

$$\Phi_f \colon S^{\mathbb{Z}^k} \longrightarrow S^{\mathbb{Z}^k}, \quad \Phi_f((x))_v = f(x[v]). \tag{2.1}$$

---

[2] For cellular automata a system state $x = (x_v)_v$ is usually called a configuration.

In other words, the cellular automaton dynamics results from the synchronous or parallel application of the maps f to the cell states x_v.

We can also construct CA over finite lattices. One standard way to do this is by imposing *periodic boundary conditions*. In one-dimension we can achieve this by identifying vertices i and $i + n$ in $\mathbb{Z}$ for some $n > 1$. This effectively creates a CA over $\mathbb{Z}/n\mathbb{Z}$. Naturally we can extend this to higher dimensions, in which case we would consider k-dimensional tori.

Another way to construct a CA over a finite structure is through *zero boundary conditions*. In one-dimension this means we would use the line graph Line_n as lattice and add two additional vertices at the ends and fix their states to zero; see Example 2.2.

Example 2.2 (One-dimensional CA). This example shows the three different types of graph or grid structures for one-dimensional CA that we discussed in the text. If we use the neighborhood structure given by $N = (-1, 0, 1)$, we see that to compute the new state for a cell v the map f only takes as arguments the state of the cell v and the states of the nearest neighbors of v. For this reason this class of maps is often referred to as *nearest-neighbor rules*. The corresponding lattices are shown in Figure 2.1.   ◇

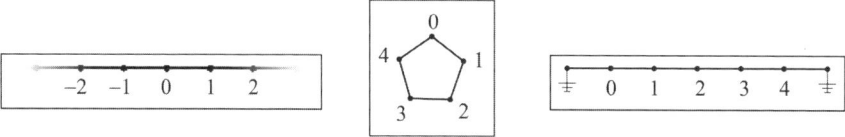

Fig. 2.1. From left to right: the lattice of a CA in the case of (a) $\mathbb{Z}$, (b) $\mathbb{Z}/5\mathbb{Z}$ with periodic boundary conditions, and (c) $\mathbb{Z}/5\mathbb{Z}$ with zero boundary conditions.

Two of the commonly used neighborhood structures N are the *von Neumann neighborhood* and the *Moore neighborhood*. These are shown in Figure 2.2. For $\mathbb{Z}^2$ the von Neumann neighborhood is

$$N = ((0,0), (-1,0), (0,-1), (1,0), (0,1)) .$$

The *radius* of a one-dimensional CA rule f with neighborhood defined by N is the norm of the largest element of N. The radius of the rule in Example 2.2 is therefore 1.

We see that the lattice and the function of a cellular automaton give us an SDS base graph Y as follows. For the vertices of Y we take all the cells. A vertex v is adjacent to all vertices v' in $n[v]$. If v itself is included in $n[v]$, we make the convention of omitting the loop $\{v, v\}$.

In analogy to SDS, one central goal of CA research is to derive as much information as possible about the global dynamics of the CA map Φ_f based on known, local properties such as the map f and the neighborhood structure.

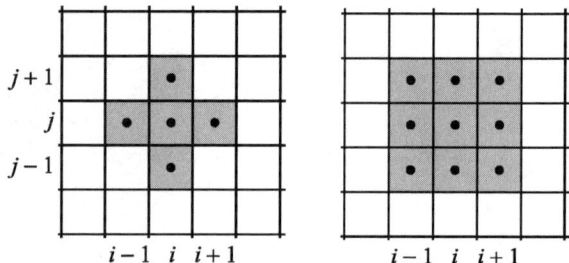

Fig. 2.2. The von Neumann neighborhood (left) and the Moore neighborhood (right) for an infinite two-dimensional CA.

The *phase space* of a CA is the directed graph with all possible configurations as vertices, and where vertices x and y are connected by a directed edge (x, y) if $\Phi_f(x) = y$. Even in the case of CA over finite lattices, it is impractical to display the whole phase space, and *space-time diagrams* (see Section 4.1) are often used to visualize certain orbits or trajectories.

Example 2.3. The CA rule f_{90} is given by $f_{90}(x_{i-1}, x_i, x_{i+1}) = x_{i-1} + x_{i+1}$ modulo 2. This *linear* function has been studied extensively in, for example, [32]. In Figure 2.3 we have shown two typical space-time diagrams for the CA with local rule f_{90} over the lattice Circ_{512}.   ⋄

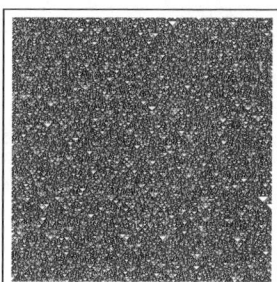

Fig. 2.3. Space-time diagrams for CA with cell function f_{90}. In the left diagram the initial configuration contains a single state that is 1. In the right diagram the initial configuration was chosen at random.

CA differ from SDS in several ways. For instance, for CA the graph Y, which is derived from the lattice and neighborhood $n[v]$, is regular and translation invariant, whereas the graph of an SDS is arbitrary, although finite. Furthermore, CA have a fixed function or rule, associated to every vertex, while SDS have a vertex-indexed family of functions. Perhaps most importantly, CA and SDS differ in their respective update schemes. As a result, CA and SDS differ significantly with respect to, for example, invertibility as we will show in the exercises.

In principle one can generalize the concept of CA and consider them over arbitrary graphs with vertex-indexed functions. One may also consider asynchronous CA. The dynamics of the latter class of CA depends critically on the particular choice of update order [48].

In the remainder of this section we will give a brief account of some basic facts and terminology on CA that will be used in the context of SDS.

2.1.3 Elementary CA Rules

A large part of the research on CA has been concerned with the finite and infinite one-dimensional cases where the lattice is $\mathbb{Z}/n\mathbb{Z}$ and $\mathbb{Z}$, respectively. An example of a phase space of a one-dimensional CA with periodic boundary conditions is shown in Figure 2.1. The typical setting uses radius-1 vertex functions with binary states. In other words, the functions are of the form $f \colon \mathbb{F}_2^3 \longrightarrow \mathbb{F}_2$ where $\mathbb{F}_2 = \{0, 1\}$ is the field with two elements. Whether the lattice is $\mathbb{Z}$ or $\mathbb{Z}/n\mathbb{Z}$, we refer to this class of functions as the *elementary CA rules* and the corresponding global CA maps as elementary CA.

Example 2.4. Let Φ_f be the CA with local rule $f \colon \mathbb{F}_2^3 \longrightarrow \mathbb{F}_2$ given by $f(x, y, z) = (1 + y)(1 + z) + (1 + xyz)$. In this case we see that the state $(1, 0, 1, 1)$ maps to $(1, 1, 1, 0)$. The phase space of Φ_f is shown in Figure 2.4.

$\diamond$

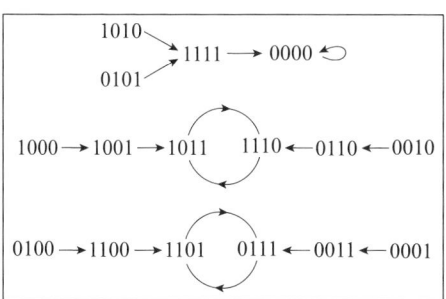

Fig. 2.4. The phase space of the elementary CA of Example 2.4.

Enumeration of Elementary CA Rules

Clearly, there are $|\mathbb{F}_2|^{|\mathbb{F}_2^3|} = 2^8 = 256$ elementary CA rules. Any such function or rule f can be specified as in Table 2.1 by the values a_0 through a_7. We identify the triple $x = (x_2, x_1, x_0) \in \mathbb{F}_2^3$ with the decimal number $k = k(x) = x_2 \cdot 2^2 + x_1 \cdot 2 + x_0$. Let the value of f at x be a_k for $0 \le k \le 7$.[3] We can then encode the map f as the decimal number $r = r(f)$ with $0 \le r \le 255$ through

---

[3] In the literature the a_i's are sometimes ordered the opposite way.

(x_{i-1}, x_i, x_{i+1})	111	110	101	100	011	010	001	000
f	a_7	a_6	a_5	a_4	a_3	a_2	a_1	a_0

Table 2.1. Specification of elementary CA rules.

$$r = r(f) = \sum_{i=0}^{7} a_i 2^i . \tag{2.2}$$

This assignment of a decimal number in $\{0, 1, 2, \ldots, 255\}$ to the rule f was popularized by S. Wolfram, and it is often referred to as the *Wolfram enumeration* of elementary CA rules [47, 49]. This enumeration procedure can be generalized to other classes of rules, and some of these are outlined in Problem 2.2.

Example 2.5. The map $\mathrm{parity}_3 \colon \mathbb{F}_2^3 \longrightarrow \mathbb{F}_2$ given by $\mathrm{parity}_3(x_1, x_2, x_3) = x_1 + x_2 + x_3$ with addition modulo 2 (i.e., in the field $\mathbb{F}_2$) can be represented by

$(x_{i-1}x_i x_{i+1})$	111	110	101	100	011	010	001	000
parity	1	0	0	1	0	1	1	0

and thus

$$r(\mathrm{parity}_3) = 2^7 + 2^4 + 2^2 + 2 = 150 . \qquad \diamond$$

A lot of work has gone into the study of this rule [32], and it is often referred to as the XOR function or the parity function. One of the reasons this rule has attracted much attention is that the induced CA is a *linear CA*. As a result all the machinery from algebra and matrices over finite fields can be put to work [33, 50].

2.1. What is the rule number of the elementary CA rule in Example 2.4?

[1]

Equivalence of Elementary CA Rules

Clearly, all the elementary CA are different as functions: For different elementary rules f_1 and f_2 we can always find a system state x such that the induced CA maps differ for x. However, as far as dynamics is concerned, many of the elementary rules induce cellular automaton maps where the phase spaces look identical modulo labels (states) on the vertices. The precise meaning of "look identical" is that their phase spaces are isomorphic, directed graphs as in Section 4.3.3. When the phase spaces are isomorphic, we refer to the corresponding CA maps as *dynamically equivalent*. Two cellular automata Φ_f and $\Phi_{f'}$ with states in $\mathbb{F}_2$ are dynamically equivalent if there exists a bijection $h \colon \mathbb{F}_2^n \longrightarrow \mathbb{F}_2^n$ such that

$$\Phi_{f'} \circ h = h \circ \Phi_f . \tag{2.3}$$

The map h is thus a one-to-one correspondence of trajectories of Φ_f and $\Phi_{f'}$. Alternatively, we may view h as a relabeling of the states in the phase space.

Example 2.6. The phase spaces of the elementary CA with local rules 124 and 193 are shown in Figure 2.5. It is easy to check that the phase spaces are isomorphic. Moreover, the phase spaces are also isomorphic to the phase space shown in Figure 2.4 for the elementary CA 110. ◇

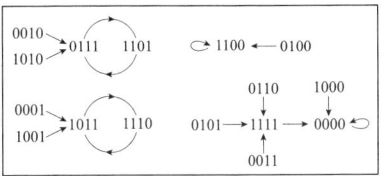 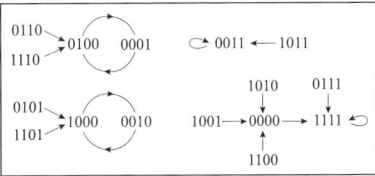

Fig. 2.5. The phase spaces of the elementary CA 124 (left) and 193 (right).

We will next show two things: (1) there at most 88 dynamically non-equivalent elementary CA, and (2) if we use a *fixed* sequential permutation update order rather than a synchronous update, then the corresponding bound for the number of dynamically non-equivalent systems is 136.

For this purpose we represent each elementary rule f by a binary 8-tuple $(a_7, \ldots, a_0)$ (see Table 2.1) and consider the set

$$R = \{(a_7, a_6, a_5, a_4, a_3, a_2, a_1, a_0) \in \mathbb{F}_2^8\} . \tag{2.4}$$

Rules that give dynamically equivalent CA are related by two types of symmetries: (1) 0/1-flip symmetries (inversion) and (2) left-right symmetries. Let $\gamma \colon R \longrightarrow R$ be the map given by

$$\gamma(r = (a_7, a_6, a_5, a_4, a_3, a_2, a_1, a_0)) = (\bar{a}_0, \bar{a}_1, \bar{a}_2, \bar{a}_3, \bar{a}_4, \bar{a}_5, \bar{a}_6, \bar{a}_7), \tag{2.5}$$

where $\bar{a}$ equals $1 + a$ computed in $\mathbb{F}_2$. With the map inv_n defined by

$$\mathsf{inv}_n \colon \mathbb{F}_2^n \longrightarrow \mathbb{F}_2^n, \quad \mathsf{inv}_n(x_1, \ldots, x_n) = (\bar{x}_1, \ldots, \bar{x}_n) \tag{2.6}$$

(note that $\mathsf{inv}_n^2 = \mathsf{id}$), a direct calculation shows that

$$\Phi_{\gamma(f)} = \mathsf{inv} \circ \Phi_f \circ \mathsf{inv}^{-1} ;$$

hence, 0/1-flip symmetry yields isomorphic phase spaces for Φ_f and $\Phi_{\gamma(f)}$.

As for left-right symmetry, we introduce the map $\delta \colon R \longrightarrow R$ given by

$$\delta(r = (a_7, a_6, a_5, a_4, a_3, a_2, a_1, a_0)) = (a_7, a_3, a_5, a_1, a_6, a_2, a_4, a_0) . \tag{2.7}$$

The Circ_n-automorphism $i \mapsto n + 1 - i$ induces in a natural way the map

$$\mathsf{rev}_n \colon \mathbb{F}_2^n \longrightarrow \mathbb{F}_2^n, \quad \mathsf{rev}_n(x_1, \ldots, x_n) = (x_n, \ldots, x_1) \tag{2.8}$$

(note $\mathsf{rev}_n^2 = \mathsf{id}$), and we have

$$\Phi_{\delta(f)} = \mathsf{rev} \circ \Phi_f \circ \mathsf{rev}^{-1} .$$

Example 2.7 (Left-right symmetry). The map defined by $f(x_1, x_2, x_3) = x_3$ induces a CA that acts as a left-shift (or counterclockwise shift if periodic boundary conditions are used). It is the rule $r = (1, 0, 1, 0, 1, 0, 1, 0)$ and it has Wolfram encoding 170. For this rule we have $\delta(r) = (1, 1, 1, 1, 0, 0, 0, 0)$, which is rule 240. We recognize this rule as the map $f(x_1, x_2, x_3) = x_1$, which is the rule that induces the "right-shift CA" as you probably expected.   ◇

In order to compute the number of non-equivalent elementary CA, we consider the group $G = \langle \gamma, \delta \rangle$. Since $\gamma \circ \delta = \delta \circ \gamma$ and $\delta^2 = \gamma^2 = 1$, we have $G = \{1, \gamma, \delta, \gamma \circ \delta\}$ and G acts on R. The number of non-equivalent rules is bounded above by the number of orbits in R under the action of G and there are 88 such orbits.

Proposition 2.8. *For $n \geq 3$ there are at most 88 non-equivalent phase spaces for elementary cellular automata.*

Proof. By the discussion above the number of orbits in R under the action of G is an upper bound for the number of non-equivalent CA phase spaces. By the Frobenius lemma [see (3.18)], this number is given by

$$N = \frac{1}{4} \sum_{\eta \in G} |\mathsf{Fix}(g)| = \frac{1}{4} (|\mathsf{Fix}(1)| + |\mathsf{Fix}(\gamma)| + |\mathsf{Fix}(\delta)| + |\mathsf{Fix}(\gamma \circ \delta)|) . \quad (2.9)$$

We leave the remaining computations to the reader as Problem 2.2.   □

2.2. Compute the terms $|\mathsf{Fix}(1)|$, $|\mathsf{Fix}(\gamma)|$, $|\mathsf{Fix}(\delta)|$, and $|\mathsf{Fix}(\gamma \circ \delta)|$ in (2.9) and verify that you get $N = 88$.   [1]

Note that we have not shown that the bound 88 is a sharp bound. That is another exercise — it may take some patience.

2.3. Is the bound 88 for the number of dynamically non-equivalent elementary CA sharp? That is, if f and g are representative rules for different orbits in R under G, then are the phase spaces of Φ_f and Φ_g non-isomorphic as directed graphs?   [3]

Example 2.9. Consider the elementary CA rule numbered 14 and represented as $r = (0, 0, 0, 0, 1, 1, 1, 0)$. In this case we have $G(r) = \{r, \gamma(r), \delta(r), \gamma \circ \delta(r)\} = \{r_{14}, r_{143}, r_{84}, r_{214}\}$ using the Wolfram encoding.   ◇

2.4. (a) What is R^G (the set of elements in R fixed by all $g \in G$) for the action of G on the elementary CA rules R in (2.4)?
(b) Do left-right symmetric elementary rules induce equivalent permutation-SDS? That is, for a fixed sequential permutation update order π, do we get equivalent global update maps? What happens if we drop the requirement of a fixed permanent updates order?
(c) What is the corresponding transformation group G' acting on elementary rules in the case of SDS with a fixed update order π? How many orbits are there in this case?
(d) Show that $R^G = R^{G'}$.   [2-C]

Other Classes of CA Rules

In addition to elementary CA rules, the following particular classes of CA rules are studied in the literature: the *symmetric rules*, the *totalistic rules*, and the radius-2 rules. Recall that a function $f: K^n \longrightarrow K$ is symmetric if for every permutation $\sigma \in S_n$ we have $f(\sigma \cdot x) = f(x)$ where $\sigma \cdot (x_1, \ldots, x_n) = (x_{\sigma^{-1}(1)}, \ldots, x_{\sigma^{-1}(n)})$. Thus, a symmetric rule f does not depend on the order of its argument. A totalistic function is a function that only depends on $(x_1, \ldots, x_n)$ through the sum $\sum x_i$ (taken in $\mathbb{N}$). Of course, over $\mathbb{F}_2$ symmetric and totalistic rules coincide. The radius-2 rules are the rules of the form $f: K^5 \longrightarrow K$ that are used to map $(x_{i-2}, x_{i-1}, x_i, x_{i+1}, x_{i+2})$ to the new state x_i' of cell i.

In some cases it may be natural or required that we handle the state of a vertex v differently than the states of its neighbor vertices when we update the state x_v. If the map f used to update the state v is symmetric in the arguments corresponding to the neighbor states of cell v, we call f_v *outer-symmetric*.

The classes of *linear CA over finite fields* and general linear maps over finite fields have been analyzed extensively in, e.g., [32,33,50,51]. Let K be a field. A map $f: K^n \longrightarrow K$ is linear if for all $\alpha, \beta \in K$ and all $x, y \in K^n$ we have $f(\alpha x + \beta y) = \alpha f(x) + \beta f(Y)$. A CA induced by a linear rule is itself a linear map. Linear maps over rings have been studied in [52].

Example 2.10. The elementary CA rule 90, which is given as $f_{90}(x_1, x_2, x_3) = x_1 + x_3$, is outer-symmetric but not totalistic or symmetric. The elementary CA rule $g(x_1, x_2, x_3) = (1 + x_1)(1 + x_2)(1 + x_3)$, which is rule 1, is totalistic and symmetric. Note that the first rule is linear, whereas the second rule is nonlinear. ◇

Example 2.11. A space-time diagram of a radius-2 rule is shown in Figure 2.6. By using the straightforward extension of Wolfram's encoding to this class of CA rules, we see that this particular rule has encoding 3283936144, or $(195, 188, 227, 144)$ in the notation of [53]. ◇

In the case of linear CA over $\mathbb{Z}/n\mathbb{Z}$, we can represent the CA map through a matrix $A \in K^{n \times n}$. This means we can apply algebra and finite field theory to analyze the corresponding phase spaces through normal forms of A. We will not go into details about this here — a nice overview can be found in [33]. We content ourselves with the following result.

Theorem 2.12 ([33]). *Let K be a finite field of order q and let $M \in K^{n \times n}$. If the dimension of $\ker(M)$ is k, then there is a rooted tree T of size q^k such that the phase space of the dynamical system given by the map $F(x) = Mx$ consists of q^{n-k} cycle states, each of which has an isomorphic copy of T attached at the root vertex.*

In other words, for a finite linear dynamical system over a field, all the transient structures are identical.

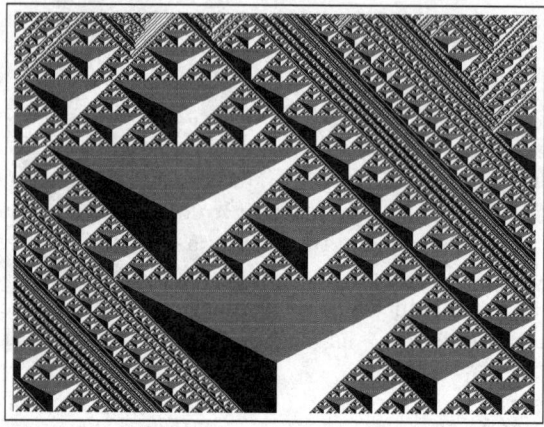

Fig. 2.6. A space-time diagram for the radius-2 CA over $\mathbb{Z}/1024\mathbb{Z}$ with rule number 3283936144 starting from a randomly chosen initial state.

2.5. Consider the finite linear dynamical system $f\colon \mathbb{F}_2^4 \longrightarrow \mathbb{F}_2^4$ with matrix (relative to standard basis)

$$M = \begin{bmatrix} 0\,1\,0\,0 \\ 0\,0\,0\,0 \\ 0\,0\,1\,1 \\ 0\,0\,1\,0 \end{bmatrix}.$$

Show that the phase space consists of one fixed point and one cycle of length three. Also show that the transient tree structures at the periodic points are all identical.   [1]

2.6. Use the elementary CA 150 over $\mathbb{Z}/n\mathbb{Z}$ to show that the question of whether or not a CA map is invertible depends on n. (As we will see in Chapter 4, this does not happen with a sequential update order.)   [1+C]

2.7. How many linear, one-dimensional, elementary CA rules of radius r are there? Give their Wolfram encoding in the case $r = 1$.   [1+]

2.8. How many elementary CA rules $f\colon \mathbb{F}_2^3 \longrightarrow \mathbb{F}_2$ satisfy the symmetry condition

$$f(x_{i-1}, x_i, x_{i+1}) = f(x_{i+1}, x_i, x_{i-1})$$

and the quiescence condition

$$f(0,0,0) = 0 \ ?$$

An analysis of the cellular automata induced by these rules can be found in, e.g., [32, 49].   [1]

2.2 Random Boolean Networks

Boolean networks (BN) were originally introduced by S. Kauffman [54] as a modeling framework for gene-regulatory networks. Since their introduction some modifications have been made, and here we present the basic setup as given in, e.g., [55–58], but see also [59].

A Boolean network has vertices or genes $V = \{v_1, \ldots, v_n\}$ and functions $F = (f_1, \ldots, f_n)$. Each gene v_i is linked or "wired" to k_i genes as specified by a map $e_i \colon \{1, \ldots, k_i\} \longrightarrow V$. The Boolean state x_{v_i} of each gene is updated as

$$x_{v_i} \mapsto f_i(x_{e_i(1)}, \ldots, x_{e_i(k_i)}) ,$$

and the whole state configuration is updated synchronously. Traditionally, the value of k_i was the same for all the vertices. A gene or vertex v that has state 1 is said to be expressed.

A *random Boolean network* (RBN) can be obtained in the following ways. First, each vertex v_i is assigned a sequence of maps $f^i = (f^i_1, \ldots, f^i_{l_i})$. At each point t in time a function f^i_t is chosen from this sequence for each vertex at random according to some distribution. The *function configuration* $(f^1_t, \ldots, f^n_t)$ that results is then used to compute the system configuration at time $t + 1$ based on the system configuration at time t. Second, we may consider for a fixed function f_i over k_i-variables the map $e_i \colon \{1, \ldots, k_i\} \longrightarrow V$ to be randomly chosen. That amounts to choosing a random directed graph in which v_i has in-degree k_i.

Since random Boolean networks are stochastic systems, they cannot be described using the traditional phase-space notion. As you may have expected, the framework of *Markov chains* is a natural way to capture their behavior. The idea behind this approach is straightforward and can be illustrated as follows.

Let $0 \le p \le 1.0$ and let $i \in \mathbb{Z}/n\mathbb{Z}$ be a vertex of an elementary CA (see the previous section) with update function f and states in $\{0, 1\}$. Let f' be some other elementary CA function. If we update vertex i using the function f with probability p and with function f' with probability $(1 - p)$ and use the function f for all other vertices states, we have a very basic random Boolean network. This stochastic system may be viewed as a weighted superposition of two deterministic cellular automata. By this we mean the following: If the state of vertex i is always updated using the map f, we obtain a phase space Γ, and if we always update the state of vertex i using the function f', we get a phase space $\tilde{\Gamma}$. The weighted sum "$p\Gamma + (1 - p)\tilde{\Gamma}$" is the directed, weighted graph with vertices all states of state space, with a directed edge from x to y if any of the two phase spaces contains this transition. The weight of the edge (x, y) is p (respectively, $1 - p$) if only Γ (respectively, $\tilde{\Gamma}$) contains this transition, and 1 if both phase spaces contain the transition. In general, the weight of the edge (x, y) is the sum of the probabilities of the configurations that has an associated phase space, which includes this transition. We may call the resulting weighted graph the *probabilistic phase space*. The evolution of the random Boolean network may therefore be viewed as a random walk on the probabilistic phase space. The corresponding weighted adjacency matrix directly and naturally encodes the associated Markov chain matrix of the RBN.

This Markov chain approach is the basis used for the framework of random Boolean networks as studied by, e.g., Shmulevich and Dougherty [55]. The following example provides a specific illustration.

Example 2.13. Let $Y = \mathsf{Circ}_3$ and, with the exception of f_0, let each function f_i be induced by $\mathrm{nor}_3 \colon \mathbb{F}_2^3 \longrightarrow \mathbb{F}_2$. For f_0 we use nor_3 with probability $p = 0.4$ and parity_3 with probability $q = 1 - p$. In the notation above we get the phase spaces Γ, $\tilde{\Gamma}$, and $p\Gamma + (1 - p)\tilde{\Gamma}$ as shown in Figure 2.7.   ◇

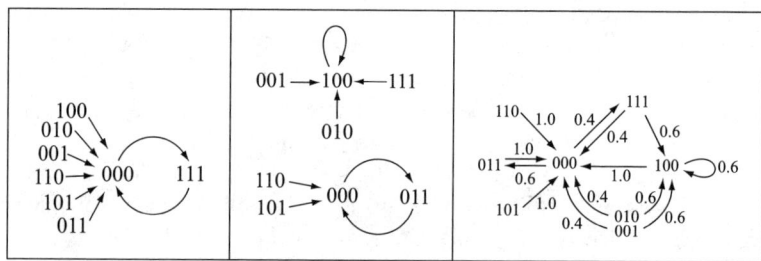

Fig. 2.7. The phase spaces Γ, $\tilde{\Gamma}$, and $p\Gamma + (1 - p)\tilde{\Gamma}$ of Example 2.13.

The concept of Boolean networks resembles several features of SDS. For instance, an analogue of the SDS dependency graph can be derived via the maps e_i. However, research on Boolean networks focuses on analyzing the functions, while for SDS the study of graph properties and update orders is of equal importance. As for sequential update schemes, we remark that aspects of asynchronous RBN have been studied in [60].

2.3 Finite-State Machines (FSMs)

Finite-state machines (FSM) [61–63] and their extensions constitute another theory and application framework. Their use ranges from tracking and response of weapon systems to dishwasher logic and all the way to the "AI-logic" of "bots" or "enemies" in computer games. Finite-state machines are not dynamical systems, but they do exhibit similarities with both SDS and cellular automata.

Definition 2.14. A finite-state machine (or a finite automaton) is a five-tuple $M = (K, \Sigma, \tau, x_0, A)$ where K is a finite set (the *states*), Σ is a finite set (the *alphabet*), $\tau \colon K \times \Sigma \longrightarrow K$ is the *transition function*, $x_0 \in K$ is the *start state*, and $A \subset K$ is the *set of accept states*.

Thus, for each state $x \in K$ and for each letter $s \in \Sigma$ there is a directed edge (x, x_s). The finite-state machine reads input from, e.g., an *input tape*. If the finite-state machine is in state x and reads the input symbol $s \in \Sigma$, it will transition to state x_s. If at the end of the input tape the current state is one of the states from A, the machine is said to accept the input tape. One therefore speaks about the set of input tapes or sequences accepted by the machine. This set of accepted input sequences is the *language accepted by M*. An FSM is often represented pictorially by its *transition diagram*, which has the states as vertices and has directed edges $(x, \tau(x, s))$ labeled by s.

If the reading of a symbol and the subsequent state transition take place every time unit, we see that each input sequence σ generates a time series of states $(M_\sigma(x_0, t))_{t=0}$. Here $M_\sigma(x_0, t)$ denotes the state at time t under the time evolution of M given the input sequence σ. The resemblance to finite dynamical systems is evident.

Example 2.15. In real applications the symbol may come in the form of events from some input system. A familiar example is traffic lights at a road intersection. The states in this case could be all permissible red–yellow–green configurations. A combination of a clock and vehicle sensors can provide events that are encoded as input symbols every second, say. The transition function implements the traffic logic, hopefully in a somewhat fair way and in accord with traffic rules. ◇

Our notion of all finite-state machine is often called a *deterministic finite-state machine* (DFSM), see, e.g., [61], where one can find in particular the equivalence of *regular languages* and finite-state machines.

Problems

2.9. Enumeration of CA rules
How many symmetric CA rules of radius 2 are there for binary states? How many outer-totalistic CA rules of radius 2 are there over $\mathbb{F}_2$? How many outer-symmetric CA rules of radius r are there with states in $\mathbb{F}_p$, the finite field with p elements (p prime)?   [1+]

2.10.
A *soliton* is, roughly speaking, a solitary localized wave that propagates without change in shape or speed even upon collisions with other solitary waves. Examples of solitons occur as solutions to several partial differential equations. In [64] it is demonstrated that somewhat similar behavior occur in *filter automata*.

The state space is $\{0, 1\}^{\mathbb{Z}}$. Let x^t denote the state at time t. For a filtered automaton with radius r and rule f the successor configuration to x^t is computed in a left-to-right (sequential) fashion as

$$x_i^{t+1} = f(x_{i-r}^{t+1}, \ldots, x_{i-1}^{t+1}, x_i^t, x_{i+1}^t, \ldots, x_{i+r}^t).$$

Argue, at least in the case of periodic boundary conditions, that a filter automaton is a particular instance of a sequential dynamical system.

Implement this system as a computer program and study orbits starting from initial states that contain a small number of states that are 1. Use the radius-3 and radius-5 functions f_3 and f_5 where $f_k \colon \mathbb{F}_2^{2k+1} \longrightarrow \mathbb{F}_2$ is given by

$$
f_k(x_{-k}, \ldots, x_{-1}, x_0, x_1, \ldots, x_k) = \begin{cases} 0 & \text{if each } x_i \text{ is zero,} \\ \displaystyle\sum_{i=-k}^{k} x_i & \text{otherwise,} \end{cases}
$$

where the summation is in $\mathbb{F}_2$. Note that these filter automata can be simulated by a CA; see [64].   [1+C]

Answers to Problems

2.1. 110.

2.2. Every rule $(a_7, \ldots, a_0)$ is fixed under the identity element, so $|\mathsf{Fix}(1)| = 256$. For a rule to be fixed under γ it must satisfy $(a_7, \ldots, a_0) = (\bar{a}_0, \ldots, \bar{a}_7)$, and there are 2^4 such rules. Likewise there are 2^6 rules fixed under δ and 2^4 rules fixed under $\gamma \circ \delta$.

2.4. (b) No. The SDS of the left-right rule is equivalent to the SDS of the original rule but with a different update order. What is the update order relation? (c) $G' = \{1, \gamma\}$. There are 136 orbits.

2.6. Derive the matrix representation of the CA and compute its determinant (in $\mathbb{F}_2$) for $n = 3$ and $n = 4$.

2.7. 2^{2r+1}.

2.8. 2^5.

2.9. (i) $2^6 = 64$. (ii) $2^5 \cdot 2^5 = 2^{10} = 1024$. (iii) $(2^{2r+1})^p$.

2.10. Some examples of orbits are shown in Figure 2.8.

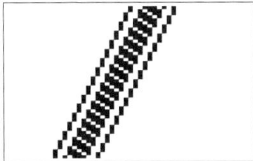 

Fig. 2.8. "Solitions" in an automata setting. In the left diagram the rule f_3 is used, while in the right diagram the rule f_5 is used.

3

Graphs, Groups, and Dynamical Systems

In this chapter we provide some basic terminology and background on the graph theory, combinatorics, and group theory required throughout the remainder of the book. A basic knowledge of group theory is assumed — a guide to introductory as well as more advanced references on the topics is given at the end of the chapter. We conclude this chapter by providing a short overview of the "classical" continuous and discrete dynamical systems. This overview is not required for what follows, but it may be helpful in order to put SDS theory into context.

3.1 Graphs

A graph Y is a four-tuple $Y = (v[Y], e[Y], \omega, \tau)$ where $v[Y]$ is the *vertex set* of Y and $e[Y]$ is the *edge set* of Y. The maps ω and τ are given by

$$\omega \colon e[Y] \longrightarrow v[Y] , \quad \tau \colon e[Y] \longrightarrow v[Y] . \tag{3.1}$$

For an edge $e \in e[Y]$ we call the vertices $\omega(e)$ and $\tau(e)$ the *origin* and *terminus* of e, respectively. The vertices $\omega(e)$ and $\tau(e)$ are the *extremities* of e. We sometimes refer to e as a *directed* edge and display this graphically as $\omega(e) \xrightarrow{\ e\ } \tau(e)$.

Two vertices v and v' are *adjacent* in Y if there exists an edge $e \in e[Y]$ such that $\{v, v'\} = \{\omega(e), \tau(e)\}$. A graph Y is *undirected* if there exists an *involution*

$$e[Y] \longrightarrow e[Y], \qquad e \mapsto \bar{e}, \tag{3.2}$$

such that $\bar{e} \neq e$ and $\tau(\bar{e}) = \omega(e)$, in which case we have $\omega(\bar{e}) = \tau(\bar{\bar{e}}) = \tau(e)$. We represent undirected graphs by diagrams — two vertices v_1 and v_2 and two edges e and $\bar{e}$ with the property $\omega(e) = v_1$ and $\tau(e) = v_2$ are represented by the diagram $v_1 \quad\rule{1cm}{0.4pt}\quad v_2$. For instance, for the four edges e_0, $\bar{e}_0$, e_1, and $\bar{e}_1$ with $\omega(e_0) = \omega(e_1)$ and $\tau(e_0) = \tau(e_1)$, we obtain the diagram

$$\omega(e_0) \overbrace{\underbrace{}_{e_0}}^{e_1} \tau(e_0) \ , \quad \text{and the diagram} \quad \circlearrowleft v$$

represents the graph with vertex $v = \omega(e) = \tau(e)$ and edges e and $\bar{e}$. *In the following, and in the rest of the book, we will assume that all graphs are undirected unless stated otherwise.*

A graph $Y' = (\mathrm{v}[Y'], \mathrm{e}[Y'], \omega', \tau')$ is a *subgraph* of Y if Y' is a graph with $\mathrm{v}[Y'] \subset \mathrm{v}[Y]$ and $\mathrm{e}[Y'] \subset \mathrm{e}[Y]$, such that the maps ω' and τ' are the restrictions of ω and τ. For any vertex $v \in \mathrm{v}[Y]$ the graph $\mathsf{Star}_Y(v)$ is the subgraph of Y given by

$$\mathrm{e}[\mathsf{Star}_Y(v)] = \{e \in \mathrm{e}[Y] \mid \omega(e) = v \text{ or } \tau(e) = v\},$$
$$\mathrm{v}[\mathsf{Star}_Y(v)] = \{v' \in \mathrm{v}[Y] \mid \exists e \in \mathrm{e}[\mathsf{Star}_Y(v)] \ : \ v' = \omega(e) \text{ or } v' = \tau(e)\} \ .$$

We denote the *ball of radius* 1 *around* $v \in \mathrm{v}[Y]$ and the *sphere of radius* 1 *around* v by

$$B_Y(v) = \mathrm{v}[\mathsf{Star}_Y(v)], \tag{3.3}$$
$$B'_Y(v) = B_Y(v) \setminus \{v\} \ , \tag{3.4}$$

respectively. A sequence of vertices and edges of the form

$$(v_1, e_1, \ldots, v_m, e_m, v_{m+1}) \quad \text{where} \quad \forall\, 1 \le i \le m, \ \omega(e_i) = v_i, \ \tau(e_i) = v_{i+1}$$

is a *walk* in Y. If the end points v_1 and v_{m+1} coincide, we obtain a *closed walk* or a *cycle* in Y. If all the vertices are distinct, the walk is a *path* in Y. Two vertices are *connected* in Y if there exists a path in Y that contains both of them. A *component* of Y is a maximal set of pairwise connected Y vertices. An edge e with $\omega(e) = \tau(e)$ is a *loop*. A graph Y is *loop-free* if its edge set contains no loops. An *independent set* of a graph Y is a subset $I \subset \mathrm{v}[Y]$ such that no two vertices v and v' of I are adjacent in Y. The set of all independent sets of a graph Y is denoted $\mathcal{I}(Y)$.

A *graph morphism*[1] $\varphi \colon Y \longrightarrow Z$ is a pair of maps $\varphi_1 \colon \mathrm{v}[Y] \longrightarrow \mathrm{v}[Z]$ and $\varphi_2 \colon \mathrm{e}[Y] \longrightarrow \mathrm{e}[Z]$ such that the diagram

$$
\begin{array}{ccc}
\mathrm{e}[Y] & \xrightarrow{\ \ \varphi_2\ \ } & \mathrm{e}[Z] \\
\downarrow{\scriptstyle \omega \times \tau} & & \downarrow{\scriptstyle \omega \times \tau} \\
\mathrm{v}[Y] \times \mathrm{v}[Y] & \xrightarrow{\ \varphi_1 \times \varphi_1\ } & \mathrm{v}[Z] \times \mathrm{v}[Z]
\end{array}
$$

commutes. A graph morphism $\varphi \colon Y \longrightarrow Z$ thus preserves adjacency.

3.1. In light of $\overline{\varphi_2(e)} = \varphi_2(\bar{e})$, show that if Y is an undirected graph, then so is the image graph $\varphi(Y)$. [1]

---

[1] Graph morphisms are also referred to as *graph homomorphisms* in the literature.

A bijective graph morphism of the form $\varphi\colon Y \longrightarrow Y$ is an *automorphism* of Y. The automorphisms of Y form a group under function composition. This is the automorphism group of Y, and it is denoted $\mathsf{Aut}(Y)$.

Let Y and Z be undirected graphs and let $\varphi\colon Y \longrightarrow Z$ be a graph morphism. We call φ *locally surjective* or *locally injective*, respectively, if all the restriction maps

$$\varphi|_{\mathsf{Star}_Y(v)}\colon \mathsf{Star}_Y(v) \longrightarrow \mathsf{Star}_Z(\varphi(v)) \tag{3.5}$$

are all surjective or all injective, respectively. A graph morphism that is both locally surjective and locally injective is called a *local isomorphism* or a *covering*.

Example 3.1. The graph morphism $\varphi\colon Y \longrightarrow Z$ shown in Figure 3.1 is surjective but not locally surjective. ◇

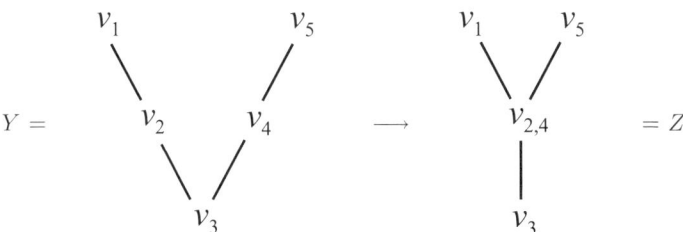

Fig. 3.1. The graph morphism φ of Example 3.1.

3.1.1 Simple Graphs and Combinatorial Graphs

An undirected graph Y is a *simple* graph if the mapping $\{e, \bar{e}\} \mapsto \{\omega(e), \tau(e)\}$ is injective. Accordingly, a simple graph has no multiple edges but may contain loops. Thus, the graph

$$Y' = \bigcirc v \underline{\hspace{1cm}} v'$$

is a simple graph. An undirected graph Y is a *combinatorial* graph if

$$\omega \times \tau\colon e[Y] \longrightarrow v[Y] \times v[Y], \qquad e \mapsto (\omega(e), \tau(e)), \tag{3.6}$$

is injective. Thus, an undirected graph is a combinatorial graph if and only if it is simple and loop-free. In fact, we have [65]:

Lemma 3.2. *An undirected graph Y is combinatorial if and only if Y contains no cycle of length ≤ 2.*

3.2. Prove Lemma 3.2. [1+]

Combinatorial graphs allow one to identify the pair $\{e, \bar{e}\}$ and its set of extremities $\{\omega(e), \tau(e)\}$, which we refer to as a *geometric edge*. We denote the set of geometric edges by $\tilde{e}[Y]$, and identify $\tilde{e}[Y]$ and $e[Y]$ for combinatorial graphs.[2] Every combinatorial graph corresponds uniquely to a simplicial complex of dimension ≤ 1; see [66].

For an undirected graph Y there exists a unique combinatorial graph Y_c obtained by identifying multiple edges of Y and by removing loops, i.e.,

$$v[Y_c] = v[Y], \tag{3.7}$$

$$\tilde{e}[Y_c] = \{\{\omega(e), \tau(e)\} \mid e \in e[Y], \omega(e) \neq \tau(e)\}. \tag{3.8}$$

Equivalently, we have a well-defined mapping $Y \mapsto Y_c$. Suppose Y is a combinatorial graph and $\varphi \colon Y \longrightarrow Z$ is a graph morphism. Then, in general, $\varphi(Y)$ is not a combinatorial graph; see Example 3.5.

Example 3.3. Figure 3.2 shows two graphs. The graph on the left is directed and has two edges e_1 and e_2 such that $\omega(e_1) = \omega(e_2) = 1$ and $\tau(e_1) = \tau(e_2) = 2$. It also has a loop at vertex 1. The graph on the right is the Peterson graph, a combinatorial graph that has provided counterexamples for many conjectures. ◇

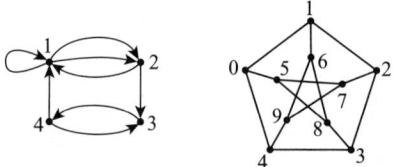

Fig. 3.2. The graphs of Example 3.3.

The *vertex join* of a combinatorial graph Y and a vertex v is the combinatorial graph, $Y \oplus v$, defined by

$$v[Y \oplus v] = v[Y] \cup \{v\}, \tag{3.9}$$

$$e[Y \oplus v] = e[Y] \cup \{\{v, v'\} \mid v' \in v[Y]\}.$$

The vertex join operation is a special case of the more general *graph join* operation [12].

Example 3.4 (Some common graph classes). The *line graph* Line_n of order n is the combinatorial graph with vertex set $\{1, 2, \ldots, n\}$ and edge set $\{\{i, i+1\} \mid i = 1, \ldots, n-1\}$. It can be depicted as

---

[2] Graph theory literature has no standard notation for the various graph classes. The graphs in Definition (3.1) are oftentimes called directed multigraphs. Refer to [12] for a short summary of some of the terms used and their inconsistency!

Line_n: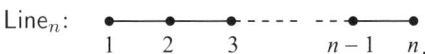
$$1 \quad 2 \quad 3 \qquad n-1 \quad n.$$

The graph Circ_n is the *circle graph* on n vertices $\{0, 1, \dots, n-1\}$ where two vertices i and j are connected if $i - j \equiv \pm 1 \bmod n$.

Circ_n: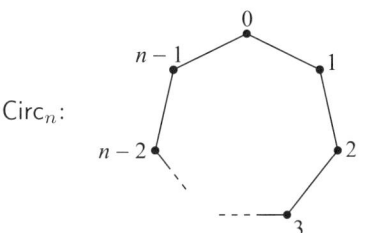

3.3. (An alternative way to define paths and cycles in graphs) Prove that for undirected graphs Y a path corresponds uniquely to a graph morphism $\mathsf{Line}_n \longrightarrow Y$ and a cycle to a graph morphism $\mathsf{Circ}_n \longrightarrow Y$.   [1+]

Example 3.5. The map $\varphi \colon \mathsf{Circ}_6 \longrightarrow \mathsf{Circ}_3$ defined by $\varphi(0) = \varphi(3) = 0$, $\varphi(1) = \varphi(4) = 1$, and $\varphi(2) = \varphi(5) = 2$ is a graph morphism. It is depicted on the left in Figure 3.3. Let C_2 be the graph with vertex set $\{0, 1\}$ and edge set $\{e_1, \bar{e}_1, e_2, \bar{e}_2\}$. The graph morphism $\psi \colon \mathsf{Circ}_4 \longrightarrow C_2$ given by $\psi(0) = \psi(2) = 0$, $\psi(1) = \psi(3) = 1$, $\psi(\{0, 1\}) = \psi(\{2, 3\}) = \{e_1, \bar{e}_1\}$, and $\psi(\{1, 2\}) = \psi(\{0, 3\}) = \{e_1, \bar{e}_1\}$ is depicted on the right in Figure 3.3.   ◇

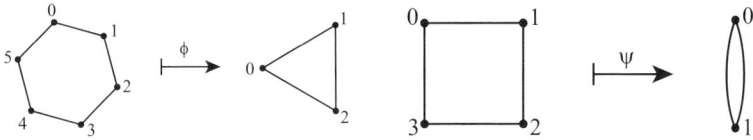

Fig. 3.3. The graph morphisms $\varphi \colon \mathsf{Circ}_6 \longrightarrow \mathsf{Circ}_3$ (left) and $\psi \colon \mathsf{Circ}_4 \longrightarrow C_2$ (right) from Example 3.5.

Using the vertex join operation we can construct other graph classes. For example, the *wheel graph*, which we write as Wheel_n, is the the vertex join of Circ_n and the vertex n so that

$$\mathrm{v}[\mathsf{Wheel}_n] = \{0, 1, \dots, n\},$$
$$\mathrm{e}[\mathsf{Wheel}_n] = \mathrm{e}[\mathsf{Circ}_n] \cup \{\{i, n\} \mid i = 0, \dots, n-1\}.$$

Wheel$_n$ can be depicted as follows:

Wheel$_n$:

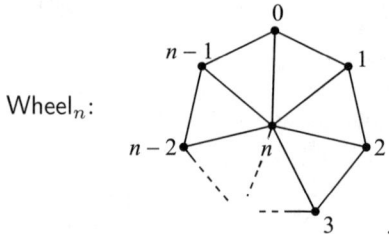

Finally, the *binary hypercube* Q_2^n is the graph where the vertices are the n-tuples over $\{0,1\}$ and where two vertices $v = (x_1,\ldots,x_n)$ and $v' = (x_1',\ldots,x_n')$ are adjacent if they differ in precisely one coordinate. Clearly, this is a graph with 2^n vertices and $(2^n \cdot n)/2 = n \cdot 2^{n-1}$ edges.   ◇

Q_2^3:

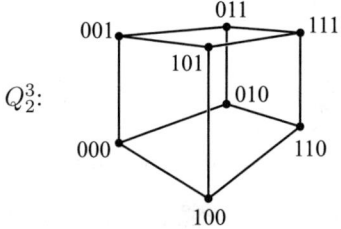

3.1.2 The Adjacency Matrix of a Graph

Let Y be a simple undirected graph with vertex set $\{v_1, v_2, \ldots, v_n\}$. The *adjacency matrix* A or A_Y of Y is the $n \times n$ matrix with entries $a_{i,j} \in \{0,1\}$ where the entry $a_{i,j}$ equals 1 if Y has $\{v_i, v_j\} \in \tilde{e}[Y]$ and equals zero otherwise. Clearly, since Y is undirected, the matrix A is symmetric. The adjacency matrix of a simple directed graph is defined analogously, but it is generally not symmetric.

Example 3.6. As an example take the graph $Y = \mathsf{Circ}_4$ with vertex set $\{1, 2, 3, 4\}$ shown below.

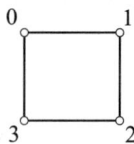

Its adjacency matrix A is given by

$$A = \begin{bmatrix} 0 & 1 & 0 & 1 \\ 1 & 0 & 1 & 0 \\ 0 & 1 & 0 & 1 \\ 1 & 0 & 1 & 0 \end{bmatrix}.$$

          ◇

The following result will be used in Chapter 5, where we enumerate fixed points of SDS.

Proposition 3.7. *Let Y be a graph with adjacency matrix A. The number of walks of length k in Y that start at vertex v_i and end at vertex v_j is $[A^k]_{i,j}$, the (i,j) entry of the kth power of A.*

The result is proved by induction. Obviously, the assertion holds for $k = 1$. Assume it is true for $k = m$. We can show that it holds for $k = m + 1$ by decomposing a walk of length $m + 1$ from vertex v_i to vertex v_j into a walk of length m from the initial vertex v_i to an intermediate vertex v_k followed by a walk of length 1 from the intermediate vertex v_k to the final vertex v_j. By the induction hypothesis, $[A^m]_{i,k}$ counts the number of walks from v_i to v_k and A counts the number of walks from v_k to v_j. By multiplying A^k and A, we sum up all these contributions for all possible intermediate vertices v_k.

Example 3.8. We compute matrix powers of A from the previous example as follows:

$$A^2 = \begin{bmatrix} 2&0&2&0 \\ 0&2&0&2 \\ 2&0&2&0 \\ 0&2&0&2 \end{bmatrix}, \quad A^3 = \begin{bmatrix} 0&4&0&4 \\ 4&0&4&0 \\ 0&4&0&4 \\ 4&0&4&0 \end{bmatrix}, \quad \text{and } A^4 = \begin{bmatrix} 8&0&8&0 \\ 0&8&0&8 \\ 8&0&8&0 \\ 0&8&0&8 \end{bmatrix}.$$

For example, there are four walks from 0 to 1 of length 3. Likewise there are eight closed cycles of length 4 starting at vertex 0. ◇

A particular consequence of this result is that the number of closed cycles of length n in Y starting at v_i is $[A^n]_{i,i}$. The *trace* of a matrix A, written $\mathsf{Tr}\ A$, is the sum of the diagonal elements of A. It follows that the total number of cycles in Y of length n is $\mathsf{Tr}\ A^n$.

The *characteristic polynomial* of an $n \times n$ matrix A is $\chi_A(x) = \det(xI - A)$, where I is the $n \times n$ identity matrix. We will use the following classical theorem in the proof of Theorem 5.3:

Theorem 3.9 (Cayley–Hamilton). *Let A be a square matrix with entries in a field and with characteristic polynomial $\chi_A(x)$. Then we have*

$$\chi_A(A) = 0 \ .$$

That is, a square matrix A satisfies its own characteristic polynomial. For a proof of the Cayley–Hamilton theorem, see [67].

Example 3.10. The characteristic polynomial of the adjacency matrix of Circ_4 is $\chi(x) = x^4 - 4x^2$, and as you can readily verify, we have

$$\chi(A) = A^4 - 4A^2 = \begin{bmatrix} 8&0&8&0 \\ 0&8&0&8 \\ 8&0&8&0 \\ 0&8&0&8 \end{bmatrix} - 4 \begin{bmatrix} 2&0&2&0 \\ 0&2&0&2 \\ 2&0&2&0 \\ 0&2&0&2 \end{bmatrix} = 0 \ ,$$

the 4×4 zero matrix. ◇

3.1.3 Acyclic Orientations

Let Y be a loop-free, undirected graph. An *orientation* of Y is a map

$$\mathcal{O}_Y : \mathrm{e}[Y] \longrightarrow \mathrm{v}[Y] \times \mathrm{v}[Y]. \tag{3.10}$$

An orientation of Y naturally induces a graph $G(\mathcal{O}_Y) = (\mathrm{v}[Y], \mathrm{e}[Y], \omega, \tau)$ where $\omega \times \tau = \mathcal{O}_Y$. The orientation $\mathcal{O}_Y$ is *acyclic* if $G(\mathcal{O}_Y)$ has no (directed) cycles. The set of all acyclic orientations of Y is denoted $\mathsf{Acyc}(Y)$. In the following we will identify an orientation $\mathcal{O}_Y$ with its induced graph $G(\mathcal{O}_Y)$.

Example 3.11. The four orientations of $Z = \; v_1 \overset{e_1}{\underset{e_2}{\frown}} v_2 \;$ are

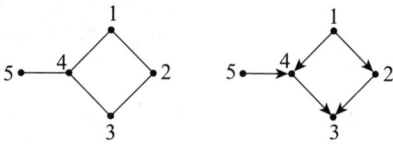

3.4.   Prove that we have a bijection

$$\beta \colon \mathsf{Acyc}(Y) \longrightarrow \mathsf{Acyc}(Y_c),$$

where Y_c is defined in Section 3.1.1, Eqs. (3.7) and (3.8).   [1+]

Let $\mathcal{O}_Y$ be an acyclic orientation of Y and let $\mathcal{P}(\mathcal{O}_Y)$ be the set of all (directed) paths π in $G(\mathcal{O}_Y)$. Furthermore, let $\Omega(\pi)$, $T(\pi)$, and $\ell(\pi)$ denote the first vertex, the last vertex, and the length of π, respectively. We consider the map $\mathsf{rnk} \colon \mathrm{v}[Y] \longrightarrow \mathbb{N}$ defined by

$$\mathsf{rnk}(v) = \max_{\pi \in \mathcal{P}(\mathcal{O}_Y)} \{ \ell(\pi) \mid T(\pi) = v \} . \tag{3.11}$$

Any acyclic orientation $\mathcal{O}_Y$ induces a *partial ordering* $\leq_{\mathcal{O}_Y}$ by setting

$$v \leq_{\mathcal{O}_Y} v' \iff [v \text{ and } v' \text{ are connected in } G(\mathcal{O}_Y) \text{ and } \mathsf{rnk}(v) \leq \mathsf{rnk}(v')] . \tag{3.12}$$

Example 3.12. On the left side in Figure 3.4 we have shown a graph Y on five vertices, and on the right side we have shown one acyclic orientation $\mathcal{O}_Y$ of Y. With this acyclic orientation we have $\mathsf{rnk}(1) = \mathsf{rnk}(5) = 0, \mathsf{rnk}(2) = \mathsf{rnk}(4) = 1$,

Fig. 3.4. A graph on five vertices (left) and an acyclic orientation of this graph depicted as a directed graph (right).

and $\mathsf{rnk}(3) = 2$. In the partial order we have $5 \leq_{\mathcal{O}_Y} 3$, while 2 and 4 are not comparable.   ◇

3.1.4 The Update Graph

Let Y be a combinatorial graph with vertex set $\{v_1, \ldots, v_n\}$, and let S_Y be the symmetric group over v$[Y]$. The identity element of S_Y is written id.

Let Y be a combinatorial graph. Two S_Y-permutations $(v_{i_1}, \ldots, v_{i_n})$ and $(v_{h_1}, \ldots, v_{h_n})$ are adjacent if there exists some index k such that (a) $v_{i_l} = v_{h_l}$, $l \neq k, k+1$, and (b) $\{v_{i_k}, v_{i_{k+1}}\} \notin e[Y]$ hold. This notion of adjacency induces a combinatorial graph over S_Y referred to as the *update graph*, and it is denoted $U(Y)$. The update graph has e$[U(Y)] = \{\{\sigma, \pi\} \mid \sigma, \pi$ are adjacent$\}$. We introduce the equivalence relation $\sim_Y$ on S_Y by

$$\pi \sim_Y \pi' \iff \pi \text{ and } \pi' \text{ are connected by a } U(Y) \text{ path.} \qquad (3.13)$$

The equivalence class of π is written $[\pi]_Y = \{\pi' \mid \pi' \sim_Y \pi\}$, and the set of all equivalence classes is denoted $S_Y / \sim_Y$. In the following we will assume that the vertices of Y are ordered according to $v_i < v_j$ if and only if $i < j$. An *inversion pair* (v_r, v_s) of a permutation $\pi \in S_Y$ is a pair of entries in π satisfying $\pi(v_i) = v_r$ and $v_s = \pi(v_k)$ with $r > s$ and $i < k$. The following lemma characterizes the component structure of $U(Y)$.

Lemma 3.13. *Let Y be a combinatorial graph and let $\pi \in S_Y$. Then there exists a $U(Y)$ path connecting π and the identity permutation* id *if and only if all inversion pairs (v_r, v_s) of π satisfy $\{v_r, v_s\} \notin e[Y]$.*

Proof. Let $\pi = (v_{i_1}, \ldots, v_{i_n}) \neq$ id and let (v_l, v_s) be an inversion pair of π. If we assume that π and id are connected, then there is a corresponding $U(Y)$ path that consists of pairwise adjacent vertices π' and π'' of the form $\pi' = (\ldots, v, v', \ldots)$ and $\pi'' = (\ldots, v', v \ldots)$. By the definition of $U(Y)$ we have $\{v, v'\} \notin e[Y]$, and in particular this holds for all inversion pairs.

Moreover, if all inversion pairs (v, v') of π satisfy $\{v, v'\} \notin e[Y]$, then it is straightforward to construct a path in $U(Y)$ connecting π and id, completing the proof of the lemma. $\square$

Example 3.14. As an example of an update graph we compute $U(\mathsf{Circ}_4)$. This graph has 14 components and is shown in Figure 3.5. We see that all the

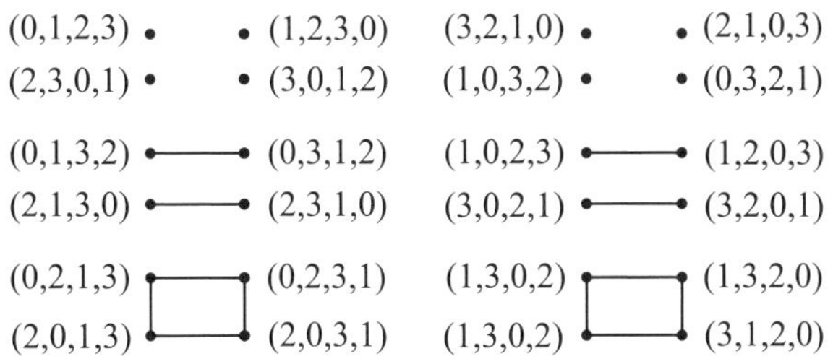

Fig. 3.5. The graph $U(\mathsf{Circ}_4)$.

isolated vertices in $U(\mathsf{Circ}_4)$ in Figure 3.5 correspond to Hamiltonian paths in Circ_4. This is true in general. Why?   ◇

3.1.5 Graphs, Permutations, and Acyclic Orientations

Any permutation $\pi = (v_{i_1}, \ldots, v_{i_n}) \in S_Y$ induces a *linear ordering* $<_\pi$ on $\{v_{i_1}, \ldots, v_{i_n}\}$ defined by $v_{i_r} <_\pi v_{i_h}$ if and only if $r < h$, where $<$ is the natural order. A permutation π of the vertices of a combinatorial graph Y induces an orientation $\mathcal{O}_Y(\pi)$ of Y by orienting each of its edges $\{v, v'\}$ as (v, v') if $v <_\pi v'$ and as (v', v), otherwise. It is clear that the orientation $\mathcal{O}_Y(\pi)$ is acyclic. For any combinatorial graph Y we therefore obtain a map

$$f'_Y \colon S_Y \longrightarrow \mathsf{Acyc}(Y), \quad \pi \mapsto \mathcal{O}_Y(\pi) \,. \tag{3.14}$$

In the following proposition, we relate permutations of the vertices of a combinatorial graph Y and the set of its acyclic orientations. The result also arises in the context of the theory of partially commutative monoids and is related to the Cartier–Foata normal form [68], but see also [69].

Proposition 3.15. *For any combinatorial graph Y there exists a bijection*

$$f_Y \colon [S_Y / \sim_Y] \longrightarrow \mathsf{Acyc}(Y) \,. \tag{3.15}$$

Proof. We have already established the map $f'_Y \colon S_Y \longrightarrow \mathsf{Acyc}(Y)$. Our first step is to show that f'_Y is constant on each equivalence class $[\pi]_Y$. To prove this it is sufficient to consider the case with two adjacent vertices π and π' in $U(Y)$. The general case will then follow by induction on the length of the path connecting π and π'. By definition, if π and π' are adjacent, they differ in exactly two consecutive entries, and the corresponding entries are not connected by an edge in Y. Consequently, we must have $f'_Y(\pi) = f'_Y(\pi')$, and we have a well-defined map

$$f_Y \colon [S_Y / \sim_Y] \longrightarrow \mathsf{Acyc}(Y) \,.$$

It remains to show that f_Y is a bijection. To this end, let $\mathcal{O}_Y$ be an acyclic orientation and consider the corresponding partition $(\mathsf{rnk}^{-1}(h))_{0 \le h \le n}$ [Section 3.1.3, Eq. (3.11)] of the vertices of Y. Let $H = \{h \mid \mathsf{rnk}^{-1}(h) \ne \varnothing\}$, where $|H| = t + 1$. We set $\mathsf{rnk}^{-1}(h) = (v_{i_h^1}, \ldots, v_{i_h^{m_h}})$ where $v_{i_h^1} <_\pi \cdots <_\pi v_{i_h^{m_h}}$ for $h \in H$. It is straightforward to verify that

$$g_Y \colon \mathsf{Acyc}(Y) \to [S_Y / \sim_Y], \quad \mathcal{O}_Y \mapsto [(v_{i_0^1}, \ldots, v_{i_0^{m_0}}, \ldots, v_{i_t^1}, \ldots, v_{i_t^{m_t}})]_Y, \tag{3.16}$$

is a well-defined map satisfying

$$g_Y \circ f_Y = \mathsf{id} \quad \text{and} \quad f_Y \circ g_Y = \mathsf{id} \,,$$

and the proof of the proposition is complete.   □

The permutation

$$\widehat{\pi} = (v_{i_0^1}, \ldots, v_{i_0^{m_0}}, \ldots, v_{i_t^1}, \ldots, v_{i_t^{m_t}}) \tag{3.17}$$

that we constructed in the above proof is called the *canonical permutation* of $[\pi]_Y$. The element $\widehat{\pi}$ is a special case of the Cartier–Foata normal form [68].

Example 3.16. Since we have $|\mathsf{Acyc}(\mathsf{Circ}_4)| = 14$, Proposition 3.15 shows that $U(\mathsf{Circ}_4)$ has 14 components (Example 3.14). To find the canonical permutation of the component containing $\pi = (2, 0, 1, 3)$, we first construct the acyclic orientation $\mathcal{O}_Y(\pi)$:

$$\mathcal{O}(\pi)(\{0,1\}) = (0,1), \qquad \mathcal{O}(\pi)(\{1,2\}) = (2,1),$$
$$\mathcal{O}(\pi)(\{2,3\}) = (2,3), \qquad \mathcal{O}(\pi)(\{0,3\}) = (0,3).$$

From this we get $\mathsf{rnk}^{-1}(0) = \{0,2\}$ and $\mathsf{rnk}^{-1}(1) = \{1,3\}$, and therefore $\widehat{\pi} = (0,2,1,3)$. ◇

The bijection f_Y allows us to count the $U(Y)$-components. In Chapter 4 we will prove that the number of components of $U(Y)$ is an upper bound for the number of functionally different sequential dynamical systems, obtained solely by varying the permutation update order. We next show how to compute this number through a recursion formula for the number of acyclic orientations of a graph.

Let e be an edge of Y. The graph Y'_e is the graph that results from Y by deleting e, and the graph Y''_e is the graph that we obtain from Y by contracting the edge e. Writing $a(Y) = |\mathsf{Acyc}(Y)|$, we now have

$$a(Y) = a(Y') + a(Y''), \tag{3.18}$$

where we have omitted the reference to the edge e. This recursion can be found in [70], where acyclic orientations of graphs are related to the chromatic polynomial χ as

$$a(Y) = (-1)^n \chi(-1).$$

3.5. Prove the recursion relation (3.18). **[2]**

Note that a graph with no edges has one acyclic orientation. Any graph map satisfying the relation (3.18) is called a *Tutte-invariant*. In Section 8.2.2 we will show how the acyclic orientations of a graph Y and the number $a(Y)$ are of significance in an area of mathematical biology.

Example 3.17. To illustrate the use of formula (3.18), we will compute the number of acyclic orientations of $Y = \mathsf{Circ}_n$ for $n \geq 3$. Pick the edge $e = \{0, n-1\}$. Then we have $Y'_e = \mathsf{Line}_n$ and $Y''_e = \mathsf{Circ}_{n-1}$, and thus

$$a(\mathsf{Circ}_n) = a(\mathsf{Line}_n) + a(\mathsf{Circ}_{n-1}) = 2^{n-1} + a(\mathsf{Circ}_{n-1}).$$

This recursion relation is straightforward to solve, and, using, for example, $a(\mathsf{Circ}_3) = 6$, we get $a(\mathsf{Circ}_n) = 2^n - 2$. This is, of course, not very surprising since there are 2^n orientations of Circ_n, two of which are cyclic. Problem 3.8 asks for a formula for $a(\mathsf{Wheel}_n)$. ◇

3.2 Group Actions

Group actions are central in the analysis of several aspects of sequential dynamical systems. Their use in the study of equivalence is one example. Recall that if X is a set and if G is a finite group, then G *acts* on X if there is a group homomorphism of G into the group of permutations of the set X, denoted S_X, in which case we call X, a G-set. If G acts on X, we have a map

$$G \times X \longrightarrow X, \quad (g, x) \mapsto gx ,$$

that satisfies $(1, x) = x$ and $(gh, x) = (g, (h, x))$ for all $g, h \in G$ and all $x \in X$.

Let $x \in X$. The *stabilizer* or *isotropy group* of x is the subgroup of G given by

$$G_x = \{g \in G \mid gx = x\} ,$$

and the G *orbit* of x is the set

$$G(x) = \{gx \mid g \in G\} .$$

For each $x \in X$ we have the bijection

$$G/G_x \longrightarrow G(x), \quad gG_x \mapsto gx , \tag{3.19}$$

which in particular implies that the size of the orbit of x equals the index of the subgroup G_x in G.

The lemma of Frobenius[3] is a classical result that relates the number of orbits N of a group action to the cardinalities of the *fixed sets*

$$\mathsf{Fix}(g) = \{x \in X \mid gx = x\} .$$

Lemma 3.18 (Frobenius).

$$N = \frac{1}{|G|} \sum_{g \in G} |\mathsf{Fix}(g)| \tag{3.20}$$

Proof. Consider the set $M = \{(g, x) \mid g \in G, x \in X; gx = x\}$. On the one hand, we may represent M as a disjoint union

$$M = \dot{\bigcup}_{g \in G} \{(g, x) \mid x \in X; \ gx = x\} ,$$

from which $|M| = \sum_g |\mathsf{Fix}(g)|$ follows. On the other hand, we can represent M as the disjoint union

$$M = \dot{\bigcup}_{x \in X} \{(g, x) \mid g \in G; \ gx = x\} ,$$

---

[3] This lemma is usually attributed to Burnside.

from which we derive $|M| = \sum_{x \in X} |G_x|$. In view of (3.19) we conclude that $|G_x| = |G|/|G(x)|$; consequently,

$$|M| = |G| \sum_{x \in X} \frac{1}{|G(x)|} = |G|N ,$$

and the proof of the lemma is complete.   □

Let X be the set $\{1, 2, \ldots, n\}$, and let G be a group acting on X and on the set K. Then the group action on X induces a natural group action on the set of all maps $f \colon \{1, 2, \ldots, n\} \longrightarrow K$ via

$$\{\rho \cdot f\}(i) = \rho\, f(\rho^{-1}(i)). \tag{3.21}$$

In particular, we may consider f as a n-tuple $x = (x_1, \ldots, x_n) = (x_j) \in K^n$. If G acts trivially on K, we obtain the following action of G on K^n:

$$\cdot \colon G \times K^n \longrightarrow K^n, \quad (\rho, (x_j)) \mapsto \rho \cdot (x_j) = (x_{\rho^{-1}(j)}). \tag{3.22}$$

It is clearly a group action: $(hg) \cdot (x_j) = (x_{g^{-1}h^{-1}(j)}) = h \cdot (g \cdot (x_j))$. The action $\cdot \colon G \times K^n \longrightarrow K^n$ on n-tuples induces a G-action on maps $\Phi \colon K^n \longrightarrow K^n$ by

$$\{\rho \bullet \Phi\}(x_j) = \rho \cdot (\Phi(\rho^{-1} \cdot (x_j))) . \tag{3.23}$$

3.2.1 Groups Acting on Graphs

Let G be a group and let Y be a combinatorial graph with automorphism group $\mathsf{Aut}(Y)$. Then G acts on Y if there exists a homomorphism from G into $\mathsf{Aut}(Y)$. Equivalently, the group G acts on Y if it acts on $\mathrm{v}[Y]$ and $\mathrm{e}[Y]$, we have the commutative diagrams

$$
\begin{array}{ccc}
\mathrm{e}[X] \xrightarrow{\ \omega\ } \mathrm{v}[X] & \qquad & \mathrm{e}[X] \xrightarrow{\ \tau\ } \mathrm{v}[X] \\
\downarrow{g} \qquad\quad \downarrow{g} & \qquad & \downarrow{g} \qquad\quad \downarrow{g} \\
\mathrm{e}[Y] \xrightarrow{\ \omega\ } \mathrm{v}[Y] & \qquad & \mathrm{e}[Y] \xrightarrow{\ \tau\ } \mathrm{v}[Y]
\end{array}
\tag{3.24}
$$

i.e., $g\omega(e) = \omega(ge)$ and $g\tau(e) = \tau(ge)$. If G acts on Y, then its action induces the *orbit graph* $G \setminus Y$ where

$$\mathrm{v}[G \setminus Y] = \{G(v) \mid v \in \mathrm{v}[Y]\}, \quad \mathrm{e}[G \setminus Y] = \{G(e) \mid e \in \mathrm{e}[Y]\},$$

and where $\omega_{G \setminus Y} \times \tau_{G \setminus Y} \colon \mathrm{e}[G \setminus Y] \longrightarrow \mathrm{v}[G \setminus Y] \times \mathrm{v}[G \setminus Y]$ is given by

$$G(e) \mapsto (G(\omega(e)), G(\tau(e))) .$$

The canonical map

$$\pi_G \colon Y \longrightarrow G \setminus Y, \quad v \mapsto G(v) \tag{3.25}$$

is then a surjective and locally surjective morphism.

3.6. Let G act on Y and let G_v be the isotropy group of vertex v. Prove that

$$G_v \setminus \mathrm{Star}_Y(v) \cong \mathrm{Star}_{G \setminus Y}(G(v)) \,.$$

[**2**]

The following example shows that the orbit graph of a combinatorial graph is not necessarily a combinatorial graph.

Example 3.19. Consider the 3-cube shown in Figure 3.6. The permutation

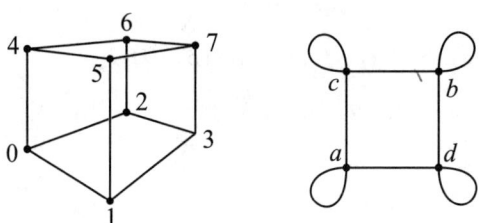

Fig. 3.6. The graph $Y = Q_2^3$ and the orbit graph $\langle(0,4)(1,5)(2,6)(3,7)\rangle \setminus Q_2^3$ shown on the left and right, respectively.

$\gamma = (0,4)(1,5)(2,6)(3,7)$ is an automorphism of Q_2^3. Of course, since the orbits of $\langle \gamma \rangle$ coincide with the cycles of γ, we see that the orbit graph $Y = \langle \gamma \rangle \setminus Q_2^3$ has four vertices. If we denote the orbits containing 0, 1, 2, and 3 by a, b, c, and d, respectively, we get the orbit graph shown on the right in Figure 3.6.

◇

3.7. Give an example of a combinatorial graph Y and a group $G < \mathsf{Aut}(Y)$ such that $G \setminus Y$ is not a simple graph. [**1+**]

3.2.2 Groups Acting on Acyclic Orientations

Let Y be an undirected, loop-free graph and let G be a group acting on Y. According to Eq. (3.22), if G acts on the graph Y, then G acts naturally on the set of acyclic orientations of Y [Section 3.1.3, Eq. (3.10)]

$$\mathcal{O}_Y : \mathrm{e}[Y] \longrightarrow \mathrm{v}[Y] \times \mathrm{v}[Y]$$

via

$$(g\mathcal{O}_Y)(e) = g(\mathcal{O}_Y(g^{-1}e)) \,, \tag{3.26}$$

where G acts on $\mathrm{v}[Y] \times \mathrm{v}[Y]$ via $g(v, v') = (g(v), g(v'))$. Furthermore, we set $G(v, v') = (G(v), G(v'))$ and

$$\mathsf{Acyc}(Y)^G = \{\mathcal{O} \in \mathsf{Acyc}(Y) \mid \forall g \in G; \ g\mathcal{O} = \mathcal{O}\} \,.$$

Suppose we have $\mathcal{O}(e) = (v, v')$. We observe that $g\mathcal{O} = \mathcal{O}$ is equivalent to

$$\forall g \in G; \qquad \mathcal{O}(ge) = g(\mathcal{O}(e)) = (gv, gv') . \qquad (3.27)$$

In particular, we note that $\mathsf{Fix}(g) = \mathsf{Acyc}(Y)^{\langle g \rangle}$. Our objective is to provide a combinatorial interpretation for the set $\mathsf{Fix}(g)$.

We first give an example.

Example 3.20. Let $g = (v_1, v_3)(v_2, v_4)$, i.e., $gv_1 = v_3$, $gv_2 = v_4$, and $g^{-1} = g$,

$$Y = \begin{array}{ccc} v_1 & \!\!\!\!\!\!\!\!- & v_2 \\ | & & | \\ v_4 & \!\!\!\!\!\!\!\!- & v_3 \end{array} \quad \text{and} \quad \mathcal{O} = \begin{array}{ccc} v_1 & \!\!\!\longrightarrow & v_2 \\ \downarrow & & \uparrow \\ v_4 & \!\!\!\longleftarrow & v_3 \end{array}.$$

Then we have $\mathcal{O} \in \mathsf{Acyc}(Y)^{\langle g \rangle}$:

$$g(\mathcal{O}(\{v_1, v_2\})) = (v_3, v_4) = \mathcal{O}(\{v_3, v_4\}) = \mathcal{O}(\{gv_1, gv_2\}),$$
$$g(\mathcal{O}(\{v_1, v_4\})) = (v_3, v_2) = \mathcal{O}(\{v_3, v_2\}) = \mathcal{O}(\{gv_1, gv_4\}) .$$

The canonical morphism $\pi_{\langle g \rangle}$ maps Y as

$$Y = \begin{array}{ccc} v_1 & \!\!\!\!\!\!\!\!- & v_2 \\ | & & | \\ v_4 & \!\!\!\!\!\!\!\!- & v_3 \end{array} \quad \longrightarrow \quad \{v_1, v_3\} \overparen{} \{v_2, v_4\} = \langle g \rangle \backslash Y,$$

and $\mathcal{O}$ induces the acyclic orientation $\{v_1, v_3\} \overrightarrow{} \{v_2, v_4\}$. $\qquad \diamond$

The example illustrates how acyclic orientations of a combinatorial graph Y fixed by a group G induce acyclic orientations of the orbit graph $G \backslash Y$ in a natural way. Let

$$\pi_G \colon Y \longrightarrow G \backslash Y, \quad v_j \mapsto G(v_j)$$

be the canonical projection. The map π_g is locally surjective, that is, for any vertex v_j of Y the restriction map

$$\pi_G|_{\mathsf{Star}_Y(v_j)} \colon \mathsf{Star}_Y(v_j) \longrightarrow \mathsf{Star}_{G \backslash Y}(G(v_j))$$

is surjective.

Theorem 3.21. *Let Y be a combinatorial graph acted upon by G. Then*
(a) If $G \backslash Y$ contains at least one loop, then $\mathsf{Acyc}(Y)^G = \varnothing$.
(b) If $G \backslash Y$ is loop-free, then we have the bijection

$$\beta \colon \mathsf{Acyc}(Y)^G \longrightarrow \mathsf{Acyc}(G \backslash Y), \quad \mathcal{O} \mapsto \mathcal{O}_G, \qquad (3.28)$$

where $\mathcal{O}_G$ is given by

$$\forall e \in \mathrm{e}[G \backslash Y]; \ \{\omega(e), \tau(e)\} = \{G(v_i), G(v_k)\}, \quad \mathcal{O}_G(e) = G(\mathcal{O}(\{v_i, v_k\})) .$$

Proof. We first note that since Y is combinatorial its orbit graph $G \setminus Y$ is undirected.

Ad (a): Suppose $G \setminus Y$ contains a loop. Then there exists a geometric edge $\{v_i, v_k\}$ such that $g'v_k = v_i$ for some $g' \in G$. We consider the subgraph X of Y with

$$e[X] = \{\{gv_i, gv_k\} \in e[Y] \mid g \in G\},$$
$$v[X] = \{v_j \in v[Y] \mid \exists v_s \in v[Y];\ \{v_j, v_s\} \in G(\{v_i, v_k\})\}\,.$$

Any acyclic orientation $\mathcal{O}$ of Y induces by restriction an acyclic orientation $\mathcal{O}'$ of X. Suppose there exists some $\mathcal{O} \in \mathsf{Acyc}(Y)^G$, i.e., $g\mathcal{O}(\{v_i, v_k\}) = \mathcal{O}(\{gv_i, gv_k\})$. Without loss of generality we can assume that v_i is an origin of the induced acyclic orientation $\mathcal{O}'$ and in particular $\mathcal{O}'(\{v_i, v_k\}) = (v_i, v_k)$. By construction, $\{g'v_i, g'v_k\}$ (note that $g\{v_i, v_k\} = \{gv_i, gv_k\}$) is a geometric X-edge, and we obtain

$$g'\mathcal{O}(\{v_i, v_k\}) = (g'v_i, g'v_k) = \mathcal{O}(\{g'v_i, g'v_k\}),$$

which contradicts the fact that v_i is an $\mathcal{O}'$-origin. Thus, we have shown that if $G \setminus Y$ contains a loop, then $\mathcal{O} \in \mathsf{Acyc}(Y)^G = \varnothing$.

Ad (b): By (a) we can assume without loss of generality that $G \setminus Y$ is loop-free. Suppose we are given some $\mathcal{O} \in \mathsf{Acyc}(Y)^G$ and that $G \setminus Y$ contains a subgraph of the form

$$Z = \quad G(v_i) \overset{\frown}{} G(v_k)\ .$$

The graph Z is the π_G-image of the subgraph X of Y given by

$$e[X] = \{\{v_r, v_t\} \in e[Y] \mid \{v_r, v_t\} \in G(\{v_i, v_k\}) \cup G(\{v_i, v_s\})\},$$
$$v[X] = \{v_j \in v[Y] \mid \exists v_s \in v[Y],\ \{v_j, v_s\} \in e[X]\}\,.$$

By construction, $\mathcal{O}$ induces a unique orientation on all orbits $G(\{v_i, v_k\})$, $\{v_i, v_k\} \in e[Y]$ [since $\mathcal{O}(\{gv_i, gv_k\}) = g\mathcal{O}(\{v_i, v_k\})$] and accordingly an orientation of Z.

Claim 1. Any $\mathcal{O} \in \mathsf{Acyc}(Y)^G$ induces exactly one of the following two acyclic orientations of Z:

$$G(v_i) \overset{\longrightarrow}{} G(v_k) \qquad G(v_i) \overset{\longleftarrow}{} G(v_k)\ . \qquad (3.29)$$

We prove the claim by contradiction. The orientation $\mathcal{O}$ induces by restriction the acyclic orientation $\mathcal{O}'$ of X. We consider

$$\pi_G|_X : X \longrightarrow G(v_i) \overset{\frown}{} G(v_k)\ .$$

If $\mathcal{O}'$ induces the orientation $\mathcal{O}_1 = G(v_i) \overset{\longleftarrow}{} G(v_k)$, then no vertex of X can be an $\mathcal{O}'$-origin since $\pi_G|_X$ is locally surjective and $G(v_i) \overset{\longleftarrow}{} G(v_k)$ is by assumption induced by $\mathcal{O}'$. This contradicts the fact that $\mathcal{O}'$ is an acyclic orientation of X and the claim follows.

According to Claim 1, we can conclude that $\mathcal{O} \in \mathsf{Acyc}(Y)^G$ induces an orientation $\mathcal{O}_G$ of $G \setminus Y$ in which all multiple edges are unidirectional.

Claim 2. We have the bijection

$$\beta \colon \mathsf{Acyc}(Y)^G \longrightarrow \mathsf{Acyc}(G \setminus Y), \quad \mathcal{O} \mapsto \mathcal{O}_G,$$

where

$$\forall\, e \in \mathsf{e}[G \setminus Y]; \ \{\omega(e), \tau(e)\} = \{G(v_i), G(v_k)\}, \quad \mathcal{O}_G(e) = G(\mathcal{O}(\{v_i, v_k\})).$$

We prove that $\mathcal{O}_G$ is acyclic by contradiction. Suppose there exists a (directed) cycle in $\mathcal{O}_G$. Then there exists a subgraph C of Y given by

$$C = (G(v_{i_1}), e_1, \ldots, e_{j-1}, G(v_{i_j}), e_j)$$

with the property

$$\mathcal{O}_G(e_r) = \begin{cases} G(\mathcal{O}(\{v_{i_r}, v_{i_{r+1}}\})) = (G(v_{i_r}), G(v_{i_{r+1}})) & \text{for } r < j, \\ G(\mathcal{O}(\{v_{i_j}, v_{i_1}\})) = (G(v_{i_j}), G(v_{i_1})) & \text{else.} \end{cases} \tag{3.30}$$

We consider C as a subgraph of $G \setminus Y$ and introduce the subgraph P of Y being the preimage of C under π_G:

$$\mathsf{e}[P] = \{\{v_r, v_s\} \in \mathsf{e}[Y] \mid G(\{v_r, v_s\}) \in \{e_h \mid h = 1, \ldots j\}\},$$
$$\mathsf{v}[P] = \{v_j \mid \exists\, v_s \in \mathsf{v}[Y] \,; \{v_j, v_s\} \in \mathsf{e}[P]\}.$$

The orientation $\mathcal{O}$ induces by restriction the acyclic orientation $\mathcal{O}_P$ of the subgraph P. Since $\pi_G|_P \colon P \longrightarrow C$ is locally surjective and

$$G(\mathcal{O}(\{v_{i_r}, v_{i_{r+1}}\})) = (G(v_{i_r}), G(v_{i_{r+1}})),$$

no vertex of P can be an $\mathcal{O}_P$-origin, which is impossible; hence, $\mathcal{O}_G$ is acyclic. This proves that β is well-defined. The map β is bijective since each $\mathcal{O} \in \mathsf{Acyc}(Y)^G$ is completely determined by its values on representatives of the edge-orbits $G(\{v_r, v_s\})$. Therefore, $\mathcal{O} \mapsto \mathcal{O}_G$ is a bijection, hence Claim 2, and the proof of the theorem is complete. $\square$

Example 3.22. As an illustration of Claim 1 we show under which conditions an orientation of the form $G(v_i) \overset{\longleftarrow}{\underset{\longrightarrow}{}} G(v_k)$ is induced. If

$$Y = \begin{array}{ccc} v_1 & \!\!\!\!\text{------}\!\!\!\! & v_2 \\ | & & | \\ v_4 & \!\!\!\!\text{------}\!\!\!\! & v_3 \end{array}, \qquad \mathcal{O} = \begin{array}{ccc} v_1 & \!\!\!\!\longrightarrow\!\!\!\! & v_2 \\ \uparrow & & \downarrow \\ v_4 & \!\!\!\!\longleftarrow\!\!\!\! & v_3 \end{array},$$

then $\mathcal{O}$ is fixed by $\langle g = (v_1, v_3)(v_2, v_4) \rangle$

$$g\mathcal{O}(\{v_1, v_2\}) = (v_3, v_4) = \mathcal{O}(\{v_3, v_4\}) = \mathcal{O}(\{gv_1, gv_2\}),$$
$$g\mathcal{O}(\{v_1, v_4\}) = (v_2, v_3) = \mathcal{O}(\{v_3, v_2\}) = \mathcal{O}(\{gv_1, gv_4\}),$$

and $\mathcal{O}$ induces the orientation $\{v_1, v_3\}$ ⟶ $\{v_2, v_4\}$. ⟵   ◇

An immediate consequence of Proposition 3.21 is the objective of this section: a combinatorial interpretation for the terms $\mathsf{Fix}(g)$ in the Frobenius lemma.

Corollary 3.23. *Let Y be a combinatorial graph acted upon by G. Then we have*

$$N = \frac{1}{|G|} \sum_{g \in G} |\mathsf{Acyc}(\langle g \rangle \backslash Y)| . \tag{3.31}$$

Example 3.24. As an illustration of the counting result (3.31), we compute N for $Y = \mathsf{Circ}_4$ and $Y = \mathsf{Circ}_5$. First we note that any element $\gamma \in \mathsf{Aut}(Y)$ such that a $\langle \gamma \rangle$-orbit contains adjacent Y-vertices does not contribute to the sum since the corresponding orbit graph will have a loop and hence does not allow for any acyclic orientations by Theorem 3.21. The automorphism group of Circ_n is the dihedral group D_n with $2n$ elements. For Circ_5 it is clear that the identity permutation id is the only automorphism that induces loop-free orbit graphs. Since $\langle \mathsf{id} \rangle \backslash Y$ is isomorphic to Y, we derive

$$N(\mathsf{Circ}_5) = \frac{1}{10}(a(\mathsf{Circ}_5)) = \frac{1}{10}(32 - 2) = 3 .$$

For Circ_4 we leave it to the reader to verify that the only automorphisms that contribute to the sum in (3.31) are id, $(0, 2)(1, 3)$, $(0)(1, 3)(2)$, and $(1)(0, 2)(3)$ and their respective orbit graphs are isomorphic to Circ_4, Line_3, Line_3, and $\bullet \frown \bullet$. Accordingly we obtain

$$N(\mathsf{Circ}_4) = \frac{1}{8}((16 - 2) + 2^2 + 2^2 + 2^1) = 3 .$$

In Chapter 4 we will show that the number N represents an upper bound for the number of dynamically nonequivalent SDS we can generate by varying the permutation update order while keeping the graph and the functions fixed. ◇

3.3 Dynamical Systems

Classical dynamical system theory is concerned with how the state of a system evolves as a function of one or more underlying variables. For the purposes of this section we will always assume that the underlying variable is time.

There are two main classes of classical dynamical systems: continuous systems where the time evolution is governed by a system of ordinary differential equations (ODEs) of the form

$$\frac{dx}{dt} = f(x), \quad x \in E \subset \mathbb{R}^n,$$

and discrete systems whose time evolution results from iterating a map

$$F : \mathbb{R}^n \longrightarrow \mathbb{R}^n .$$

We can, of course, consider more general state spaces, but we will restrict ourselves to $\mathbb{R}^n$ in the following.

Let us now describe the two main classes of dynamical systems and give basic terminology and definitions. Continuous and discrete systems differ in some significant ways. To be able to speak about time evolution of the continuous system we need to know that the ODE actually has a solution. If it has a solution, it would also be convenient to know if such a solution is unique. For a discrete system this is not a primary concern — the dynamics is obtained by iterating the map F.

In light of this, we start by presenting conditions for existence and uniqueness of solutions for systems of ODEs. We will then present a selection of theorems for both continuous and discrete dynamical systems. In addition to giving definitions and background information, the purpose of this is to illustrate differences between the classical systems and discrete, finite dynamical systems such as sequential dynamical systems (SDS), which is the main topic of this book. As we will see later, the differences manifest themselves in tools and analysis techniques and also in the nature of the questions that are being posed. In contrast to the combinatorial and algebraic techniques used to study sequential dynamical systems, the techniques used for classical dynamical systems tend to rely on continuity and differentiability.[4]

3.3.1 Classical Continuous Dynamical Systems

The classical continuous dynamical systems appear in the context of systems of ordinary differential equations of the form

$$x' = F(x), \quad x \in E , \tag{3.32}$$

where E is some open subset of $\mathbb{R}^n$ and $F: E \longrightarrow \mathbb{R}^n$ is a *vector field* on E. Unless otherwise stated, we will assume that F is at least continuously differentiable on E, which we write as $F \in C^1(E)$, or *smooth* (infinitely differentiable), which we write as $F \in C^\infty(E)$.

---

[4] Of course, algebraic theory and combinatorial theory play an important part in classical dynamical systems when analyzed through, for example, symbolic dynamics.

The vector field F gives rise to a flow $\varphi_t \colon E \longrightarrow \mathbb{R}^n$, where $\varphi_t(x) = \varphi(t, x)$ is a smooth function defined for all $x \in E$ and all $t \in I = (a, b)$ with $a < 0 < b$. The flow satisfies (3.32), that is,

$$\frac{d}{dt}\varphi(x, t)|_{t=t'} = F(\varphi(x, t')) \quad \text{for all } x \in E, \ t' \in I \ .$$

For $x \in E$ and $s, t, s + t \in I$, the flow has the properties [71]

$$\varphi_0(x) = x \quad \text{and} \quad \varphi_{t+s}(x) = \varphi_t(\varphi_s(x)) \ .$$

The system (3.32) is often augmented by an *initial condition*

$$x(0) = x_0 \in E \ .$$

In this case the solution of (3.32) — if it exists (actually it does, but more on that below) — is the map $x(t) = \varphi(x_0, t)$ satisfying $x(0) = x_0$. The map $x(t)$ defines an *orbit* or solution curve of the system (3.32) that passes through x_0. The geometric interpretation of a solution curve is as a curve in $\mathbb{R}^n$ that is everywhere tangential to F, that is, $x'(t) = F(x(t))$. The collection of all solution curves of (3.32) is the *phase space*. The image of the phase space is the *phase portrait*. Locally, the phase space and the phase portrait are given by flow maps.

Example 3.25. On the left in Figure 3.7 we have shown some of the solution curves for the two-dimensional system

$$x' = x^2 + xy, \quad y' = \tfrac{1}{2}y^2 + xy \ .$$

On the right we have shown some of the solution curves for the Hamiltonian system (see, e.g., [72])

$$x' = y, \quad y' = x + x^2 \ .$$

$\diamond$

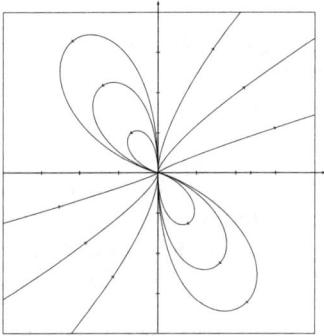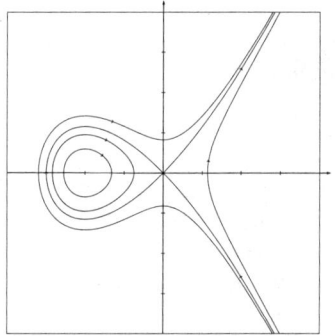

Fig. 3.7. Solution curves for the systems in Example 3.25.

It is not obvious that (3.32) has a solution or, if it does, that such a solution is unique. The following theorem summarizes the basic facts on these questions:

Theorem 3.26. *Let E be an open subset of $\mathbb{R}^n$, let $x_0 \in E$, and assume that $F \in C^1(E)$. Then*
   (i)   *There exists $a > 0$ such that the initial-value problem given by (3.32) and $x(0) = x_0$ has a unique solution for $t \in (-a, a)$.*
   (ii) *There exists a maximal open interval (α, β) for which the solution is unique.*

The standard proof of these statements is based on Picard's method and Banach's fixed-point theorem. The interested reader is referred to, e.g., [72,73]. To be fair, we should state that the condition F be continuously differentiable is somewhat stronger than what is required. It is enough that F is *locally Lipschitz* on E, i.e.,

$$|f(x) - f(y)| \le K\,|x - y|$$

for all x, y in some sufficiently small open subset of E, and where K is some finite constant (the Lipschitz constant).

So where are the dynamical systems? So far there have only been systems of ordinary differential equations and flows.

Definition 3.27 (Dynamical system). Let E be an open subset of $\mathbb{R}^n$. A dynamical system is a C^1 map satisfying

1. $\varphi(0, x) = x$ for all $x \in E$ and
2. $\varphi(t, \varphi(s, x)) = \varphi(t + s, x)$ for all $s, t \in \mathbb{R}$ and all $x \in E$.

As for flows we often write $\phi(t, x)$ as $\phi_t(x)$. It is clear that

$$F(x) = \frac{d}{dt}\varphi(t, x)\,|_{t=0}$$

defines a C^1 vector field on E and that for all $x_0 \in E$ the map $\varphi(t, x_0)$ is a solution to the initial-value problem

$$x' = F(x), \quad x(0) = x_0 \;.$$

The converse does not hold since the flow of (3.32) is generally only defined on some finite interval I and not $\mathbb{R}$. The interested reader may look up the "global existence theorem" in [72] for a way to remedy this.

3.3.2 Classical Discrete Dynamical Systems

The classical discrete dynamical systems arise from iterates of a map

$$F \colon \mathbb{R}^n \longrightarrow \mathbb{R}^n \;, \tag{3.33}$$

which is typically assumed to be continuous. Starting from an initial state x_0 we get the *forward orbit* of x_0 denoted by $O^+(x_0)$ as the sequence of points x_0, $F(x_0)$, $F^2(x_0)$, $F^3(x_0), \ldots$, that is, $O^+(x_0) = (F^k(x_0))_{k=0}^\infty$. Here $F^k(x_0)$ denotes the *k-fold composition* defined by $F^0(x_0) = x_0$ and $F^k(x_0) = F(F^{k-1}(x_0))$. If F is a homeomorphism, which means that F is continuous with a continuous inverse, we define the *backward orbit* $O^-(x_0) = (F^k(x_0))_{k=0}^{-\infty}$ and the *full orbit* as $O(x_0) = (F^k(x_0))_{k=-\infty}^\infty$.

The concept of flow is in this case captured directly in terms of the map F. If F is a homeomorphism, we define the corresponding flow as

$$\phi \colon \mathbb{R}^n \times \mathbb{Z} \longrightarrow \mathbb{R}^n, \quad \phi(x,t) = \phi_t(x) = F^t(x) . \tag{3.34}$$

Again the *phase space* of the dynamical system induced by F is the collection of all orbits.

Example 3.28. The map $F \colon \mathbb{R}^2 \longrightarrow \mathbb{R}^2$ given by

$$F(x,y) = \begin{bmatrix} a - by - x^2 \\ x \end{bmatrix} \tag{3.35}$$

is the Hénon map. It is a much-studied two-dimensional map [74] exhibiting many of the properties typically associated with chaotic dynamical systems. A part of its orbit starting at $(0,0)$ is shown in Figure 3.8. It is an approximation of its "strange attractor." ◇

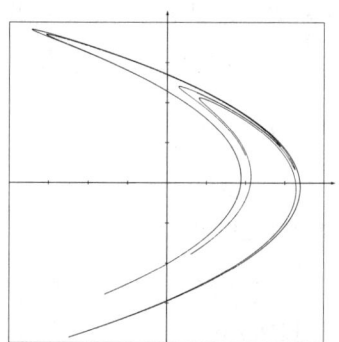

Fig. 3.8. An orbit of the Hénon map of Example 3.28.

The goal of dynamical system theory is to understand as much as possible about the orbits of (3.32) and (3.33). In practice, certain states, orbits, and phase-space features have received more research attention than others. Classical examples include fixed points, periodic points, and limit cycles.

Definition 3.29 (Fixed points, periodic points).
(i) The state x_0 of (3.32) or (3.33) is a *fixed point* if for all t we have $\phi(x_0, t) = x_0$. The set of fixed points of ϕ is denoted $\mathsf{Fix}(\phi)$.
(ii) The state x_0 is a *periodic point* if there exists $0 < t_0 < \infty$ such that $\phi(x_0, t_0) = x_0$. The smallest such value t_0 is the *prime period* of x_0. If x_0 is periodic, then the set $\Gamma(x_0) = \{\phi(x_0, t)|t\}$ is the *periodic orbit* containing x_0. The set of all periodic points of ϕ is denoted $\mathsf{Per}(\phi)$.

Fixed points and periodic orbits are examples of *limit sets*. More generally, a point p is an ω-limit point of x if there exists a sequence $(\phi_{t_i}(x))_i$ such that $\phi_{t_i}(x) \longrightarrow p$ and $t_i \to \infty$. The set of all ω-limit points of an orbit Γ is denoted $\omega(\Gamma)$. The notion of α-limit points is analogous, the only difference being that $t_i \to -\infty$. The set $\omega(\Gamma) \cup \alpha(\Gamma)$ is the *limit set* of Γ. Thus, a periodic orbit is its own ω-limit set and α-limit set.

A subset E of $\mathbb{R}^n$ is a *forward invariant set* (*backward invariant set*) if for all $x \in E$ we have $\phi_t(x) \in E$ for $t \geq 0$ ($t \leq 0$).

The notion of invariant sets naturally extends to sequential dynamical systems. This is not the case for the concept of *stability*. We say that a periodic orbit Γ of (3.32) is stable if for each $\varepsilon > 0$ there exists a neighborhood U of Γ such that for all $x \in U$ the distance[5] $d(\phi_t(x), \Gamma) < \varepsilon$ for all $t > 0$. If we additionally have $\lim_{t\to\infty} d(\phi_t(x), \Gamma) = 0$, then Γ is *asymptotically stable*. An asymptotically stable periodic orbit is often referred to as a *limit cycle*. Asymptotically stable fixed points are defined in the same manner although they could, of course, be viewed as a special case of a periodic orbit.

3.3.3 Linear and Nonlinear Systems

Whenever the right-hand side in (3.32) or the map (3.33) is a linear function, we refer to the system as *linear*. A system that is not linear is *nonlinear*. Using matrix notation, linear systems of the form (3.32) and (3.33) can be written as

$$\frac{dx}{dt} = Ax \tag{3.36}$$

and

$$F(x) = Ax \ .$$

These systems are well-understood. An extensive account of the continuous linear systems is given in [71]. For a description of linear maps over finite fields, see [33, 50].

Of course, interesting systems are usually nonlinear, so a natural question is why one should study linear systems. One reason is the celebrated Hartman–Grobman theorem, which states that, subject to rather mild conditions, a nonlinear system can locally be represented by a linear system — the two systems are locally equivalent. However, before we present the details we first need to clarify what we mean by equivalence.

---

[5] For definitions see, for example, [72, 73].

Definition 3.30 (Topological equivalence). Two maps $F, G\colon \mathbb{R}^n \longrightarrow \mathbb{R}^n$ are topologically equivalent if there exists a homeomorphism $h\colon \mathbb{R}^n \longrightarrow \mathbb{R}^n$ such that

$$G \circ h = h \circ F. \tag{3.37}$$

We close this chapter with the Hartman–Grobman theorem stated for discrete dynamical systems.

Theorem 3.31 (Hartman–Grobman). *Let $F\colon \mathbb{R}^n \longrightarrow \mathbb{R}^n$ be a C^1 map, and let x_0 be a fixed point of F such that the Jacobian $DF(x_0)$ has no eigenvalues of absolute value 1. Then there exists a homeomorphism h defined on some neighborhood U of x_0 such that for all $x \in U$*

$$h \circ F = DF(x_0) \circ h.$$

In other words, under the condition of the theorem the phase space of the linear system and that of the nonlinear system are equivalent in some neighborhood U of x_0. A standard application of the Hartman–Grobman theorem is to determine stability properties of fixed points. The problems at the end of this chapter elaborates some more on these concepts and the use of Theorem 3.31.

In Chapter 4 we will address the same question of equivalence in the context of sequential dynamical systems. As will become clear, the lack of continuity and derivatives will make things a lot different.

References

The following is a list of references for the material presented in this chapter that can be used for further study.

Algebra. There are many good introductory books to this area. Examples include the books by Fraleigh [75] and Bhattacharya [76], where the latter is somewhat more advanced. The books by Jacobson [77] and Hungerford [78] are classical texts, but they are typically considered more demanding. Van der Waerden's two volumes [79, 80] based on the lectures of E. Artin and E. Noether are highly recommended.

Combinatorics and Graph Theory. It can be hard to find good texts on graph theory. Although written for an entirely different purpose, Serre's book on trees [66] contains an excellent section on graphs acted upon by groups. Dicks' book [81] is another nice reference on graphs and groups. Diestel's book [82] and Godsil and Royle's book [83] are good choices. In combinatorics many like Riordan's book [84]. We have not used this book, but we can recommend van Lint and Wilson's book [85]. Stanley's book [21] is a demanding but excellent introductory combinatorics text that you should open at least once.

Dynamical Systems. For continuous dynamical systems, Hirsch and Smale's book [71] is a classic that we recommend. The book by Perko [72] provides an alternative introduction to continuous dynamical systems. These two books provide the necessary background for more advanced texts like the ones by Guckenheimer and Holmes [86] and Coddington and Levinson [87]. Devaney's book [88] provides an introduction to discrete dynamical system, and the work on one-dimensional dynamics presented by de Melo and van Strien [89] can serve as an advanced followup text.

Problems

3.8. Compute $a(\mathsf{Wheel}_n)$.   [1+]

3.9. Compute $a(Q_2^3)$.   [2-]

3.10. Characterize $U(K_n)$ and $U(E_n)$, where E_n is the empty graph on n vertices.   [2-]

3.11. Show that different solution curves of (3.32) cannot cross. Can a solution curve of (3.32) cross itself?   [2]

3.12. The *logistic map* is the map $F_\mu \colon \mathbb{R} \longrightarrow \mathbb{R}$ given by

$$F_\mu(x) = \mu x(1 - x) , \tag{3.38}$$

with $\mu > 0$. It is also referred to as the *quadratic family*. Depending on the value of μ, the associated discrete dynamical system can exhibit fascinating dynamics; see, e.g., [88]. In this problem we will see how to use Theorem 3.31 to study the stability properties of this dynamical system near its fixed points.

Show that the dynamical system has fixed points $x_0 = 0$ and $x_\mu = 1 - 1/\mu$. The linearization of the dynamical system at x_0 is given by

$$x_{n+1} = \Big(\frac{dF}{dx}|_{x=0}\Big)x = \mu x . \tag{3.39}$$

Use Theorem 3.31 and the linear system (3.39) to discuss the behavior of the nonlinear dynamical system determined by F_μ around $x = 0$ as a function of μ. What is $\frac{dF_\mu}{dx}|_{x=x_\mu}$? Use this to show that x_μ is an attracting fixed point for $1 < \mu < 3$.   [2]

3.13. In this problem we will see how to apply Theorem 3.31 to the two-dimensional discrete dynamical system from Example 3.28 (the Hénon map). Recall that the map is given by (3.35):

$$F(x, y) = \begin{bmatrix} a - by - x^2 \\ x \end{bmatrix},$$

with $F\colon \mathbb{R}^2 \longrightarrow \mathbb{R}^2$ and $a, b > 0$. What are the fixed points of this system? The linearization of this map at (x_0, y_0) is given by

$$G(x, y) = J(x_0, y_0) \begin{bmatrix} x \\ y \end{bmatrix} = \begin{bmatrix} -2x_0 & -b \\ 1 & 0 \end{bmatrix} \begin{bmatrix} x \\ y \end{bmatrix} . \tag{3.40}$$

What are the eigenvalues of the matrix J in this case? Use this to determine the stability properties of the fixed points for the original Hénon map as a function of a and b. **[2]**

3.14. This problem illustrates the use of the Hartman–Grobman theorem for two-dimensional continuous systems. We will elaborate on Example 3.25 and consider the dynamical system given by

$$x' = f(x, y) = y, \quad y' = g(x, y) = x + x^2 . \tag{3.41}$$

An equilibrium point for this system is a point (x_0, y_0) where f and g are simultaneously zero. What are the equilibrium points for (3.41)?

The linearization of (3.41) at a point (x_0, y_0) is

$$\begin{bmatrix} x' \\ y' \end{bmatrix} = J(x_0, y_0) = \begin{bmatrix} \frac{\partial f}{\partial x} & \frac{\partial f}{\partial y} \\ \frac{\partial g}{\partial x} & \frac{\partial g}{\partial y} \end{bmatrix}_{(x_0, y_0)} . \tag{3.42}$$

What is the Jacobian matrix J of (3.41) at a general point (x, y)? Compute its value for the two equilibrium points you just found.

By an extension of Theorem 3.31 (see [71]) to the flow map of (3.41) we have that the nonlinear system and its linearization at a point (x_0, y_0) are topologically equivalent in a neighborhood of (x_0, y) if the matrix $J(x_0, y_0)$ has no eigenvalues where the real part is zero. Find the eigenvalues of the Jacobian matrix for both equilibrium points.

The linear system can be diagonalized. Use this to determine the stability properties of the equilibrium point $(0, 0)$. **[2]**

3.15. Consider the system of ordinary differential equations given by

$$\begin{aligned} x' &= -2y + yz, \\ y' &= 2x - 2xz, \\ z' &= xy . \end{aligned} \tag{3.43}$$

It is clear that $(0, 0, 0)$ is an equilibrium point of the dynamical system, but since

$$J(0, 0, 0) = \begin{bmatrix} 0 & -2 & 0 \\ 2 & 0 & 0 \\ 0 & 0 & 0 \end{bmatrix} \tag{3.44}$$

has eigenvalues 0 and $\pm 2i$, we cannot apply the extension of Theorem 3.31 as in the previous problem.

Let F be the vector field associated with (3.43), and define the function $V \colon \mathbb{R}^3 \longrightarrow \mathbb{R}$ by $V(x, y, z) = x^2 + y^2 + z^2$. A key observation is that $V(x, y, z) > 0$ for $(x, y, z) \neq (0, 0, 0)$ and $V(0, 0, 0) = 0$. Moreover, the inner product

$$\dot{V} = \operatorname{grad} V \cdot F = 2x(-2y + yz) + 2y(2x - 2xz) + 2z(xy) = 0 \,.$$

What can you conclude from this? The function V is an example of a *Liapunov function*. [2]

Answers to Problems

3.2. Proof of Lemma 3.2. Suppose we are given the two edges $e, \bar{e}$, where $v = \omega(e)$, i.e., 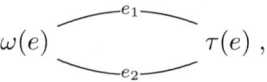 v. Then we have $e \mapsto (v, v)$ and $\bar{e} \mapsto (v, v)$, and Y is not combinatorial. For a cycle of length 2,

$$\omega(e) \overset{\overset{e_1}{\frown}}{\underset{e_2}{\smile}} \tau(e) ,$$

we have the two different edges e_1, e_2 such that $e_1 \mapsto (\omega(e_1), \tau(e_1))$ and $e_2 \mapsto (\omega(e_2), \tau(e_2))$, i.e., Y is not combinatorial. Hence, if Y is combinatorial, it contains no cycle of length ≤ 2. Suppose Y contains no cycle of length ≤ 2. Then Y cannot contain multiple edges and has no loops, from which it follows that $\omega \times \tau : e[Y] \longrightarrow v[Y] \times v[Y]$ is injective.

3.5. The relation (3.18) can be proved as follows: Consider an acyclic orientation $\mathcal{O}$ of Y'. We observe that $\mathcal{O}$ induces at least one and at most two acyclic orientations of Y. In the case it induces two acyclic orientations we can conclude that it induces one acyclic orientation of Y'', and Eq. (3.18) follows.

3.6. First we observe that G_v acts on $\mathsf{Star}_Y(v)$ and consider the map

$$f : G_v \setminus \mathsf{Star}_Y(v) \longrightarrow \mathsf{Star}_{G \setminus Y}(G(v)), \quad G_v(v') \mapsto G(v') .$$

By construction, f is a surjective graph morphism. We show that f is injective. Let e and e' be two edges of Y with $\omega(e) = \omega(e') = v$, and suppose $G(e) = G(e')$. Then there exists some $g \in G$ such that $ge = e'$ holds. We obtain $\omega(ge) = g\omega(e) = \omega(e') = v$, and as a result $gv = v$, i.e., $g \in G_v$. The case of two edges e, e' with $\tau(e) = \tau(e') = v$ is completely analogous. Hence, we have proved the following: For any two edges e, e' of $\mathsf{Star}_Y(v)$, $G(e) = G(e')$ implies $G_v(e) = G_v(e')$; hence, f is injective.

3.7. An example is $Y = \mathsf{Circ}_4$ with G the subgroup of $\mathsf{Aut}(Y)$ generated by $(0, 2)(1, 3)$ (cycle form).

3.8. $3^n - 3$.

3.9. 1862.

3.11. Solution curves cannot cross — this would violate the uniqueness of solution property.

3.12. By, for example, Banach's fixed-point theorem [73] we see that the fixed point $x_0 = 0$ is an attracting fixed point for $0 < \mu < 1$. It is a repelling fixed point for $\mu > 1$. One can also show that it is an attracting fixed point for $\mu = 1$, but Theorem 3.31 does not apply in this situation.

Here $\frac{dF_\mu}{dx}|_{x=x_\mu} = 2 - \mu$. For $1 < \mu < 3$ we have $-1 < 2 - \mu < 1$, and by Banach's fixed-point theorem, it follows that x_μ is an attracting fixed point in this parameter range.

3.13. Solving the equation for the fixed points gives

$$x_0 = y_0 = (-(1+b) \pm \sqrt{(1+b)^2 + 4a})/2 \ .$$

Since $a > 0$ there are two fixed points. You may want to refer to [88] for more information on the Hénon map.

3.14. Here $f(x, y) = y$ and $g(x, y) = x + x^2$ are simultaneously zero at $(0, 0)$ and $(-1, 0)$. The Jacobian matrix of the system is

$$J(x, y) = \begin{bmatrix} 0 & 1 \\ 1 + 2x & 0 \end{bmatrix} . \tag{3.45}$$

Here $J(0, 0) = \begin{bmatrix} 0 & 1 \\ 1 & 0 \end{bmatrix}$ and $J(-1, 0) = \begin{bmatrix} 0 & 1 \\ -1 & 0 \end{bmatrix}$. The matrix $J(0, 0)$ has eigenvalues $\lambda = -1$ and $\lambda = 1$ and $J(0, 0)$ has eigenvalues $\lambda = -i$ and $\lambda = i$. The point $(0, 0)$ is therefore an unstable equilibrium point for (3.41). It is an example of a *saddle point*, which is also suggested by Figure 3.7. We cannot apply the Hartman–Grobman theorem to the point $(-1, 0)$, but a symmetry argument can be used to conclude that this is a *center*.

3.15. Since $\dot{V} = 0$, the solution curves to the system of ordinary differential equations are tangential to the level surfaces of the function V. The origin is a stable equilibrium point for this system. If we had $\dot{V} < 0$ for $(x, y, z) \neq 0$, we could have concluded that the origin would also be asymptotically stable and thus an attracting equilibrium point. See, for example, [71].

Sequential Dynamical Systems over Permutations

In this chapter we will give the formal definition of sequential dynamical systems (SDS). We will study SDS where the update order is a permutation of the vertex set of the underlying graph. In Chapter 7 we will extend our analysis to update orders that are words over the vertex set, that is, systems where vertices can be updated multiple times within a system update. Since most graphs in this chapter are combinatorial graphs (Section 3.1.1), we will, by abuse of terminology, refer to combinatorial graphs simply as graphs unless ambiguity may arise.

4.1 Definitions and Terminology

4.1.1 States, Vertex Functions, and Local Maps

Let Y be a (combinatorial) graph with vertex set $v[Y] = \{v_1, \ldots, v_n\}$ and let $d(v)$ denote the degree of vertex v. We can order the vertices of $B_Y(v)$ using the natural order of their indices, i.e., we set $v_j < v_k$ if and only if $j < k$ and consequently obtain the $(d(v) + 1)$-tuple

$$\left(v_{j_1}, \ldots, v_{j_{d(v)+1}}\right).$$

We can represent the $(d(v) + 1)$-tuple $(v_{j_1}, \ldots, v_{j_{d(v)+1}})$ via the map

$$n[v] \colon \{1, 2, \ldots, d(v) + 1\} \longrightarrow v[Y], \quad i \mapsto v_{j_i}. \tag{4.1}$$

For instance, if vertex v_2 has neighbors v_1 and v_5, we obtain

$$n[v_2] = (n[v_2](1), n[v_2](2), n[v_2](3)) = (v_1, v_2, v_5).$$

We let K denote a finite set and assign a *vertex state* $x_v \in K$ to each vertex $v \in v[Y]$. In many cases we will assume that K has the structure of a finite field. For $K = \mathbb{F}_2$ we refer to states as binary states. The choice of binary states of course represents the minimal number of states we can have, but it is also a common choice in, for example, the study of cellular automata.

The n-tuple of vertex states $(x_{v_1}, \ldots, x_{v_n})$ is called a *system state*. We will use x, y, z, and so on to denote system states. When it is clear from the context whether we mean vertex state or system state, we may omit "vertex" or "system." The family of vertex states associated with the vertices in $B_Y(v)$ [Eq. (3.3)] induced by $n[v]$ is denoted $x[v]$, that is,

$$x[v] = (x_{n[v](1)}, \ldots, x_{n[v](d(v)+1)}) . \tag{4.2}$$

When necessary, we will reference the underlying graph Y explicitly and write $n[v; Y]$ and $x[v; Y]$, respectively. In analogy with our notation $B_Y(v)$ and $B'_Y(v)$ [Eqs. (3.3) and (3.4)], we will write $n'[v; Y]$ and $x'[v; Y]$ for the corresponding tuples in which v and x_v are omitted, i.e.,

$$n'[v; Y] = (v_{j_1}, \ldots, \hat{v}, \ldots, v_{j_{d(v)+1}}), \tag{4.3}$$

$$x'[v; Y] = (x_{n[v](1)}, \ldots, \hat{x}_v, \ldots, x_{n[v](d(v)+1)}), \tag{4.4}$$

where $\hat{v}, \hat{x}_v$ means that the corresponding coordinate is omitted.

Example 4.1. Let $Y = \mathsf{Circ}_4$, which has vertex set $v[\mathsf{Circ}_4] = \{0, 1, 2, 3\}$ and edges as shown in Figure 4.1. In this case we simply use the natural order on

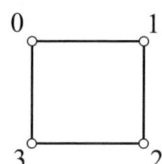

Fig. 4.1. The graph Circ_4.

$v[Y]$ and obtain, for instance, $n[0] = (0, 1, 3)$ and $n[1] = (0, 1, 2)$.   ◇

For each vertex v of Y the *vertex function* is the map

$$f_v \colon K^{d(v)+1} \longrightarrow K .$$

We define the *local function* $F_v \colon K^n \longrightarrow K^n$ by

$$F_{v_i}(x) = (x_{v_1}, \ldots, x_{v_{i-1}}, f_{v_i}(x[v_i]), x_{v_{i+1}}, \ldots, x_{v_n}) . \tag{4.5}$$

Thus, F_{v_i} maps all variables $x_{v_j} \neq x_{v_i}$ identically, and the v_ith coordinate only depends on the variables x_{v_j} with $v_j \in B_Y(v_i)$. When we want to emphasize the graph Y, we refer to a local map as $F_{v,Y}$. Finally, we set $\mathbf{F}_Y = (F_v)_v$.

4.1.2 Sequential Dynamical Systems

As in Section 3.1.4, we let S_Y denote the symmetric group over the vertex set of Y. We will use Greek letters, e.g., π and σ, for the elements of S_Y. A permutation $\pi = (\pi_1, \ldots, \pi_n) \in S_Y$ naturally induces an order of the vertices in Y through $\pi_i < \pi_j$ if $i < j$.

Throughout this book we will use the term *family* to specify an *indexed set*. A family where the index set is the integers is a *sequence*.

Definition 4.2 (Sequential dynamical system). Let $Y = (\mathrm{v}[Y], \mathrm{e}[Y], \omega, \tau)$ be an undirected graph (Section 3.1), let $(f_v)_{v \in \mathrm{v}[Y]}$ be a family of vertex functions, and let $\pi \in S_Y$. The sequential dynamical system (SDS) is the triple $(Y, (F_v)_v, \pi)$. Its associated SDS-map is $[\mathbf{F}_Y, \pi] \colon K^n \longrightarrow K^n$ defined by

$$[\mathbf{F}_Y, \pi] = F_{\pi_n} \circ F_{\pi_{n-1}} \circ \cdots \circ F_{\pi_1} . \tag{4.6}$$

It is important to note that SDS are defined over undirected graphs and not over combinatorial graphs. The main reason for this is the concept of SDS morphisms, which involves graph morphisms. Graph morphisms generally do not map combinatorial graphs into combinatorial graphs (see Section 3.1.1). However, local maps are defined using the concept of adjacency, which is independent of the existence of multiple edges, and we therefore obtain

$$(Y, (F_v)_v, \pi) = (Y_c, (F_v)_v, \pi) .$$

Accordingly, we postulate Y to be undirected for technical reasons arising from the notion of SDS morphisms, and we may always replace Y by Y_c.

The graph Y of an SDS is referred to as the *base graph*. The application of the Y-local map F_v is the *update of vertex v*, and the application of $[\mathbf{F}_Y, \pi]$ is a *system update*. We will occasionally write $\prod_{v=\pi_1}^{\pi_n} F_v$ for the right-hand side of (4.6), where $\prod$ denotes the composition product of maps as in (4.6).

In Chapter 7, and in some propositions and problems, we also consider SDS where the update order is a *word* $w = (w_1, \ldots, w_k)$ over $\mathrm{v}[Y]$, that is, a sequence of Y-vertices. For future reference, we therefore define an SDS over a word w as the triple $(Y, (F_v)_v, w)$, where its associated SDS-map is $[\mathbf{F}_Y, w] \colon K^n \longrightarrow K^n$ defined by

$$[\mathbf{F}_Y, w] = F_{w_k} \circ F_{w_{k-1}} \circ \cdots \circ F_{w_1} . \tag{4.7}$$

In this context we use the terminology *permutation-SDS* and *word-SDS* to emphasize this point as appropriate.

Example 4.3. Continuing Example 4.1 with the graph $Y = \mathsf{Circ}_4$, we let each vertex function be the function $f \colon \mathbb{F}_2^3 \longrightarrow \mathbb{F}_2$ that returns the sum of its arguments in $\mathbb{F}_2$. Thus, x_1 is mapped to $f(x_0, x_1, x_2) = x_0 + x_1 + x_2$. The corresponding local map $F_1 \colon \mathbb{F}_2^4 \longrightarrow \mathbb{F}_2^4$ is given by

$$F_1(x_0, x_1, x_2, x_3) = (x_0, x_0 + x_1 + x_2, x_2, x_3) .$$

$\diamond$

Let K be a finite field. For a system state $x \in K^n$ we sometimes need to compute the sum of the vertex states in $\mathbb{N}$. Note that we include 0 in the natural numbers so that $\mathbb{N} = \{0, 1, 2, \dots\}$. This is done to distinguish this sum from sums taken in the respective finite field K. We set

$$\mathsf{sum}_l \colon K^l \longrightarrow \mathbb{N}, \quad \mathsf{sum}_l(x_1, \dots, x_l) = x_1 + \cdots + x_l \quad \text{(computed in } \mathbb{N}) . \quad (4.8)$$

Below is a list of vertex functions that will be used throughout the rest of the book. In these definitions we set $x = (x_1, \dots, x_k)$.

$$\mathsf{nor}_k \colon \mathbb{F}_2^k \longrightarrow \mathbb{F}_2, \quad \mathsf{nor}_k(x) = (1 + x_1) \cdots (1 + x_k) \quad (4.9)$$

$$\mathsf{nand}_k \colon \mathbb{F}_2^k \longrightarrow \mathbb{F}_2, \quad \mathsf{nand}_k(x) = 1 + x_1 \cdots x_k \quad (4.10)$$

$$\mathsf{parity} \colon \mathbb{F}_2^k \longrightarrow \mathbb{F}_2, \quad \mathsf{parity}_k(x) = x_1 + \cdots + x_k \quad (4.11)$$

$$\mathsf{or}_k \colon \mathbb{F}_2^k \longrightarrow \mathbb{F}_2, \quad \mathsf{or}_k(x) = \begin{cases} 1, & \mathsf{sum}_k(x) > 0 \\ 0, & \text{otherwise} \end{cases} \quad (4.12)$$

$$\mathsf{and}_k \colon \mathbb{F}_2^k \longrightarrow \mathbb{F}_2, \quad \mathsf{and}_k(x) = x_1 \cdots x_k \quad (4.13)$$

$$\mathsf{minority}_k \colon \mathbb{F}_2^k \longrightarrow \mathbb{F}_2, \quad \mathsf{minority}_k(x) = \begin{cases} 1, & \mathsf{sum}_k(x) \leq \lfloor k/2 \rfloor \\ 0, & \text{otherwise} \end{cases} \quad (4.14)$$

$$\mathsf{majority}_k \colon \mathbb{F}_2^k \longrightarrow \mathbb{F}_2, \quad \mathsf{majority}_k(x) = \begin{cases} 1, & \mathsf{sum}_k(x) \geq \lceil k/2 \rceil \\ 0, & \text{otherwise} \end{cases} \quad (4.15)$$

Note that all these functions are symmetric and Boolean. A function $f \colon K^l \longrightarrow K$ is a *symmetric function* if and only if $f(\sigma \cdot x) = f(x)$ for all $x \in K^l$ and all $\sigma \in S_l$ with $\sigma \cdot x$ defined in (3.22). This is a natural class to study in the context of SDS since they induce SDS, which allow for the action of graph automorphisms.

Example 4.4. Let $Y = \mathsf{Circ}_4$ as in Example 4.1. For each vertex we use the vertex function $\mathsf{nor}_3 \colon \mathbb{F}_2^3 \longrightarrow \mathbb{F}_2$ defined in (4.9) with corresponding Y-local maps

$$F_0(x) = (\mathsf{nor}(x_0, x_1, x_3), x_1, x_2, x_3) ,$$
$$F_1(x) = (x_0, \mathsf{nor}(x_0, x_1, x_2), x_2, x_3) ,$$
$$F_2(x) = (x_0, x_1, \mathsf{nor}(x_1, x_2, x_3), x_3) ,$$
$$F_3(x) = (x_0, x_1, x_2, \mathsf{nor}(x_0, x_2, x_3)) .$$

Consider the system state $x = (0, 0, 0, 0)$. Using the update order $\pi = (0, 1, 2, 3)$, we compute in order

$$F_0(0, 0, 0, 0) = (1, 0, 0, 0) ,$$
$$F_1 \circ F_0(0, 0, 0, 0) = (1, 0, 0, 0) ,$$
$$F_2 \circ F_1 \circ F_0(0, 0, 0, 0) = (1, 0, 1, 0) ,$$
$$F_3 \circ F_2 \circ F_1 \circ F_0(0, 0, 0, 0) = (1, 0, 1, 0) .$$

Thus, we have $(F_3 \circ F_2 \circ F_1 \circ F_0)(0, 0, 0, 0) = (1, 0, 1, 0)$. In other words: The map of the SDS over the graph Circ_4 with nor_3 as vertex functions and the update order $(0, 1, 2, 3)$ applied to the system state $(0, 0, 0, 0)$ gives the new system state $(1, 0, 1, 0)$. We write this as

$$[\mathbf{Nor}_{\mathsf{Circ}_4}, (0, 1, 2, 3)](0, 0, 0, 0) = (1, 0, 1, 0) \ .$$

Repeated applications of $(F_3 \circ F_2 \circ F_1 \circ F_0)$ yield the system states $(0, 0, 0, 1)$, $(0, 1, 0, 0)$, $(0, 0, 1, 0)$, $(1, 0, 0, 0)$, $(0, 1, 0, 1)$, and $(0, 0, 0, 0)$ again. These system states constitute a *periodic orbit*, a concept we will define below.

The crucial point to notice here is the importance of the particular order in which the local maps F_v are applied. This distinguishes SDS from, for example, generalized cellular automata where the maps F_v are applied synchronously.

◇

Let $(f_v)_{v \in \mathrm{v}[Y]}$ be a family of vertex functions for some graph Y. If all maps are induced by a particular function, e.g., nor functions, and only vary in their respective arity, we refer to the corresponding SDS-map as $[\mathbf{Nor}_Y, \pi]$.

A sequence $(g_l)_{l=1}^n$ with $g_l \colon K^l \longrightarrow K$ *induces* a family of vertex functions $(f_v)_{v \in \mathrm{v}[Y]}$ by setting $f_v = g_{d(v)+1}$. The resulting SDS is then called the SDS over Y *induced by the sequence* $(g_l)_{l=1}^n$. Accordingly, an SDS is induced if all vertices of Y of the same degree l have identical vertex functions induced by g_l. For instance, the SDS in Example 4.4 is induced by the function $\mathrm{nor}_3 \colon \mathbb{F}_2^3 \longrightarrow \mathbb{F}_2$.

In this book we use the following conventions:

- vertex functions are all denoted in lowercase, e.g., nor_3,
- local maps have the first letter in uppercase and the remaining letters in lowercase, e.g., Nor_v,
- the vertex-indexed family of local maps is written in bold where the first letter is in uppercase and the remaining letters in lowercase, e.g., $\mathbf{Nor}_Y = (\mathrm{Nor}_v)_v$.

4.1.3 The Phase Space of an SDS

Let x be a system state. As in Section 3.3.2 the *forward orbit* of x under $[\mathbf{F}_Y, \pi]$ is the sequence $O^+(x)$ given by

$$O^+(x) = (x, [\mathbf{F}_Y, \pi](x), [\mathbf{F}_Y, \pi]^2(x), [\mathbf{F}_Y, \pi]^3(x), \dots) \ .$$

If the SDS-map $[\mathbf{F}_Y, \pi]$ is bijective, we have the sequence

$$O(x) = ([\mathbf{F}_Y, \pi]^l(x))_{l \in \mathbb{Z}} \ .$$

The orbit $O^+(x)$ is often referred to as a *time series*. Since we only consider finite graphs and the states are taken from finite sets, all orbits are finite.

In the case of binary states we can represent an orbit or time series as a *space-time diagram*. A vertex state that is zero is represented as a white square and a vertex state that is one is represented as a black square. A system state $x = (x_1, x_2, \ldots, x_n)$ is displayed using the black-and-white box representations of its vertex states and is laid out in a left-to-right manner. Starting from the initial configuration each successive configuration is displayed below its predecessor.

Example 4.5. In Figure 4.2 we have shown an example of a space-time diagram. You may want to verify that $[\mathbf{Nor}_{\mathsf{Circ}_5}, (0, 1, 2, 3, 4)]$ is an SDS-map that generates this space-time diagram.   ◇

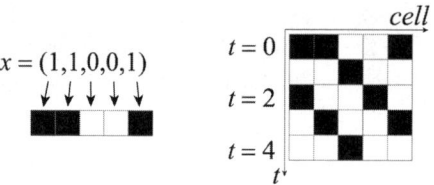

Fig. 4.2. An example of a space-time diagram.

Example 4.6. A space-time diagram for an SDS over Circ_{512} induced by $(\mathrm{parity}_k)_k$ is shown in Figure 4.3.   ◇

Fig. 4.3. A space-time diagram for the SDS map $[(\mathrm{Parity}_k)_k, \pi]$ starting from a randomly chosen initial state $x \in \mathbb{F}_2^{512}$. The update order is $\pi = (0, 1, \ldots, 511)$.

The *phase space* of an SDS-map $[\mathbf{F}_Y, \pi]$ is the directed graph $\Gamma = \Gamma([\mathbf{F}_Y, \pi])$ defined by

$$\mathrm{v}[\Gamma] = K^n, \quad \mathrm{e}[\Gamma] = \{(x, y) \mid x, y \in K^n, \ y = [\mathbf{F}_Y, \pi](x)\}, \quad (4.16)$$
$$\omega \times \tau : \mathrm{e}[\Gamma] \longrightarrow \mathrm{v}[\Gamma] \times \mathrm{v}[\Gamma], \quad (x, [\mathbf{F}_Y, \pi](x)) \mapsto (x, [\mathbf{F}_Y, \pi](x)) .$$

The map $\omega \times \tau$ is injective by construction. As a result we do not have to reference the maps ω and τ explicitly. As for combinatorial graphs, Γ is completely specified by its vertex and edge sets. By abuse of terminology, we will sometimes speak about the phase space of an SDS $(Y, \mathbf{F}_Y, \pi)$, in which case it is understood that we refer to its SDS-map.

In view of the definition of orbits and periodic points in Section 3.3.2, we observe that Γ-vertices contained in cycles are precisely the *periodic points* of the SDS-map $[\mathbf{F}_Y, \pi]$. The set of periodic points of $[\mathbf{F}_Y, \pi]$ is denoted $\mathsf{Per}([\mathbf{F}_Y, \pi])$. Likewise, the subset of Γ-vertices contained in cycles of length 1 are the fixed points of $[\mathbf{F}_Y, \pi]$, denoted $\mathsf{Fix}([\mathbf{F}_Y, \pi])$. The remaining Γ-vertices are *transient system states*. By abuse of terminology, we will also speak about the periodic points and fixed points of an SDS.

Example 4.7. In Figure 4.4 we have shown all the system state transitions for the SDS-map

$$[\mathbf{Nor}_{\mathsf{Circ}_4}, \pi] = \mathrm{Nor}_{\pi_3} \circ \mathrm{Nor}_{\pi_2} \circ \mathrm{Nor}_{\pi_1} \circ \mathrm{Nor}_{\pi_0} : \mathbb{F}_2^4 \longrightarrow \mathbb{F}_2^4 \;,$$

in the case of $\pi = (0, 1, 2, 3)$ and $\pi = (0, 2, 1, 3)$. It is easy to see that changing the permutation update order can lead to SDS with entirely different phase spaces. We will analyze this in detail in Section 4.3.   ◇

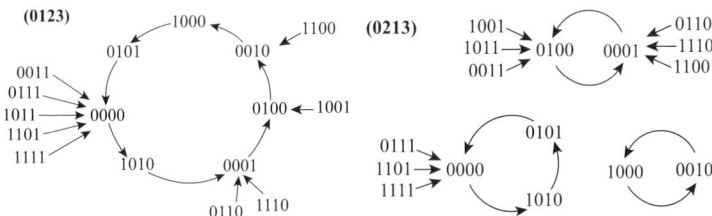

Fig. 4.4. The phase spaces for the SDS-maps of $(\mathsf{Circ}_4, \mathbf{Nor}_{\mathsf{Circ}_4}, (0, 1, 2, 3))$ and $(\mathsf{Circ}_4, \mathbf{Nor}_{\mathsf{Circ}_4}, (0, 2, 1, 3))$ on the left and right, respectively. Clearly, the phase spaces are different. We also note that the phase spaces are not isomorphic as directed graphs.

As for presenting phases spaces, it is convenient to encode a binary n-tuple $x = (x_1, x_2, \ldots, x_n)$ as the decimal number by

$$k = \sum_{i=1}^{n} x_i \cdot 2^{i-1} \;. \tag{4.17}$$

Example 4.8. In Figure 4.5 we have shown the phase space of the SDS $(\mathsf{Circ}_4, \mathbf{Nor}_{\mathsf{Circ}_4}, (0, 1, 2, 3))$ using the regular binary n-tuple labeling and the corresponding base-10 encoding given by (4.17).   ◇

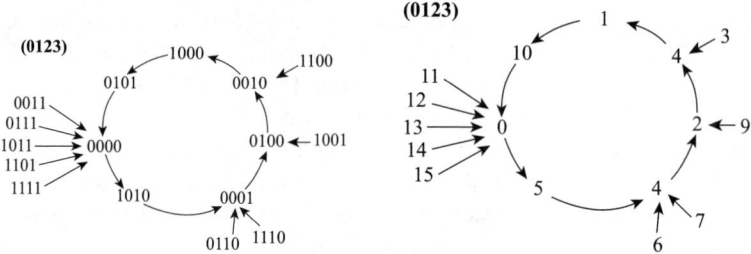

Fig. 4.5. The phase space of $(\text{Circ}_4, \textbf{Nor}_{\text{Circ}_4}, (0,1,2,3))$ with binary states (left) and base 10 encoded states (right).

4.1. Using the programming language of your choice, write functions that convert a binary n-tuple to its decimal representation given by (4.17) and a matching function that converts a decimal number to its corresponding n-tuple. Are there limitations on this method, and, if so, is it a problem in practice?

Assume the vertex states are from the finite set $K = \{0, 1, \ldots, q\}$. We can view the corresponding n-tuples as $(q+1)$-ary numbers. Write functions that convert between n-tuples with entries in K and their base-10 representations. For example, if $q = 2$, then the 4-tuple $(2, 1, 0, 1)$ has decimal representation $1 \cdot 3^3 + 0 \cdot 3^2 + 1 \cdot 3^1 + 2 \cdot 3^0 = 27 + 3 + 2 = 32$. What is the decimal representation of $(3, 1, 2, 0)$ assuming $q = 3$? Assuming $n = 6$ and $q = 3$, find the base-4, 6-tuple representation of 1234.   [1C]

4.1.4 SDS Analysis — A Note on Approach and Comments

SDS analysis is about understanding and characterizing the phase-space structure of an SDS. Since SDS have SDS-maps that are finite dynamical systems, we could in principle obtain the entire phase space by exhaustive computations. However, even small or moderately sized SDS with binary states over graphs that have 100 vertices, say, would have $2^{100} > 10^{30}$ states. As a result the main theme of SDS research is to derive phase-space information based on the structure of the base graph Y, the local maps, and the update order.

Let $K = \{0, 1\}$ and $\mathrm{v}[Y] = \{v_1, \ldots, v_n\}$. First, any SDS-map $[\mathbf{F}_Y, \pi]$ is a map from K^n to K^n. So why not study general maps $f : K^n \longrightarrow K^n$? The reason is, of course, that SDS exhibit an additional structure that allows for interesting analysis and results. In light of this, a natural question is therefore: When does a map $f : K^n \longrightarrow K^n$ allow for an SDS representation? A characterization of this class even for the subset of linear maps would be of interest.

Let us revisit the definition of an SDS. Suppose we did not postulate the graph Y explicitly. We can then obtain the base graph Y as follows: As a vertex set takes $\{v_1, \ldots, v_n\}$, and as edges take all ordered pairs (v, v') for which the vertex function f_v depends on the vertex state $x_{v'}$ where $v \neq v'$. As such, the graph Y is a directed graph, but we can, of course, obtain a combinatorial graph; see [90]. In other words, for a given family of local maps $(F_v)_v$ there exists a unique minimal graph Y that could serve as the base graph, and in this sense the graph may be viewed as redundant in the definition of SDS. We chose to explicitly reference the base graph in the definition since this allows us to consider varying families of local maps over a given combinatorial structure. In a way this is also why we did not try to define an SDS as just a map but as a triple. In principle one could also speculate replacing the local maps by an algebraic structure, like a ring or monoid, which would result in a combinatorial version of a scheme [91].

4.2. What is meant by an SDS being induced? For the graph Circ_6, what is $n[5]$? How is the function Nor_5 defined in this case?   [1]

4.3. Compute the phase space of $[\mathbf{Majority}_{\mathsf{Line}_3}, (2, 1, 3)]$.   [1]

4.2 Basic Properties

In this section we present some elementary properties of SDS.

4.2.1 Decomposition of SDS

As a lead-in to answer the question of SDS decomposition, we pose some slightly more general questions. How does an SDS-map $\phi = [\mathbf{F}_Y, \pi]$ depend on the update order π, and under which conditions does $[\mathbf{F}_Y, \pi] = [\mathbf{F}_Y, \pi']$ hold? In other words, if we fix Y and the family of local maps $(F_v)_v$, then when do two permutations give rise to the same SDS-map?

Clearly, the answer depends on both the local maps and the structure of the graph Y. If the local maps are all trivial, it does not matter what order we use in the composition, and the same holds if we have a graph with no edges. Here is a key observation: If we have two non-adjacent vertices v and v' in a graph Y, then we always have the commutation relation

$$F_v \circ F_{v'} = F_{v'} \circ F_v . \tag{4.18}$$

Equation (4.18) holds for any choice of vertex functions and for any choice of K. Extending this observation, we see that if we have two permutations π and π' that only differ in two adjacent positions, that is,

$$\pi = (\pi_1, \ldots, \pi_{i-1}, \pi_i, \pi_{i+1}, \pi_{i+2}, \ldots, \pi_n) \quad \text{and}$$
$$\pi' = (\pi_1, \ldots, \pi_{i-1}, \pi_{i+1}, \pi_i, \pi_{i+2}, \ldots, \pi_n) ,$$

and such that $\{\pi_i, \pi_{i+1}\}$ is not an edge in Y, then we have the identity of SDS-maps $[\mathbf{F}_Y, \pi] = [\mathbf{F}_Y, \pi']$. Thus, recalling the definition of the equivalence relation $\sim_Y$ from Section 3.1.3, we conclude that $\pi \sim_Y \pi'$ implies $[\mathbf{F}_Y, \pi] = [\mathbf{F}_Y, \pi']$. This justifies the construction of the update graph $U(Y)$ of a graph Y in Section 3.1.3. Accordingly, we have proved:

Proposition 4.9. *Let Y be a graph and let $(F_v)_v$ be a family of Y-local maps. Then we have*

$$\pi \sim_Y \pi' \Longrightarrow [F_Y, \pi] = [F_Y, \pi'].$$

It is now clear how to decompose an SDS-map in the case when the base graph Y is not connected.

Proposition 4.10. *Let Y be the the disjoint union of the graphs Y_1 and Y_2 and let π_Y be an update order for Y. We have*

$$[F_{Y_2}, \pi_{Y_2}] \circ [F_{Y_1}, \pi_{Y_1}] = [F_Y, \pi_Y] = [F_{Y_1}, \pi_{Y_1}] \circ [F_{Y_2}, \pi_{Y_2}], \tag{4.19}$$

where π_{Y_i} is the update order of Y_i induced by π_Y for $i = 1, 2$.

Proof. Let $(\pi_{Y_1} | \pi_{Y_2})$ denote the concatenation of the two update orders over Y_1 and Y_2. Clearly, $\pi_Y \sim_Y (\pi_{Y_1} | \pi_{Y_2}) \sim_Y (\pi_{Y_2} | \pi_{Y_1})$, and by Proposition 4.9 we have equality. $\square$

Note that an immediate corollary of Proposition 4.10 is that $[F_Y, \pi_Y]^k = [F_{Y_1}, \pi_{Y_1}]^k \circ [F_{Y_2}, \pi_{Y_2}]^k$. Thus, the dynamics of the two subsystems is entirely decoupled. As a result we may without loss of generality always assume that the base graph of an SDS is connected.

4.4. Let Y_1 and Y_2 be graphs and let Γ_1 and Γ_2 be phase spaces of two SDS-maps ϕ_1 and ϕ_2 over Y_1 and Y_2, respectively. The product of these two dynamical systems is a new dynamical system $\phi \colon \mathrm{v}[\Gamma_1] \times \mathrm{v}[\Gamma_2] \longrightarrow \mathrm{v}[\Gamma_1] \times \mathrm{v}[\Gamma_2]$ where $\phi(x, y) = (\phi_1(x), \phi_2(y))$. Characterize the dynamics of the product in terms of the dynamics of the two SDS ϕ_1 and ϕ_2.   [2]

4.2.2 Fixed Points

Fixed points of SDS are the simplest type of periodic orbits. These states have the property that they do not depend on the particular choice of permutation update order:

Proposition 4.11. *Let Y be a graph and let $(F_v)_v$ be Y-local functions. Then for any $\pi, \pi' \in S_Y$ we have*

$$\mathsf{Fix}([\mathbf{F}_Y, \pi]) = \mathsf{Fix}([\mathbf{F}_Y, \pi']). \tag{4.20}$$

Proof. If $x \in K^n$ is a fixed point of the permutation SDS-map $[\mathbf{F}_Y, \pi]$, then by the structure of the Y-local maps we necessarily have $F_v(x) = x$ for all $v \in \mathrm{v}[Y]$. It is therefore clear that x is fixed under $[\mathbf{F}_Y, \pi']$ for any permutation update order π'. $\square$

4.5. In Proposition 4.11 we insisted on permutation update orders. What happens to Proposition 4.11 if the update order is a *word* over v[Y]?   [1+]

It is clear that we obtain the same set of fixed points whether we update our system synchronously or asynchronously. Why? In Chapter 5 we will revisit fixed points and show that they can be fully characterized for certain graphs classes such as, for example, Circ_n.

You may have noticed already that the Nor-SDS encountered so far never had any fixed point, and you may even have shown that this true in general: A Nor-SDS with a permutation update order has no fixed points. The same holds for Nand-SDS, which are *dynamically equivalent* to Nor-SDS; see Section 4.3.3. If we restrict ourselves to symmetric functions, it turns out that $(\text{nor}_k)_k$ and $(\text{nand}_k)_k$ are the only sequences of functions $(g_k)_k$ that induce fixed-point-free SDS for any choices of base graph. For any other sequence of symmetric functions there exists a graph such that the corresponding SDS has at least one fixed point.

Theorem 4.12. *Let* $(g_k)_k$ *with* $g_k \colon \mathbb{F}_2^k \longrightarrow \mathbb{F}_2$ *be a sequence of symmetric functions such that the induced permutation* SDS*-map* $[\mathbf{F}_Y, \pi]$ *has no fixed points for any choice of base graph* Y. *Then we have*

$$(g_k)_k = (\text{nor}_k)_k \quad or \quad (g_k)_k = (\text{nand}_k)_k . \tag{4.21}$$

Proof. We prove this in two steps: First, we show that each map $f_v = g_{d(v)+1}$ has to be either nor_v or nand_v. In the second step we show that if the sequence $(g_k)_k$ contains both nor functions and nand functions, then we can construct an SDS that has at least one fixed point. For the proof we may assume that the graphs Y are connected and thus that every vertex has degree at least 1. Recall that since we are considering induced SDS, all vertices of the same degree d have local functions induced by the same map g_{d+1}. By a slight abuse of notation we will write $f_v = \text{nor}_v$ instead of $f_v = \text{nor}_{d(v)+1}$.

Step 1. For each $k = 1, 2, 3, \ldots$ we have either $g_k = \text{nor}_k$ or $g_k = \text{nand}_k$. It is easy to see that the statement holds for $k = 1$. Consider the case of $k = 2$. It is clear that for the SDS to be fixed-point-free the symmetric function g_2 must satisfy $g_2(0,0) = 1$ and $g_2(1,1) = 0$, since we would otherwise have a fixed point over the graph K_2. Moreover, since the g_k's are symmetric, either we have $g_2(0,1) = g_2(1,0) = 1$ so that $g_2 = \text{nand}_2$, or we have $g_2(0,1) = g_2(1,0) = 0$, in which case $g_2 = \text{nor}_2$. This settles the case where $k = 2$.

Assume next that $k > 2$, and suppose that $g_k \neq \text{nor}_k$ and $g_k \neq \text{nand}_k$. Then there must exist two k-tuples $x = (x_1, \ldots, x_k)$ and $y = (y_1, \ldots, y_k)$ with $l = |\{i \mid x_i = 1\}|$ and $l' = |\{i \mid y_i = 1\}|$ such that $0 < l, l' < k$ and $g_k(x) = 1$, $g_k(y) = 0$. There are two cases to consider: We have either *(i)* $g_2(0,1) = 0$ or *(ii)* $g_2(0,1) = 1$. In case *(i)* we take $Y(l, k-l)$ to be the graph with $l(k-l)$ vertices and $\binom{l}{2} + l(k-l)$ edges constructed from K_l as follows: For each vertex v of K_l we add $k-l$ new vertices and join these with an edge to vertex v. The graph $Y(4,3)$ is shown in Figure 4.6. The state we obtain by assigning 1 to each vertex state of the K_l subgraph and 0 to the remaining vertex states is clearly a fixed point. In case (ii) we use the graph $Y(k - l', l')$. We construct a fixed point by assigning 0 to each $K_{k-l'}$ vertex state and by assigning 1 to the remaining vertex states. We have thus shown that the only possible vertex functions are nor_v and nand_v. It remains to show that they cannot occur simultaneously.

Step 2. We will show that either $(g_k)_k = (\text{nor}_k)_k$ or $(g_k)_k = (\text{nand}_k)_k$. Suppose that $g_l = \text{nor}_l$ and $g_{l'} = \text{nand}_{l'}$. Let Y be the graph union of the empty graphs $Y_1 = E_{l-1}$ and $Y_2 = E_{l'-1}$. That is, Y has vertex set $\text{v}[Y_1] \cup \text{v}[Y_2]$ and edge set $\text{e}[Y] = \{\{v, v'\} \mid v \in Y_1; v' \in Y_2\}$. Using $\text{nand}_{l'}$ as a function for each vertex v' in Y_1 and nor_l for each vertex v in Y_2, we construct a fixed point by assigning 0 to each vertex state in Y_2 and 1 to each vertex state in Y_1, and the proof is complete. □

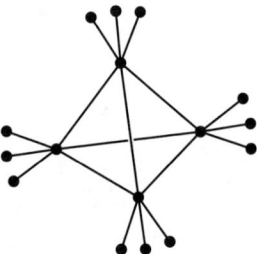

Fig. 4.6. The graph $Y(m, n)$ for $m = 4$ and $n = 3$ used in the proof of Theorem 4.12.

4.2.3 Reversible Dynamics and Invertibility

In this section we study SDS with bijective SDS-maps. An SDS for which the SDS-map is a bijection is an *invertible* SDS. From a dynamical systems point of view, having an invertible SDS-map means we can go backwards in time in a unique, well-defined way. For this reason such SDS are sometimes referred to as *reversible* and we say that they have *reversible dynamics*.

4.6. Describe the phase space of an invertible sequential dynamical system.

[1+]

The goal of this section is to derive criteria for an SDS to be invertible. We first characterize the SDS that are reversible over $K = \mathbb{F}_2$. For this purpose we introduce the maps $\text{id}_k, \text{inv}_k \colon \mathbb{F}_2^k \longrightarrow \mathbb{F}_2^k$ defined by

$$\text{inv}_k(x_1, \ldots, x_k) = (1 + x_1, \ldots, 1 + x_k), \tag{4.22}$$

$$\text{id}_k(x_1, \ldots, x_k) = (x_1, \ldots, x_k). \tag{4.23}$$

For the following proposition recall the definitions of $x[v]$ and $x'[v]$ [eqs. (4.2) and (4.4)].

Proposition 4.13. *Let* $(Y, \mathbf{F}_Y, \pi)$ *be an* SDS *with map* $[\mathbf{F}_Y, \pi]$. *Then* $(Y, \mathbf{F}_Y, \pi)$ *is invertible if and only if for each vertex* $v_i \in \mathrm{v}[Y]$ *and each* $x'[v_i]$ *the map*

$$g_{x'[v_i]} \colon \mathbb{F}_2 \longrightarrow \mathbb{F}_2, \qquad g_{x'[v_i]}(x_{v_i}) = f_{v_i}(x[v_i]), \tag{4.24}$$

is a bijection. If the SDS-*map* $[\mathbf{F}_Y, \pi]$ *is bijective where* $\pi = (\pi_1, \pi_2, \dots, \pi_n)$, *then its inverse is an* SDS-*map and is given by*

$$[\mathbf{F}_Y, \pi]^{-1} = [\mathbf{F}_Y, \pi^*], \tag{4.25}$$

where $\pi^* = (\pi_n, \pi_{n-1}, \dots, \pi_2, \pi_1)$.

Proof. Consider first the map $[\mathbf{F}_Y, \pi]$, i.e.,

$$F_{\pi_n} \circ F_{\pi_n} \circ \cdots \circ F_{\pi_1} . \tag{4.26}$$

As a finite product of maps, $F_{\pi_n} \circ F_{\pi_n} \circ \cdots \circ F_{\pi_1}$ is invertible if and only if each map F_{v_i} is. (Why?) By definition of F_{v_i} we have

$$F_{v_i}(x) = (x_{v_1}, \dots, x_{v_{i-1}}, f_{v_i}(x[v_i]), x_{v_{i+1}}, \dots, x_{v_n}) .$$

This map is bijective if and only if the map $g_{x'[v_i]}(x_{v_i}) = f_{v_i}(x[v_i])$ is bijective for any fixed choice of $x'[v_i]$. The only two such maps are inv_1 (the inversion map) and id_1 (the identity map), establishing the first assertion.

In both cases, that is, if $g_{x'[v_i]}(x_{v_i}) = f_{v_i}(x[v_i])$ is the inversion map or the identity map, we obtain that $F_{v_i}^2$ is the identity. From

$$[\mathbf{F}_Y, \pi^*] \circ [\mathbf{F}_Y, \pi] = F_{\pi(1)} \circ \cdots \circ F_{\pi(n-1)} \circ \underbrace{F_{\pi(n)} \circ F_{\pi(n)}} \circ F_{\pi(n-1)} \circ \cdots \circ F_{\pi(1)}$$

and $F_{v_i}^2 = 1$ we can conclude that $[\mathbf{F}_Y, \pi^*]$ is the inverse map of $[\mathbf{F}_Y, \pi]$, and the proof is complete. $\qquad\square$

Example 4.14. In this example we will consider the SDS over Circ_n where all functions f_v are induced by parity_3. We claim that the corresponding SDS are invertible.

Consider the vertex i and fix $x'[i] = (x_{i-1}, x_{i+1})$. The map $g_{x'[i]} \colon \mathbb{F}_2 \longrightarrow \mathbb{F}_2$ is given by

$$g_{x'[i]}(x_i) = f_i(x_{i-1}, x_i, x_{i+1}) = x_i + x_{i-1} + x_{i+1}.$$

If $x_{i-1} + x_{i+1}$ equals 0, then $g_{x'[i]}$ is the identity map. On the other hand, if $x_{i-1} + x_{i+1}$ equals 1, then $g_{x'[i]}$ is the inversion map and Proposition 4.13 guarantees that the corresponding SDS are invertible. In Figure 4.7 we have shown the phase spaces of $[\mathbf{Parity}_{\mathsf{Circ}_4}, (0, 1, 2, 3)]$ and $[\mathbf{Parity}_{\mathsf{Circ}_4}, (0, 1, 2, 3)]^{-1} = [\mathbf{Parity}_{\mathsf{Circ}_4}, (3, 2, 1, 0)]$. $\qquad\diamond$

The following example illustrates how to use the above proposition in order to show that a certain map f fails to induce an invertible SDS.

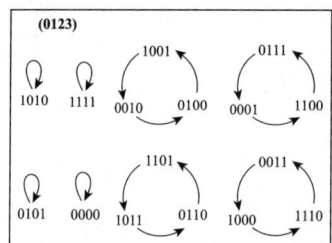 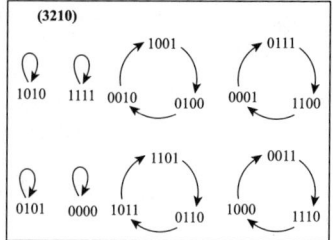

Fig. 4.7. The phase spaces of $[\mathbf{Parity}_{\mathsf{Circ}_4}, (0, 1, 2, 3)]$ and its inverse SDS-map $[\mathbf{Parity}_{\mathsf{Circ}_4}, (0, 1, 2, 3)]^{-1} = [\mathbf{Parity}_{\mathsf{Circ}_4}, (3, 2, 1, 0)]$.

Example 4.15. We claim that SDS over Circ_n induced by rule 110 (see Section 2.1.3) are not invertible. The first thing we need to do is to "decode" rule 110. Since

$$110 = 0 \cdot 2^7 + 1 \cdot 2^6 + 1 \cdot 2^5 + 0 \cdot 2^4 + 1 \cdot 2^3 + 1 \cdot 2^2 + 1 \cdot 2^1 + 0 \cdot 2^0,$$

we obtain the following table for rule 110:

(x_{i-1}, x_i, x_{i+1})	111	110	101	100	011	010	001	000
f_{110}	0	1	1	0	1	1	1	0

Here is the key observation: Since $(0, 0, 1)$ and $(0, 1, 1)$ both map to 1 under f_{110}, F_i is not injective and f does not induce an invertible SDS.   ◇

4.7. Identify all maps $f: \mathbb{F}_2^3 \longrightarrow \mathbb{F}_2$ that induce invertible SDS over $Y = \mathsf{Circ}_n$. From the previous two examples you see that parity_3 is one such map while f_{110} does not qualify. Find the remaining maps. How many such maps are there? What are the rule numbers of these maps?   **[2]**

4.8. So far the examples and problems have mainly dealt with the graph $Y = \mathsf{Circ}_n$. Building on Example 4.14, show that any SDS where the vertex functions f_v are induced by $(\mathrm{parity}_k)_k$ is invertible.   **[1+]**

Note: It may be clear already, but we point out that the question of whether $[\mathbf{F}_Y, \pi]$ is invertible does not depend on the update order π. Note, however, that different update orders generally give different periodic orbit structures, as the organization of the particular system states on the cycles will vary.

The generalization of Proposition 4.13 from $\mathbb{F}_2$ to an arbitrary finite set is straightforward. Note, however, that the inversion formula in Eq. (4.25) is only valid for $\mathbb{F}_2$. The inversion formula in the case of $K = \mathbb{F}_p$ is addressed in Problem 4.10.

4.9. How many vertex functions for a vertex v of degree d induce invertible SDS in the case of (1) $\mathbb{F}_2$ and (2) $\mathbb{F}_p$?   **[1+]**

4.10. Generalize the inversion formula to the case with vertex states in $\mathbb{F}_p$.   **[2]**

So far we have considered SDS with arbitrary vertex functions $(f_v)_v$. If we restrict ourselves to symmetric vertex functions, we obtain the following:

Proposition 4.16. *Let* $(Y, \mathbf{F}_Y, \pi)$ *be an invertible* SDS *with symmetric vertex functions* $(f_v)_v$. *Then* f_v *is either (a)* $\mathrm{parity}_{d(v)+1}$ *or (b)* $1 + \mathrm{parity}_{d(v)+1}$.

Before we prove the proposition, we introduce the notion of an *H-class*: The set $H_k = \{x \in \mathbb{F}_2^n \mid \mathrm{sum}_n(x) = k\}$ is called *H*-class k. In the case of $\mathbb{F}_2^n$ there are $n + 1$ such *H*-classes.

Proof. Let v be a vertex of degree $d_v = k - 1$ and associated symmetric vertex function f_v. We will use induction over the *H*-classes $0, 1, \ldots$ in order to show that f_v is completely determined by its value on the state (0).
Induction basis: The value $f_v(0)$ determines the value of f_v on *H*-class 1. To prove this assume $f_v(0) = y_0$. Then by Proposition 4.13 we know that the value of f_v on $(0, 0, 0, \ldots, 0)$ and the representative $(1, 0, 0, \ldots, 0)$ from *H*-class 1 must differ and thus

$$f_v(0, 0, \ldots, 0) = y_0 \quad \Longrightarrow \quad f_v(1, 0, \ldots, 0) = 1 + y_0 . \tag{4.27}$$

Induction step: The value of f_v on H_l determines the value of f_v on H_{l+1}. Let $x_l = (0, 1, 1, \ldots, 1, 0, 0, \ldots, 0) \in H_l$ and assume $f_v(x_l) = y_l$. Then in complete analogy to our argument for the induction basis we derive $((1, 1, \ldots, 1, 0, 0, \ldots, 0) \in H_{l+1})$:

$$f_k(0, 1, 1, \ldots, 1, 0, 0, \ldots, 0) = y_l \implies f_k(1, 1, 1, \ldots, 1, 0, 0, \ldots, 0) = 1 + y_l, \tag{4.28}$$

completing the induction step. If $y_0 = 0$, we obtain $f_v = \mathrm{parity}_v$, and if $y_0 = 1$, we obtain $f_v = 1 + \mathrm{parity}_v$, and the proof is complete. $\square$

The following result addresses the dynamics of SDS restricted to their periodic points. We will use this later in Section 5.3 when we characterize the periodic points of threshold systems such as $[\mathbf{Majority}_Y, \pi]$. It can be viewed as a generalization of Proposition 4.13.

Proposition 4.17. *Let* Y *be a graph and let* $(Y, \mathbf{F}_Y, \pi)$ *be an* SDS *over* $\mathbb{F}_2$ *with* SDS-*map* $\phi = [\mathbf{F}_Y, \pi]$. *Let* ψ *be the restriction of* ϕ *to* $\mathrm{Per}(\phi)$, *i.e.,* $\psi = \phi|_{\mathrm{Per}(\phi)}$. *Then* ψ *is invertible with inverse* ψ^*.

Proof. We immediately observe that the argument in the proof of Proposition 4.13 holds when restricted to periodic points. $\square$

From a computational point of view it is desirable to have efficient criteria for determining if a point is periodic. Proposition 4.17 provides the following necessary (but not sufficient) condition:

Corollary 4.18. *Let* $(Y, \mathbf{F}_Y, \pi)$ *be an* SDS *over* $\mathbb{F}_2^n$. *Then a necessary condition for* $x \in \mathbf{F}_2^n$ *to be a periodic point under* $[\mathbf{F}_Y, \pi]$ *is* $[\mathbf{F}_Y, \pi^*] \circ [\mathbf{F}_Y, \pi](x) = x$.

In light of our previous results, the proof is obvious. Thus, if we have $[\mathbf{F}_Y, \pi^*] \circ [\mathbf{F}_Y, \pi](x) \neq x$, we can conclude that x is not a periodic point. To derive a sufficient criterion for periodicity is much more subtle. In fact, we will show later that periodicity in general depends on the particular choice of permutation or word.

4.2.4 Invertible SDS with Symmetric Functions over Finite Fields

We conclude this section with a characterization of invertible SDS with symmetric vertex function over finite fields [93]. In the following we will show how to explicitly construct invertible (word)-SDS for any choice of graph Y and word w. To set the stage let $[\mathbf{F}_Y, \pi]$ be such an SDS-map.

A *vertex coloring*[1] of a (combinatorial) graph Y is a map

$$c \colon \mathrm{v}[Y] \longrightarrow C \,,$$

where C is a finite set (the set of colors) such that for any $\{v, v'\} \in \mathrm{e}[Y]$ we have $c(v) \neq c(v')$. When we want to emphasize the color set C, we refer to c as a C-coloring of Y.

Generalizing Proposition 4.13 to arbitrary finite fields K, we observe that $F_{v,Y}$ (with vertex function $f_v \colon K^m \longrightarrow K$) is bijective if and only if the function

$$g_{x'[v]} \colon K \longrightarrow K, \quad (x_v) \mapsto f_v(x[v]) \tag{4.29}$$

is a bijection for all $x'[v] \in K^{m-1}$. Consider a generalized m-cube, Q_κ^m, whose vertices are m-tuples $(x_1, \ldots, x_m)$ with $x_i \in K$ and where K is a finite field of cardinality κ. Two vertices in Q_κ^m are adjacent if they differ in exactly one coordinate. The adjacency concept in Q_κ^m reflects Eq. (4.29), as only varying one particular coordinate in Q_κ^m produces specific Q_κ^m-neighbors. This is the intuition behind the fact that the local map $F_{v,Y}$ is bijective if and only if its vertex function f_v induces a coloring of an orbit graph (Section 3.2.1) of Q_κ^m. The corresponding group inducing this orbit graph arises naturally from specific properties of the vertex function such as it being symmetric.

Example 4.19. Let $Y = Q_3^3$. Here S_3 acts on Y via

$$\sigma(v_1, v_2, v_3) = \left(v_{\sigma^{-1}(1)}, v_{\sigma^{-1}(2)}, v_{\sigma^{-1}(3)}\right) \,.$$

The orbit graph $S_3 \setminus Q_3^3$ of this action is given in Figure 4.8.   ◇

Let W_Y denote the set of words $w = (w_1, \ldots, w_q)$ over $\mathrm{v}[Y]$. In Theorem 4.20 we will show that for arbitrary Y and word $w \in W_Y$ there always exists an invertible SDS. Furthermore, we will give a combinatorial interpretation of invertible SDS via κ-colorings of the orbit graphs $S_{d_Y(v)+1} \setminus Q_\kappa^{d_Y(v)+1}$. This not only generalizes Proposition 4.16 (see also [94]) but allows for a new combinatorial perspective.

---

[1] Note that what we call a vertex coloring some refer to as a *proper* vertex coloring; see [83].

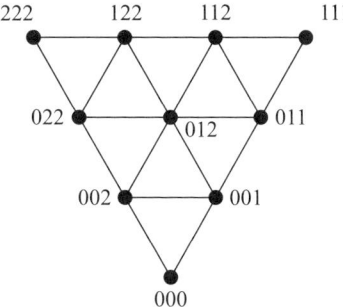

Fig. 4.8. The orbit graph $S_3 \setminus Q_3^3$ of Example 4.19.

Theorem 4.20. *Let Y be a combinatorial graph, K a finite field with $\kappa = |K|$, $m = d(v) + 1$, $w \in W_Y$, and $(Y, \mathbf{F}_Y, w)$ a word-SDS induced by symmetric vertex functions. Then for any $v \in \mathrm{v}[Y]$ we have the bijection*

$$\alpha \colon \{F_v \mid F_v \text{ is bijective}\} \longrightarrow \{c_v \mid c_v \text{ is a } \kappa\text{-coloring of } S_m \setminus Q_\kappa^m\} . \quad (4.30)$$

In particular, for arbitrary Y and w there always exists a family $\mathbf{F}_Y$ such that the SDS $(Y, \mathbf{F}_Y, w)$ is invertible.

Proof. We first observe

$$[\mathbf{F}_Y, w] = \prod_{i=1}^{k} F_{w_i, Y} \text{ is bijective} \quad \Longleftrightarrow \quad \forall w_i; \ F_{w_i} \text{ is bijective.} \quad (4.31)$$

Let $F_{v,Y}$ be a bijective Y-local map induced by the symmetric vertex function $f_v \colon K^m \longrightarrow K$. Without loss of generality we may assume that Y is connected and thus $m \geq 2$. From

$$F_{v,Y}(x_{v_1}, \ldots, x_{v_n}) = (x_{v_1}, \ldots, x_{v_{i-1}}, f_v(x[v]), x_{v_{i+1}}, \ldots, x_{v_n})$$

we conclude that the map

$$g_{x'[v]} \colon K \longrightarrow K, \quad x_v \mapsto f_v(x[v]) \quad (4.32)$$

is a bijection for arbitrary $x'[v]$ (Proposition 4.13).

Let $x[v] = (x_{v_{j_1}}, \ldots, x_{v_{j_m}})$. We consider the graph Q_κ^m with vertices $x = (x_{v_{j_1}}, \ldots, x_{v_{j_m}})$, where $j_i < j_{i+1}$ for $1 \leq i \leq m-1$. Two vertices are adjacent in Q_κ^m if they differ in exactly one coordinate. The graph Q_κ^m is acted upon by S_m through

$$\sigma(x_{v_{j_i}})_{1 \leq i \leq m} = (x_{\sigma^{-1}(v_{j_i})})_{1 \leq i \leq m} . \quad (4.33)$$

Since $S_m < \mathsf{Aut}(Q_\kappa^m)$, the above S_m-action induces the orbit graph $S_m \setminus Q_\kappa^m$. We note that $S_m \setminus Q_\kappa^m$ contains a subgraph isomorphic to the complete graph of size κ (why?); hence, each coloring of $S_m \setminus Q_\kappa^m$ requires at least κ colors.

Claim 1. The map f_v uniquely corresponds to a κ-coloring of the orbit graph $S_m \setminus Q_\kappa^m$.

By abuse of terminology we identify f_v with its induced map

$$\tilde{f}_v \colon S_m \setminus Q_\kappa^m \longrightarrow K, \quad S_m(x_{v_{j_1}}, \ldots, x_{v_{j_m}}) \mapsto f_v(x_{v_{j_1}}, \ldots, x_{v_{j_m}}),$$

which is well-defined since the $S_m \setminus Q_\kappa^m$-vertices are by definition S_m-orbits, and f_v is a symmetric function. To show that f_v is a coloring we use the local surjectivity of $\pi_{S_m} \colon Q_\kappa^m \longrightarrow S_m \setminus Q_\kappa^m$. Without loss of generality we can assume that two adjacent $S_m \setminus Q_\kappa^m$-vertices $S_m(x)$ and $S_m(x')$ have representatives $y[v]$ and $z[v]$ that differ exactly in their vth coordinate, that is,

$$y[v] = (x_{v_{j_1}}, \ldots, y_v, \ldots, x_{v_{j_m}}), \quad z[v] = (x_{v_{j_1}}, \ldots, z_v, \ldots, x_{v_{j_m}}) \, .$$

Since $g_{x'[v]} \colon K \longrightarrow K$, $x_v \mapsto f_v(x[v])$ is bijective for any $x'[v]$ [Eq. (4.32)], we have

$$g_{x'[v]}(y_v) = f_v(S_m(y[v])) \neq f_v(S_m(z[v])) = g_{x'[v]}(z_v) \, ,$$

that is, f_v is a coloring of $S_m \setminus Q_\kappa^m$. Furthermore, the bijectivity of $g_{x'[v]}$ and the fact that f_v is defined over $S_m \setminus Q_\kappa^m$ imply that f_v is a κ-coloring of $S_m \setminus Q_\kappa^m$ and Claim 1 follows.

Accordingly, we have a map

$$\alpha \colon \{F_v \mid F_v \text{ is bijective}\} \longrightarrow \{c_v \mid c_v \text{ is a } \kappa\text{-coloring of } S_m \setminus Q_\kappa^m\} \, . \quad (4.34)$$

We proceed by proving that α is a bijection. We can conclude from Claim 1 that α is an injection. To prove surjectivity we show that $S_m \setminus Q_\kappa^m$ contains a specific subgraph isomorphic to a complete graph over κ vertices. Consider the mapping

$$\vartheta \colon S_m \setminus Q_\kappa^m \longrightarrow P(K), \quad \vartheta(S_m(x)) = \{x_{v_{j_i}} \mid 1 \leq i \leq m\} \, , \quad (4.35)$$

where $P(K)$ denotes the power set of K. For any $x_{v_{j_i}} \in \vartheta(S_m(x))$ there are $\kappa - 1$ different neighbors of the form $S_m(x_k)$, where

$$\forall \, k \in K \setminus x_{v_{j_i}}; \quad x_k = (x_{v_{j_1}}, \ldots, x_{v_{j_{i-1}}}, k, x_{v_{j_{i+1}}}, \ldots, x_{v_{j_m}}) \, . \quad (4.36)$$

We denote this set by $N(x_{v_{j_i}}) = \{S_m(x_k) \mid k \neq x_{v_i}\}$. By the definition of $S_m \setminus Q_\kappa^m$, any two different vertices $S_m(x_k)$ and $S_m(x_{k'})$ are adjacent. Accordingly, the complete graph over $N(x_{v_{j_i}}) \cup \{S_m(x)\}$ is a subgraph of $S_m \setminus Q_\kappa^m$. As a result any κ coloring induces a symmetric vertex map f_v with the property that

$$g_{x'[v]} \colon K \longrightarrow K, \quad x_v \mapsto f_v(x[v])$$

is a bijection for arbitrary $x'[v]$; hence, α is surjective and the proof of Eq. (4.30) is complete.

Claim 2. For any $m \in \mathbb{N}$ and a finite field K with $|K| = \kappa$, there exists a κ-coloring of $S_m \setminus Q_\kappa^m$.

To prove Claim 2, we consider

$$\mathsf{s}_m : S_m \setminus Q_\kappa^m \longrightarrow K, \quad \mathsf{s}_m(S_m(x)) = \sum_{i=1}^{m} x_{v_{j_i}} . \tag{4.37}$$

Since s_m is a symmetric function, it is a well-defined map from $S_m \setminus Q_\kappa^m$ to K. In order to prove that s_m is a coloring, we use once more local surjectivity of the canonical projection

$$\pi_{S_m} : Q_\kappa^m \longrightarrow S_m \setminus Q_\kappa^m .$$

Accordingly, for any two $S_m(x)$-neighbors $S_m(\xi)$ and $S_m(\xi')$ we can find representatives $\tilde{\xi}$ and $\tilde{\xi}'$ in Q_κ^m that differ in exactly one coordinate. We then have

$$\mathsf{s}_m(S_m(\xi)) = \sum_{i=1}^{m} \tilde{\xi}_{v_{j_i}} \neq \sum_{i=1}^{m} \tilde{\xi}'_{v_{j_i}} = \mathsf{s}_m(S_m(\xi')) . \tag{4.38}$$

We conclude from the fact that s_m is a mapping over $S_m \setminus Q_\kappa^m$ and Eq. (4.38) that $\mathsf{s}_m : S_m \setminus Q_\kappa^m \longrightarrow K$ is a κ-coloring of $S_m \setminus Q_\kappa^m$.

Let Y be a graph and let w be a finite word over $\mathrm{v}[Y]$. Using Claim 2 and the bijection α of Eq. (4.34) for every w_i of w, we conclude that there exists at least one invertible SDS $(Y, \mathbf{F}_Y, w)$, completing the proof of Theorem 4.20. □

4.11. Show that the degree of a vertex $S_m(x)$ in $S_m \setminus Q_\kappa^m$ can be expressed as

$$d_{S_m \setminus Q_\kappa^m}(S_m(x)) = (\kappa - 1) \, |\vartheta(S_m(x))| . \tag{4.39}$$

[1+]

4.12. Construct the graph $G = S_3 \setminus Q_3^3$ from Example 4.19. How many $K = \mathbb{F}_3$ colorings does G admit? [1+]

4.13. Let K be the field with four elements. How many vertices does the graph $G = S_3 \setminus Q_4^3$ have? Sketch the graph G. How many K colorings does G admit? [2]

4.14. How many vertices does the graph $G = S_m \setminus Q_\alpha^m$ have? [1+]

Example 4.21. Let $K = \mathbb{F}_3$ and let $v \in \mathrm{v}[Y]$ be a vertex of degree 2. According to Theorem 4.20, there exists a bijective local map $F_{v,Y}$ that corresponds to the proper 3-coloring of the orbit graph $S_3 \setminus Q_3^3$;

$$\mathsf{s}_3 : S_3 \setminus Q_3^3 \longrightarrow \mathbb{F}_3, \quad \mathsf{s}_3(S_3(x)) = x_1 + x_2 + x_3.$$

We can display the s_3-3-coloring of $S_3 \setminus Q_3^3$ as follows:

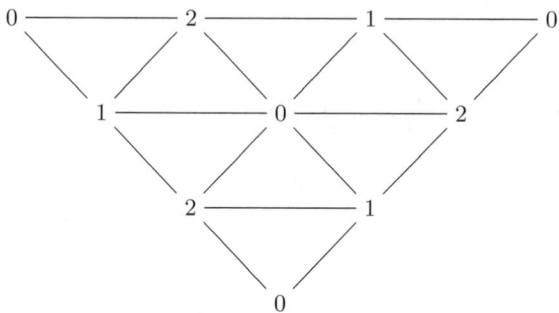

$\diamond$

When $K = \mathbb{F}_2$, Theorem 4.20 yields:

Corollary 4.22. *Let $K = \mathbb{F}_2$. Then a word-SDS $(Y, \mathbf{F}_Y, w)$ is invertible if and only if for all w_i the Y-local map F_{w_i} is induced by either* parity *or* $1 +$ parity.

Proof. For $K = \mathbb{F}_2$ the orbit graph $S_m \setminus Q_2^m$ is a line graph of size $m + 1$, that is, $S_m \setminus Q_2^m \cong \mathsf{Line}_{m+1}$.

$$(0, \ldots, 0) \text{———} (0, \ldots, 0, 1) \text{———} \cdots \cdots \text{———} (0, 1, \ldots, 1) \text{———} (1, \ldots, 1)$$

Each 2-coloring of Line_{m+1} is uniquely determined by its value on $(0, \ldots, 0)$ and there are two possible choices. Mapping $(0, \ldots, 0)$ to 0 yields the parity function, and mapping $(0, \ldots, 0)$ to 1 yields the function $1 +$ parity, and Corollary 4.22 follows. $\qquad\square$

4.3 Equivalence

Equivalence is a fundamental notion in all of mathematics. In this section we will analyze equivalence concepts of SDS. We begin our study of equivalence by asking under which conditions are two SDS maps $[\mathbf{F}_Y, \pi]$ and $[\mathbf{G}_Z, \sigma]$ identical as functions? We refer to this as *functional equivalence* and address this in Section 4.3.1.

Example 4.23. In this example we once more consider SDS over the graph Circ_4 where the vertex functions are induced by $\mathsf{nor}_3 \colon \{0, 1\}^3 \longrightarrow \{0, 1\}$. The four SDS-maps we consider are $[\mathbf{Nor}_{\mathsf{Circ}_4}, (0, 1, 2, 3)]$, $[\mathbf{Nor}_{\mathsf{Circ}_4}, (3, 2, 1, 0)]$, $[\mathbf{Nor}_{\mathsf{Circ}_4}, (0, 1, 3, 2)]$, and $[\mathbf{Nor}_{\mathsf{Circ}_4}, (0, 3, 1, 2)]$, and they are all shown in Figure 4.9. The two phase spaces at the bottom in the figure are identical. The SDS-maps $[\mathbf{Nor}_{\mathsf{Circ}_4}, (0, 1, 3, 2)]$ and $[\mathbf{Nor}_{\mathsf{Circ}_4}, (0, 3, 1, 2)]$ are functionally equivalent. The top two phase spaces are not identical, but closer inspection shows that they are *isomorphic*: If we disregard the states/labels, we see that they are identical as unlabeled graphs. $\diamond$

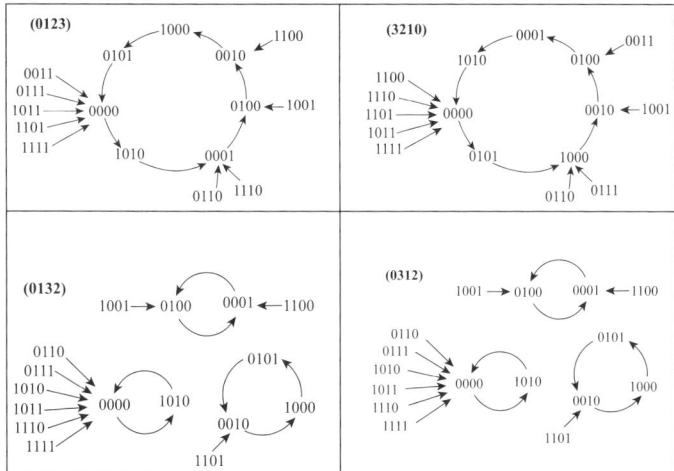

Fig. 4.9. Top left: the phase space of $[\mathbf{Nor}_{\mathsf{Circ}_4}, (0,1,2,3)]$. Top right: the phase space of $[\mathbf{Nor}_{\mathsf{Circ}_4}, (3,2,1,0)]$. Bottom left: the phase space of $[\mathbf{Nor}_{\mathsf{Circ}_4}, (0,1,3,2)]$. Bottom right: the phase space of $[\mathbf{Nor}_{\mathsf{Circ}_4}, (0,3,1,2)]$.

If two SDS phase spaces are isomorphic as (directed) graphs, we call the two SDS *dynamically equivalent*. We will analyze this type of equivalence in Section 4.3.3.

There are other concepts of equivalences and isomorphisms as well. For example, [90] considers *stable isomorphism*: Two finite dynamical systems are stably isomorphic if there is a digraph isomorphism between their periodic orbits. In other words, two finite dynamical systems are stably isomorphic if their multisets of orbit sizes coincide. We refer to Proposition 5.43 in Chapter 5, where we elaborate some more on this notion. The following example serves to illustrate the concept.

Example 4.24. Figure 4.10 shows the phase spaces of the two SDS-maps $[\mathbf{Nor}_{\mathsf{Circ}_4}, (0,1,2,3)]$ and $[(\mathbf{1}+\mathbf{Nor}+\mathbf{Nand})_{\mathsf{Circ}_4}, (0,1,2,3)]$. By omitting the system state $(1,1,1,1)$, it is easy to see that these dynamical systems have precisely the same periodic orbits and are thus stably isomorphic. ◇

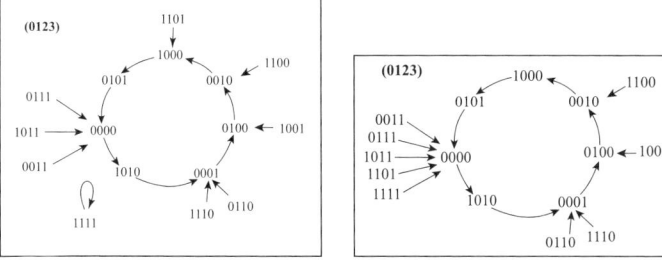

Fig. 4.10. The phase space of $(\mathsf{Circ}_4, \mathbf{Nor}_{\mathsf{Circ}_4}, (0,1,2,3))$ (right) and the phase space of $(\mathsf{Circ}_4, (\mathbf{1}+\mathbf{Nor}+\mathbf{Nand})_{\mathsf{Circ}_4}, (0,1,2,3))$ (left).

The natural framework for studying equivalence is *category theory*. Consider categories whose *objects* are SDS phase spaces. Different choices of *morphisms* between SDS phase spaces yield particular categories and are tantamount to different notions of equivalence. If we, for instance, only consider the identity as morphism, we arrive at the notion of functional equivalence. If we consider as morphisms all digraph isomorphisms, we obtain dynamical equivalence. A systematic, category theory-based approach is beyond the scope of this book, but the interested reader may want to explore this area further [95].

4.3.1 Functional Equivalence of SDS

In Section 4.2 we already encountered the situation where two SDS-maps $[\mathbf{F}_Y, \pi]$ and $[\mathbf{G}_Z, \sigma]$ are identical. There we considered the cases $Y = Z$ and $\mathbf{F}_Y = \mathbf{G}_Z$ and showed that

$$\pi \sim_Y \pi' \implies [\mathbf{F}_Y, \pi] = [\mathbf{F}_Y, \pi'] . \tag{4.40}$$

In this section we will continue this analysis assuming a fixed base graph Y and family of Y-local functions $(F_v)_v$.

A particular consequence of Eq. (4.40) is that the number of components in the update graph $U(Y)$ of Y (Section 3.1.4) is an upper bound for the number of functionally different SDS that can be generated by only varying the permutation. In Section 3.1.5 we established that there is a bijection

$$f_Y \colon [S_Y / \sim_Y] \longrightarrow \mathsf{Acyc}(Y) .$$

This shows us that $[\mathbf{F}_Y, \pi]$, viewed as a function of the update order π, only depends on the acyclic orientation $\mathcal{O}_Y(\pi)$. We can now state

Proposition 4.25. *For any combinatorial graph Y and any family of Y-local functions $(F_v)_v$ we have*

$$|\{[\mathbf{F}_Y, \pi] \mid \pi \in S_Y\}| \leq |\mathsf{Acyc}(Y)| , \tag{4.41}$$

and the bound is sharp.

Proof. The inequality (4.41) is clear from (4.40) and the bijection f_Y. It remains to show that the bound is sharp. To this end we prove the implication

$$[\pi]_Y \neq [\sigma]_Y \implies [\mathbf{Nor}_Y, \pi] \neq [\mathbf{Nor}_Y, \sigma] . \tag{4.42}$$

Without loss of generality we may assume that $\pi = \mathsf{id}$, and Lemma 3.13 guarantees the existence of a pair of Y-vertices v and v' with $\{v, v'\} \in e[Y]$ such that

$$\pi = (\ldots, v, \ldots, v', \ldots) \quad \text{and} \quad \sigma = (\ldots, v', \ldots, v, \ldots) .$$

We set $B_Y^{<\sigma}(v) = \{w \mid w \in B_Y(v) \ \wedge \ w <_\sigma v\}$. Let

$$x = (x_u)_u, \qquad x_u = \begin{cases} 1 & \text{if } u \in B_Y^{<\sigma}(v), \\ 0 & \text{otherwise.} \end{cases} \tag{4.43}$$

Obviously, $[\mathbf{Nor}_Y, \pi](x)_v = 0$ since $v' <_\sigma v$ and $x_{v'} = 1$. But clearly we have $[\mathbf{Nor}_Y, \sigma](x)_v = 1$; hence, $[\mathbf{Nor}_Y, \pi] \neq [\mathbf{Nor}_Y, \sigma]$ and

$$|\{[\mathbf{Nor}_Y, \pi] \mid \pi \in S_Y\}| = |\mathsf{Acyc}(Y)|,$$

and the proof is complete.   □

We remark that Eq. (4.40) and the bound in (4.41) are valid for vertex functions over, e.g., $\mathbb{R}^n$ and $\mathbb{C}^n$, and there are no restrictions on the vertex functions f_v.

4.3.2 Computing Equivalence Classes

In this section we give some remarks on computational issues related to SDS. Through the bijection f_Y we can bound the number of functionally nonequivalent SDS by computing $a(Y) = |\mathsf{Acyc}(Y)|$. For the computation of $a(Y)$ we have from Section 3.1.3 the recursion relation $a(Y) = a(Y'_e) + a(Y''_e)$. However, the computation of $a(Y)$ is in general of equal complexity as the computation of the chromatic number of Y.

There are various approaches to bound $a(Y)$. Let $\alpha(Y)$ be the (vertex) independence number of Y. By definition, there are at most $\alpha(Y)$ independent vertices, and clearly we have at most $n!$ linear orderings. From this we immediately deduce that $n!/\alpha(Y)^n \leq a(Y)$. In [96] a bound is derived in terms of the *degree-sequence* of Y: $a(Y) \geq \prod_{i=1}^{n}(\delta_i + 1)!^{1/\delta_i + 1}$ For graphs with $\binom{\ell}{2} + h$ edges, it is shown in [97] that for $0 \leq h < \ell$ the inequality $a(Y) \geq \ell!\,(h+1)$ holds. In [98, 99] the following upper bound for the number of acyclic orientations is given: $a(Y) \leq \prod_{i=1}^{n}(\delta_i + 1)$. In [96] an upper bound is given in terms of the number of spanning trees of Y.

Example 4.26. In Example 3.17 we saw that $a(\mathsf{Circ}_n) = 2^n - 2$. Thus, for the graph Circ_n and fixed vertex functions $(f_v)_v$ we can generate at most $2^n - 2$ functionally nonequivalent SDS by varying the permutation update order.   ◇

4.15. Derive a formula for $a(\mathsf{Wheel}_n)$.   [1+]

4.16. For a fixed sequence of vertex functions over Q_2^3 show that we can have at most 1862 functionally nonequivalent permutation SDS.   [2]

How sequential is a sequential dynamical system? This may sound like a strange question. However, if we implement an SDS on a "modern"[2]

---

[2] Of course, it is dangerous to say "modern" computer in any written work — after 10 years most things in that business are hopelessly dated!

computer with multiple processors, this question is relevant for efficient implementations.

In fact, we already encountered this question for permutation SDS in some form. Consider an SDS over a graph Y with update order π. We call a vertex of $\mathcal{O}(\pi)$ [identified with the graph $G(\mathcal{O}(\pi))$, Section 3.1.3] with the property

$$\nexists e \in e[G(\mathcal{O}(\pi))]; \quad \tau(e) = v,$$

a *source*. We can now compute the *rank layer sets* as follows:

Set $G = G(\mathcal{O}_Y(\pi))$, let $G'_0 = G$, and let $k = 0$.

- While $v[G'_k] \neq \varnothing$ repeat:
  - Let L_k be the set of sources in G'_k.
  - Let G'_{k+1} be the graph obtained by deleting all vertices in L_k from G'_k along with their incident edges.
  - Increment k by 1.

Notice that $L_k = \mathsf{rnk}^{-1}(k)$ and that this is also a practical way to construct the canonical permutation $\widehat{\pi}$ [Eq. (3.17)] associated with a given acyclic orientation. Here is the key fact: All the vertices in the layer set L_k can have their states updated simultaneously. This follows since L_k is necessarily an independent set of Y. From this it is clear that the smallest number of processor cycles we need to compute one full update pass of the SDS equals the number of layers, and this is given by $1 + \min_{k \geq 0}\{\mathsf{rnk}^{-1}(k) \neq \varnothing\}$. In general, this is the best possible result.

Example 4.27. Let $Y = \mathsf{Wheel}_6$ and let $\pi = (4, 2, 3, 5, 1, 0, 6)$. We will compute the induced acyclic orientation $\mathcal{O}_Y(\pi)$, find the layer sets (relative to $\mathcal{O}_Y(\pi)$), and compute $\widehat{\pi}$.

The directed graph representation of the induced acyclic orientation is shown in Figure 4.11. Here $\mathsf{rnk}(2) = \mathsf{rnk}(4) = 0$, $\mathsf{rnk}(1) = \mathsf{rnk}(3) = \mathsf{rnk}(5) = 1$, $\mathsf{rnk}(0) = 2$, and $\mathsf{rnk}(6) = 3$. Thus,

$$\widehat{\pi} = (4, 2, \overbrace{3, 5, 1}, 0, 6) = (\ \underbrace{2, 4}_{\mathsf{rnk}^{-1}(0)}\ ,\ \underbrace{1, 3, 5}_{\mathsf{rnk}^{-1}(1)}\ ,\ \underbrace{0}_{\mathsf{rnk}^{-1}(2)}\ ,\ \underbrace{6}_{\mathsf{rnk}^{-1}(3)}\ )\ .$$

What is the smallest number of processor cycles we would need to compute $[\mathbf{F}_Y, \pi](x)$? Since the maximal rank is 3, we see that we would need at least $3 + 1 = 4$ cycles to compute $[\mathbf{F}_Y, \pi](x)$ on a parallel multiprocessor machine.

$\diamond$

4.17. Let $Y = E_n$ be the graph on $2n$ vertices given by

$$v[E_n] = \{0, 1, \ldots, 2n - 1\}$$
$$e[E_n] = \{\{i, i+1\}, \{i, i+n-1\}, \{i, i+n\} \mid 0 \leq i < n\}\ ,$$

where all indices/vertices are computed modulo $2n$. The graph E_5 is shown in Figure 4.12.

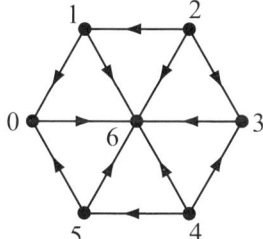

Fig. 4.11. The acyclic orientation induced by $(4, 2, 3, 5, 1, 0, 6)$ over Wheel_6.

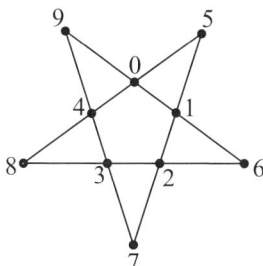

Fig. 4.12. The graph E_5 of Problem 4.17.

(*i*) Find the canonical permutation $\widehat{\pi}$ of $\pi = (0, 1, 2, 3, 4, 5, 6, 7, 8, 9)$. (*ii*) For $(Y, \mathbf{F}_{E_5}, \pi)$, what is the smallest number of computation cycles needed to evaluate the SDS-map $[\mathbf{F}_{E_5}, \pi]$ at some state x on a parallel computer with at least 10 processors? Here we assume that each processor can evaluate one vertex function per computation cycle. (*iii*) For a fixed sequence of vertex functions $(f_v)_v$ how many *functionally different* permutation SDS can we have over $Y = E_n$?   [1+]

4.3.3 Dynamical Equivalence

Functional equivalence of SDS distinguishes phase-space graphs as labeled graphs. Here we may want to classify phase spaces according to the structure of transients and periodic orbits irrespective of the particular labeling of the vertices. Accordingly, we call two SDS *dynamically equivalent* if their phase spaces are isomorphic as graphs. For finite dynamical systems we have the following:

Definition 4.28 (Dynamical equivalence). Let E be a finite set. Two finite dynamical systems given by map $H, G \colon E \longrightarrow E$ are dynamically equivalent if there exists a bijection $\phi \colon E \longrightarrow E$ such that

$$G \circ \phi = \phi \circ H . \tag{4.44}$$

We note that dynamical equivalence becomes a special case of topological conjugation if we use the discrete topology on E; see Section 3.3.3.

It is worth spending a moment to reflect on Eq. (4.44). We observe that the bijection ϕ maps the phase space of the dynamical system of H into the phase space of the dynamical system of G. For instance, assume that x is a fixed point under H so that $H(x) = x$. Then $\phi(x)$ is a fixed point for G since by Eq. (4.44) we have

$$G(\phi(x)) = \phi(H(x)) = \phi(x) .$$

We can generalize this to periodic orbits. Let y be a periodic point of period 2 under H. Since (4.44) implies

$$G^2 \circ \phi = G \circ (G \circ \phi) = G \circ \phi \circ H = g \circ H \circ H = \phi \circ H^2 ,$$

and in general $G^k \circ \phi = \phi \circ H^k$ for $k \geq 1$, we obtain that $\phi(y)$ is a periodic point of period 2 under G. If x is mapped to y under H, we see that $G(\phi(x)) = \phi(H(x)) = \phi(y)$. In other words, G maps $\phi(x)$ to $\phi(y)$, and accordingly, the phase spaces of G and H can be identified modulo the labeling of system states.

Example 4.29. In Figure 4.13 we have the isomorphic phase spaces of the SDS $(\mathsf{Circ}_4, \mathbf{Nor}_{\mathsf{Circ}_4}, (0,1,3,2))$ and $(\mathsf{Circ}_4, \mathbf{Nand}_{\mathsf{Circ}_4}, (0,1,3,2))$. The map $\mathsf{inv}_4 \colon \{0,1\}^4 \longrightarrow \{0,1\}^4$ given by

$$\mathsf{inv}(x_0, x_1, x_2, x_3) = (1 + x_0, 1 + x_1, 1 + x_2, 1 + x_3)$$

provides the bijection of Eq. (4.44) in Definition 4.28. $\diamond$

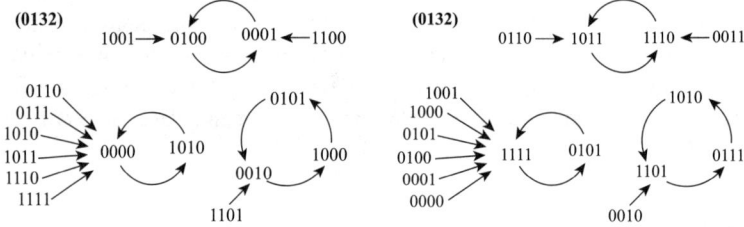

Fig. 4.13. Two isomorphic phase spaces. The phase space of $[\mathbf{Nor}_{\mathsf{Circ}_4}, (0,1,3,2)]$ (left) is mapped to the the phase space of $[\mathbf{Nand}_{\mathsf{Circ}_4}, (0,1,3,2)]$ (right) by the map inv_4.

Recall that if a group G acts on $\mathsf{v}[Y]$, then its action induces an action on system states $x = (x_{v_1}, \ldots, x_{v_n}) \in K^n$ by $gx = (x_{g^{-1}(v_1)}, \ldots, x_{g^{-1}(v_n)})$. In particular, this holds for $\mathsf{Aut}(Y)$ acting on $\mathsf{v}[Y]$.

Proposition 4.30. *Let Y be a combinatorial graph, let $\pi \in S_Y$, and let $(g_k)_{k=1}^n$ be a sequence of symmetric functions. Then we have for the SDS $(Y, \mathbf{F}_Y, \pi)$ and $(Y, \mathbf{F}_Y, \gamma\pi)$ induced by $(g_k)_{k=1}^n$*

$$\forall \gamma \in \mathsf{Aut}(Y); \quad [\mathbf{F}_Y, \gamma\pi] \circ \gamma = \gamma \circ [\mathbf{F}_Y, \pi] , \tag{4.45}$$

where $\gamma(x_{v_1}, \ldots, x_{v_n}) = (x_{\gamma^{-1}(v_1)}, \ldots, x_{\gamma^{-1}(v_n)})$.

Thus, for any sequence of symmetric functions $(g_k)_k$ the two induced SDS $(Y, \mathbf{F}_Y, \gamma\pi)$ and $(Y, \mathbf{F}_Y, \pi)$ are dynamically equivalent.

Proof. Since the SDS are induced, we have $f_v = g_{d(v)+1}$ for all vertices. We can rewrite Eq. (4.45) as

$$[\mathbf{F}_Y, \gamma\pi] = \gamma \circ [\mathbf{F}_Y, \pi] \circ \gamma^{-1} .$$

To prove this statement it is sufficient to show that for all $v \in v[Y]$ we have

$$F_{(\gamma\pi)(v)} = \gamma \circ F_{\pi(v)} \circ \gamma^{-1} .$$

The result then follows by composition. For the left-hand side we obtain

$$F_{(\gamma\pi)(v_i)}(x) = (x_{v_1}, \dots, \underbrace{f_{(\gamma\pi)(v_i)}(x[\gamma\pi(v_i)])}_{\text{pos. } (\gamma\pi)(v_i)}, \dots, x_{v_n}) .$$

Similarly, for the right side we derive

$$\gamma \circ F_{\pi(v_i), Y} \circ \gamma^{-1}(x) = \gamma \circ F_{\pi(v_i), Y}(x_{\gamma(v_1)}, \dots, x_{\gamma(v_n)})$$

$$= \gamma(x_{\gamma(v_1)}, \dots, \underbrace{f_{\pi(v_i)}(x_w \mid w \in \gamma B_Y(\pi(v_i)))}_{\text{pos. } \pi(v_i)}, \dots, x_{\gamma(v_n)})$$

$$= (x_{v_1}, \dots, \underbrace{f_{\pi(v_i)}(x_w \mid w \in \gamma B_Y(\pi(v_i)))}_{\text{pos. } \gamma\pi(v_i)}, \dots, x_{v_n}) .$$

Equality now follows since for $\gamma \in \mathsf{Aut}(Y)$ we have $\gamma B_Y(\pi(v_i)) = B_Y(\gamma\pi(v_i))$, and from $f_{\pi(v_i)} = f_{\gamma\pi(v_i)}$ since the SDS are induced and automorphisms preserve vertex degrees. $\square$

As noted in the proof we may rewrite Eq. (4.45) as

$$[F_Y, \gamma\pi] = \gamma \circ [F_Y, \pi] \circ \gamma^{-1} .$$

Clearly, this equation gives rise to a natural conjugation action of $\mathsf{Aut}(Y)$ on SDS.

4.18. In Proposition 4.30 we made some assumptions that were stronger than what we needed. Do we need symmetric functions? Do we need to only consider induced SDS? Does the proposition hold for word-SDS?   [2]

Example 4.31. We have already seen simple examples of the relation (4.45). To be specific take $\phi = [\mathbf{Nor}_{\mathsf{Circ}_4}, (0, 1, 2, 3)]$ and $\psi = [\mathbf{Nor}_{\mathsf{Circ}_4}, (3, 2, 1, 0)]$.[3] The automorphism group of the Circ_4 is D_4. We see that $\gamma = (0, 3)(1, 2)$ (cycle form) is an automorphism of Circ_4 and that $(3, 2, 1, 0) = \gamma(0, 1, 2, 3)$. Without any computations we therefore conclude by Proposition 4.30 that the SDS-maps ϕ and ψ are dynamically equivalent. Their phase spaces are shown in Figure 4.14. $\diamond$

---

[3] Again, when nothing else is said all permutations are written using the *standard form* as opposed to *cycle form*.

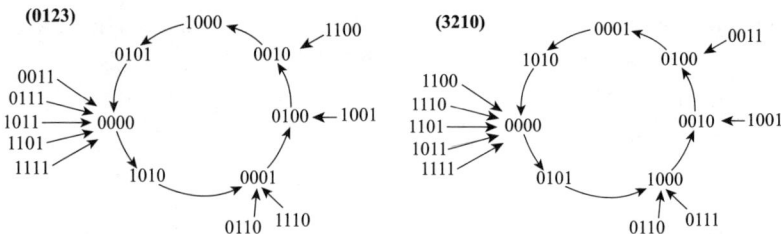

Fig. 4.14. The phase spaces of the dynamically equivalent SDS $(\mathsf{Circ}_4, \mathbf{Nor}_{\mathsf{Circ}_4}, (0, 1, 2, 3))$ (left) and $(\mathsf{Circ}_4, \mathbf{Nor}_{\mathsf{Circ}_4}, (3, 2, 1, 0))$ (right).

In light of Proposition 4.30, it is natural to consider group actions to characterize dynamically equivalent SDS. This will also allow us to derive bounds for the number of nonequivalent SDS that we can obtain by varying the update order while keeping the local functions and the graph fixed. Recall that $g_Y \colon \mathsf{Acyc}(Y) \longrightarrow S_n/\sim_Y$ is the inverse of the bijection f_Y in Proposition 3.15 of Chapter 3.

Lemma 4.32 (Aut(Y)-actions). *Let Y be a combinatorial graph. We have $\mathsf{Aut}(Y)$-actions on the sets (i) $S_n/\sim_Y$, (ii) $\mathsf{Acyc}(Y)$, and (iii) $\mathcal{F} = \{[\mathbf{F}_Y, \pi] \mid \pi \in S_Y\}$ given by*

$$(\gamma, [\pi]_Y) \mapsto \gamma[\pi]_Y := [\gamma\pi]_Y, \tag{4.46}$$

$$(\gamma, \mathcal{O}_Y) \mapsto \gamma\mathcal{O}_Y = \gamma \circ \mathcal{O} \circ \gamma^{-1}, \tag{4.47}$$

$$(\gamma, [\mathbf{F}_Y, \sigma]) \mapsto \gamma \bullet [\mathbf{F}_Y, \sigma] := [\mathbf{F}_Y, \gamma\sigma], \tag{4.48}$$

respectively. Furthermore, the actions on $S_n/\sim_Y$ and $\mathsf{Acyc}(Y)$ are compatible, i.e., we have

$$f_Y(\gamma[\pi]_Y) = \gamma f_Y([\pi]_Y) \tag{4.49}$$

and

$$h \colon \mathsf{Acyc}(Y) \longrightarrow \mathcal{F}, \quad h(\mathcal{O}_Y) = [\mathbf{F}_Y, \pi], \quad \pi \in g_Y(\mathcal{O}_Y) \tag{4.50}$$

is an $\mathsf{Aut}(Y)$-map.

Proof. We first note that the action in (4.46) is well-defined since we have

$$\pi \sim_Y \sigma \Longrightarrow \gamma\pi \sim_Y \gamma\sigma,$$

and hence $[\sigma]_Y = [\pi]_Y$ implies $[\gamma\sigma]_Y = [\gamma\pi]_Y$. It is clear that we have a group action. The maps (4.47) and (4.48) are clearly group actions, but see Problem 4.19.

Let γ be a graph automorphism and $\mathcal{O}_Y(\gamma\pi)$, $\mathcal{O}_Y(\pi)$ be the acyclic orientations induced by the permutations $\gamma\pi$ and π, respectively. Then we have

$$\gamma\mathcal{O}_Y(\pi) = \mathcal{O}_Y(\gamma\pi); \tag{4.51}$$

see Problem 4.19. From this we conclude

$$f_Y(\gamma[\pi]_Y) = f_Y([\gamma\pi]_Y) = \mathcal{O}_Y(\gamma\pi) = \gamma\mathcal{O}_Y(\pi) = \gamma f_Y([\pi]_Y) \ .$$

Using $\gamma\mathcal{O}_Y(\pi) = \mathcal{O}_Y(\gamma\pi)$, we can easily verify that h is an $\mathsf{Aut}(Y)$-map:

$$
\begin{aligned}
h(\gamma\mathcal{O}_Y(\pi)) &= h(\mathcal{O}_Y(\gamma\pi)) \\
&= [\mathbf{F}_Y, \gamma\pi] \\
&= \gamma \bullet [\mathbf{F}_Y, \pi] \qquad \text{(Proposition 4.30)} \\
&= \gamma \bullet h(\mathcal{O}_Y(\pi)) \ ,
\end{aligned}
$$

and the proof of the lemma is complete. □

4.19. Prove that (4.47) defines a group action, and establish the identity (4.51). [1+]

The results so far only address the update order aspect of dynamical equivalence — local maps and the base graph are identical for both SDS. Before we proceed by analyzing the number of dynamically nonequivalent SDS that can be generated by varying the update order, we remark that two SDS with identical base graphs, but different vertex functions can also be dynamically equivalent. For instance, for an arbitrary SDS $(Y, \mathbf{F}_Y, \pi)$ with vertex states in $K = \mathbb{F}_2$ we obtain a dynamically equivalent SDS where the Y-local functions are

$$\mathsf{inv}_n \circ F_{v,Y} \circ \mathsf{inv}_n \ ,$$

where inv_n is the inversion map, $(x_{v_i}) \mapsto (x_{v_i} + 1)$. In particular, it follows that the SDS $(Y, \mathbf{Nor}_Y, \pi)$ and $(Y, \mathbf{Nand}_Y, \pi)$ are dynamically equivalent. See also Theorem 4.12 and Example 4.29.

4.20. Are the SDS induced by the sequence of vertex functions $(\mathsf{parity}_k)_k$ and the sequence $(1 + \mathsf{parity}_k)_k$ dynamically equivalent? [1+]

4.3.4 Enumeration of Dynamically Nonequivalent SDS

How many dynamically nonequivalent SDS can be generated for a fixed graph Y and fixed family of induced local functions $\mathbf{F}_Y$ by varying the permutation update order? We denote this number by $\Delta(\mathbf{F}_Y)$. From Eq. (4.45) it is clear that $\Delta(\mathbf{F}_Y)$ cannot exceed the number of orbits in $S_Y/\sim_Y$ under $\mathsf{Aut}(Y)$. This quantity depends only on Y and is denoted by $\Delta(Y)$. Writing $a(Y) = |\mathsf{Acyc}(Y)|$, we have:

Theorem 4.33. *Let Y be a combinatorial graph, and let $\mathbf{F}_Y$ be a family of Y-local functions induced by symmetric functions. Then*

$$\Delta(\mathbf{F}_Y) \le \Delta(Y) = \frac{1}{|\mathsf{Aut}(Y)|} \sum_{\gamma \in \mathsf{Aut}(Y)} a(\langle\gamma\rangle \setminus Y) \ . \qquad (4.52)$$

Proof. Since $f_Y(\gamma[\pi]_Y) = \gamma f_Y([\pi]_Y)$, the number of orbits $\Delta(Y)$ in $S_n/\sim_Y$ induced by the $\mathsf{Aut}(Y)$-action equals the number of $\mathsf{Aut}(Y)$-orbits in $\mathsf{Acyc}(Y)$, and by Frobenius' lemma (Lemma 3.18) we have

$$\Delta(Y) = \frac{1}{|\mathsf{Aut}(Y)|} \sum_{\gamma \in \mathsf{Aut}(Y)} |(\mathsf{Fix}(\gamma)_{\mathsf{Acyc}(Y)})|.$$

The inequality (4.52) now follows from Theorem 3.21, which provides a combinatorial interpretation of the $\mathsf{Fix}(g)$ terms in Frobenius' lemma via the bijection

$$\beta\colon \mathsf{Acyc}(Y)^G \longrightarrow \mathsf{Acyc}(G \setminus Y), \quad \mathcal{O} \mapsto \mathcal{O}_G,$$

which implied Eq. (3.31): $N = \frac{1}{|G|} \sum_{g \in G} |\mathsf{Acyc}(\langle g \rangle \setminus Y)|.$ □

Accordingly, Theorem 4.33 follows from Theorem 3.21 in Chapter 3 and Lemma 4.32. Example 3.24 from Chapter 3 illustrates how this can be applied to circle graphs. We will derive a formula for $\Delta(\mathsf{Circ}_n)$ below.

4.21. Compute the bound $\Delta(Y)$ for $Y = K_n$, $n \geq 1$. *Hint.* Using the formula (4.52) is going completely overboard in this case. Think about what the bound $\Delta(Y)$ represents, and give your answers in no more than three lines!   [1+]

4.22. Compute the bound $\Delta(Y)$ for $Y = \mathsf{Wheel}_4$.   [1+]

4.23. In Example 4.35 we found that $\Delta(Y) = 3$. Is this bound sharp? *Hint.* What can you do to test if the bound is sharp?   [2-]

4.24. Is $\Delta(\mathbf{Parity}_{\mathsf{Circ}_4}) = \Delta(\mathsf{Circ}_4)$?   [2-]

4.25. How many possible permutation update orders are there for the graph $Y = Q_2^3$? How many functionally nonequivalent SDS can we obtain over Q_2^3 by only varying the update order? How many dynamically nonequivalent induced SDS can we obtain over Q_2^3 by varying only the update order? Is the bound $\Delta(Q_2^3)$ sharp?   [3-C]

Using formula (4.52), we can now compute the upper bound $\Delta(Y)$ for various classes of graphs.

Proposition 4.34 ($\Delta(\mathsf{Circ}_n)$ and $\Delta(\mathsf{Wheel}_n)$). *Let ϕ be the Euler ϕ-function. For $n \geq 3$ we have*

$$\Delta(\mathsf{Circ}_n) = \begin{cases} \frac{1}{2n} \sum_{d|n} \phi(d) \left(2^{n/d} - 2\right) + 2^{n/2}\big/4, & n \text{ even}, \\ \frac{1}{2n} \sum_{d|n} \phi(d) \left(2^{n/d} - 2\right), & n \text{ odd}, \end{cases} \quad (4.53)$$

$$\Delta(\mathsf{Wheel}_n) = \begin{cases} \frac{1}{2n} \sum_{d|n} \phi(d) \left(3^{n/d} - 3\right) + 3^{n/2}\big/2, & n \text{ even}, \\ \frac{1}{2n} \sum_{d|n} \phi(d) \left(3^{n/d} - 3\right), & n \text{ odd}. \end{cases} \quad (4.54)$$

Proof. First recall that

$$a(\mathsf{Circ}_n) = 2^n - 2 \quad \text{and} \quad a(\mathsf{Wheel}_n) = 3^n - 3 \qquad (4.55)$$

and that $\mathsf{Aut}(\mathsf{Circ}_n) = D_n$. This group is given by $\{\tau^m \sigma^k \mid m = 0, 1, \; k = 0, 1, \ldots, n-1\}$, where, using cycle notation, $\sigma = (0, 1, 2, \ldots, n-1)$ and $\tau = \prod_{i=1}^{\lceil (n-1)/2 \rceil}(i, n-i)$. By Theorem 3.21 we need to compute $a(\langle \gamma \rangle \setminus Y)$ for all $\gamma \in \mathsf{Aut}(Y)$. We start by looking at the rotations.

(*i*) If σ^k has order n, then the orbit graph $\langle \sigma^k \rangle \setminus \mathsf{Circ}_n$ consists of one single vertex with a loop attached, and therefore [Theorem 3.21, (a)] we have $\mathsf{Fix}(\sigma^k) = \varnothing$. Note that there are $\phi(n)$ automorphisms of order n.

(*ii*) If the order of σ^k is $n/2$, then the orbit graph $\langle \sigma^k \rangle \setminus \mathsf{Circ}_n$ is a graph with two vertices connected by two edges and we obtain (Theorem 3.21, Claim 1) $a(\langle \sigma^k \rangle \setminus \mathsf{Circ}_n) = 2 = 2^{n/n/2} - 2$. There are $\phi(n/2)$ such automorphisms.

(*iii*) In the case where σ^k has order n/d with $d > 2$, we have that $\langle \sigma^k \rangle \setminus \mathsf{Circ}_n \cong \mathsf{Circ}_d$ and thus $a(\langle \sigma^k \rangle \setminus \mathsf{Circ}_n) = 2^d - 2$. There are $\phi(d)$ such automorphisms.

(*iv*) Finally, it is seen that the only case in which $\langle \tau \sigma^k \rangle \setminus \mathsf{Circ}_n$ does not contain loops [Theorem 3.21, (a)] is when both n and k are even, and in this case $\langle \tau \sigma^k \rangle \setminus \mathsf{Circ}_n \cong \mathsf{Line}_{n/2+1}$ and $a(\langle \tau \sigma^k \rangle \setminus Y) = 2^{n/2}$ for all such k. There are $n/2$ automorphisms of this form.

Thus, for odd n we have

$$\Delta(\mathsf{Circ}_n) = \frac{1}{2n} \sum_{d \mid n} \phi(n/d)\, a(\mathsf{Circ}_d) = \frac{1}{2n} \sum_{d \mid n} \phi(d) \left(2^{n/d} - 2\right),$$

and for n even we will have to include the additional contribution from automorphisms $\tau \sigma^k$, which is $(1/2n)(n/2) a(\mathsf{Line}_{n/2+1}) = 2^{n/2}/4$, completing the proof for $\Delta(\mathsf{Circ}_n)$.

Now consider Wheel_n. Clearly we also have that $\mathsf{Aut}(\mathsf{Wheel}_n)$ is isomorphic to D_n. The calculation of $\Delta(\mathsf{Wheel}_n)$ now follows from what we did above and the following observation. If Y has no vertices of maximal degree (that would be $n-1$ for a graph on n vertices), then $\mathsf{Aut}(Y)$ and $\mathsf{Aut}(Y \oplus v)$ are isomorphic and $G \setminus (Y \oplus v)$ is isomorphic to $(G \setminus Y) \oplus v'$. This observation will allow us to use our calculations in case of Circ_n for Wheel_n for $n > 3$.

(*i*) By the same argument as above, we have that $\langle \sigma^k \rangle \setminus \mathsf{Wheel}_n$ contains a loop whenever σ^k has order n.

(*ii*) When σ^k has order $n/2$, then $\langle \sigma^k \rangle \setminus \mathsf{Wheel}_n \cong \mathsf{Circ}_3$ and thus the number of acyclic orientations of the orbit graph is $6 = 3^{n/(n/2)} - 3$.

(*iii*) When the order of σ^k is n/d with $d > 2$, we obtain $\langle \sigma^k \rangle \setminus \mathsf{Wheel}_n \cong \mathsf{Wheel}_d$, and $a(\langle \sigma^k \rangle \setminus \mathsf{Wheel}_n) = 3^d - 3$.

(*iv*) We only get contributions from automorphisms of the form $\tau \sigma^k$ when n and k are both even. In this case $\langle \tau \sigma^k \rangle \setminus \mathsf{Wheel}_n \cong W_{n/2+1}$, where W_n is the graph obtained from Wheel_n by deleting the edge $\{0, n-1\}$. We leave it as an exercise to conclude that $a(W_n) = 2 \cdot 3^{n-1}$ and consequently $a(\langle \tau \sigma^k \rangle \setminus \mathsf{Wheel}_n) = 2 \cdot 3^{n/2}$.

Adding up the terms as before produces the given formula, and the proof is complete. □

Example 4.35. In Example 3.24 we calculated the bound (4.52) for $Y = \mathsf{Circ}_4$ and $Y = \mathsf{Circ}_5$ directly. Here we will calculate the bound $\Delta(Y)$ for $Y = \mathsf{Circ}_6$ and $Y = \mathsf{Circ}_7$ using the formula in (4.53).

$$\Delta(\mathsf{Circ}_6) = \frac{1}{12}(\phi(1)(2^6 - 2) + \phi(2)(2^3 - 2) + \phi(3)(2^2 - 2)) + 2^{6/2}/4$$

$$= \frac{1}{12}(62 + 6 + 2 \cdot 2) + 2 = 6 + 2 = 8 \ .$$

We also get

$$\Delta(\mathsf{Circ}_7) = \frac{1}{14}(\phi(1)(2^7 - 2)) = \frac{1}{14}(126) = 9 \ .$$

◇

4.26. Compute $\Delta(\mathsf{Circ}_p)$ for p a prime with $p > 2$.   [1]

We derived a combinatorial upper bound for the number of dynamically nonequivalent SDS through the orbits of the $\mathsf{Aut}(Y)$-action on $\mathsf{Acyc}(Y)$. It is natural to ask for which graphs and for which families of local functions $\mathbf{F}_Y$ this bound is sharp, that is, when do we have $\Delta(\mathbf{F}_Y) = \Delta(Y)$ (see Problem 4.25)?

Conjecture 4.36. For any combinatorial graph Y and permutation-SDS induced by $(\mathrm{nor}_k)_k$, the bound $\Delta(Y)$ is sharp, i.e.,

$$\Delta(\mathbf{Nor}_Y) = \Delta(Y) \ . \tag{4.56}$$

In the following proposition we study the particular case of the star graph, denoted by Star_n. The star graph is the combinatorial graph given by $\mathrm{v}[\mathsf{Star}_n] = \{0, 1, 2, \ldots, n\}$ and $\mathrm{e}[\mathsf{Star}_n] = \{\{0, i\} \mid 1 \le i \le n\}$.

Proposition 4.37. *We have*

$$\Delta(\mathsf{Star}_n) = \Delta(\mathbf{Nor}_{\mathsf{Star}_n}) \ , \quad n \ge 2 \ .$$

Proof. The proof is done by considering all $\mathsf{Aut}(\mathsf{Star}_n)$-orbits of $S_n/\sim_{\mathsf{Star}_n}$ and by demonstrating that each orbit gives rise to an SDS with unique phase-space features.

It is clear that a graph automorphism must fix the center vertex 0. However, any permutation of the "outer" vertices corresponds to an automorphism. Therefore, the automorphism group of Star_n is isomorphic to S_n. Moreover, each class $[\pi]_{\mathsf{Star}_n}$ is characterized by the position of 0. Assume $\pi(j) = 0$. Then we have $[\pi]_{\mathsf{Star}_n} = \{\pi' \in S_{\mathsf{Star}_n} \mid \pi'(j) = 0\}$. We write this equivalence class as $[\pi]^j_{\mathsf{Star}_n}$. It now follows that

$$[S_{\mathsf{Star}_n}/\sim_{\mathsf{Star}_n}] = \bigcup_{j=1}^{n+1} [\pi]^j_{\mathsf{Star}_n} .$$

It is sufficient to prove that the SDS $(Y, \mathbf{Nor}_{\mathsf{Star}_n}, \pi_j)$ for $j = 1, \ldots, n+1$ have pairwise non-isomorphic phase spaces $\Gamma(\mathsf{Star}_n, \mathbf{Nor}_{\mathsf{Star}_n}, \pi_j)$. To this end let $\pi \in S_{n+1}$ be a permutation with $\pi(i) = 0$. We also set $x = (x_{\pi(1)}, \ldots, x_{\pi(i-1)})$ and $y = (x_{\pi(i+1)}, \ldots, x_{\pi(n+1)})$. If $i \neq 1, n+1$, we obtain the following orbits in phase space where underline denotes vectors and overbars denote logical complements.

$$(x1y) \longmapsto (00\bar{y}) \Longleftarrow (10y), \qquad y \neq \underline{0}, \qquad (4.57)$$

$$(x1\underline{0}) \longmapsto (\underline{001}) \longmapsto (100), \qquad x \neq \underline{0} \qquad (4.58)$$

$$(\underline{010})$$

$$(x0y) \Longleftarrow (\bar{x}0\bar{y}) \qquad x \neq \underline{0}, \underline{1} \qquad (4.59)$$

In the case $i = 1$ we obtain

$$(1y) \longmapsto (0\bar{y}) \Longleftarrow (0y) \longleftarrow (1\bar{y}), \qquad y \neq \underline{0}, \underline{1}, \qquad (4.60)$$

$$(1\underline{1}) \longmapsto (00) \longmapsto (10), \qquad (4.61)$$

$$(0\underline{1})$$

and in the case $i = n+1$ we have

$$(x0) \Longleftarrow (\bar{x}0), \qquad x \neq \underline{0}, \underline{1}, \qquad (4.62)$$

$$(x1) \longmapsto (00) \longmapsto (10) \qquad x \neq \underline{0}. \qquad (4.63)$$

$$(\underline{01})$$

It is clear from the above diagrams that for any Star_n vertex i the associated digraph has a unique component containing a 3-cycle and on this cycle there is a unique state v_i with indegree$(v_i) > 1$. In the first case indegree$(v_i) = 2^{i-1}$, in the second case indegree$(v_i) = 2$, and in the third case indegree$(v_i) = 2^n$. The only case in which these numbers are not all different is for $i = 2$. But in this case we can use, e.g., the structure in (4.60) to distinguish the corresponding digraphs. It follows that if $i \neq j$ the digraphs $\Gamma(\mathsf{Star}_n, \mathbf{Nor}_{\mathsf{Star}_n}, \pi_i)$ and $\Gamma(\mathsf{Star}_n, \mathbf{Nor}_{\mathsf{Star}_n}, \pi_j)$ are non-isomorphic, and we have shown that

$$\Delta(\mathbf{Nor}_{\mathsf{Star}_n}) = \Delta(\mathsf{Star}_n) ,$$

completing the proof of the theorem. $\qquad \qquad \square$

The reason why the sharpness proof was fairly clean for $Y = \mathsf{Star}_n$ is the large automorphism group of this graph and the clear-cut characterization of $\mathsf{Star}_n / \sim_{\mathsf{Star}_n}$. For Circ_n, for instance, the situation becomes a lot more involved.

Let $\mathsf{Star}_{l,m}$ denote the combinatorial graph derived from K_l by attaching precisely m new vertices to each vertex of K_l:

$$v[\mathsf{Star}_{l,m}] = v[K_l] \cup \bigcup_{i=1}^{l} \{i_r \mid 1 \le r \le m\}, \qquad (4.64)$$

$$e[\mathsf{Star}_{l,m}] = e[K_l] \cup \bigcup_{i=1}^{l} \{\{i, i_r\} \mid 1 \le r \le m\}.$$

The graph $\mathsf{Star}_{3,2}$ is shown in Figure 4.15.

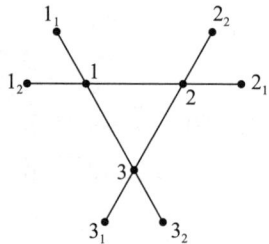

Fig. 4.15. The graph $\mathsf{Star}_{3,2}$.

Proposition 4.38. *For* $\mathsf{Star}_{2,m}$ *we have*

$$\Delta(\mathbf{Nor}_{\mathsf{Star}_{2,m}}) = \Delta(\mathsf{Star}_{2,m}) . \qquad (4.65)$$

Each permutation SDS $(\mathsf{Star}_{2,m}, \mathbf{Nor}_{\mathsf{Star}_{2,k}}, \pi)$ *has precisely one periodic orbit of length* 3.

The proof of this result goes along the same lines as the proof for Star_n, but it is rather cumbersome. If you feel up to it you may check the details in [100]. We contend ourselves with the following two results that are of independent interest.

Lemma 4.39. *Let* $m, l \ge 2$. *We have*

$$\mathsf{Aut}(\mathsf{Star}_{l,m}) \cong S_m^l \rtimes S_l . \qquad (4.66)$$

4.27. Prove Lemma 4.39.   [2]

For the graph $\mathsf{Star}_{l,m}$ it turns out we can also compute the bound $\Delta(\mathsf{Star}_{l,m})$ directly:

Proposition 4.40. *Let $m, l \geq 2$. We have*

$$\Delta(\mathsf{Star}_{l,m}) = (m+1)^l . \tag{4.67}$$

4.28. Verify the bound (4.67) in Proposition 4.40.   [2]

4.29. Settle conjecture 4.36.   [5]

4.4 SDS Morphisms and Reductions

It is natural to ask for structure-preserving maps between SDS. For dynamical systems the standard way to relate two systems is through phase-space relations as we did when studying dynamical equivalence. However, SDS exhibit additional structure, and it seems natural also to have morphisms relate the SDS base graphs, vertex functions, and update orders.

What should these structure-preserving maps be? Using the language of category theory, we are looking for the morphisms in a category where the objects are SDS. There are choices in this process, and we will be using Definition 4.41 below [101]. For an alternative approach we refer to [102].

Definition 4.41 (SDS morphism). Let $(Y, \mathbf{F}_Y, \pi)$ and $(Z, \mathbf{G}_Z, \sigma)$ be two SDS. An SDS-*morphism* between $(Y, \mathbf{F}_Y, \pi)$ and $(Z, \mathbf{G}_Z, \sigma)$ is a triple

$$(\varphi, \eta, \Phi) \colon (Y, \mathbf{F}_Y, \pi) \longrightarrow (Z, \mathbf{G}_Z, \sigma) ,$$

where $\varphi \colon Y \longrightarrow Z$ is a graph morphism, $\eta \colon S_Z \longrightarrow S_Y$ is a map that satisfies $\eta(\sigma) = \pi$, and Φ is a digraph morphism of phase spaces

$$\Phi \colon \Gamma(Z, \mathbf{G}_Z, \sigma) \longrightarrow \Gamma(Y, \mathbf{F}_Y, \pi) .$$

If all three maps φ, η, and Φ are bijections, we call (φ, η, Φ) an SDS-*isomorphism*.

A priori it is not clear that there are any SDS morphisms. The following example gives an example of an SDS morphism and also illustrates key elements of the theory developed in this section.

Example 4.42. The map $\varphi \colon Q_2^3 \longrightarrow K_4$ defined by $\varphi(0) = \varphi(7) = 1$, $\varphi(1) = \varphi(6) = 2$, $\varphi(2) = \varphi(5) = 3$, and $\varphi(3) = \varphi(4) = 4$ is a graph morphism. It identifies vertices on spatial diagonals and is depicted in Figure 4.16. Let $\sigma = (1, 3, 2, 4) \in S_Z$, let $\pi = (0, 7, 2, 5, 1, 6, 3, 4) \in S_Y$, and let $\eta \colon S_Z \longrightarrow S_Y$ be a map with $\eta(\sigma) = \pi$. Moreover, we define $\chi \colon \mathbb{F}_2^4 \longrightarrow \mathbb{F}_2^8$ by $\chi(x_1, x_2, x_3, x_4) = (x_1, x_2, x_3, x_4, x_4, x_3, x_2, x_1)$. If we take $x = (0, 1, 0, 0)$, we get the following commutative diagram:

$$
\begin{array}{ccc}
(0,1,0,0) & \xrightarrow{\;[\mathbf{Nor}_{K_4}, \sigma]\;} & (0,0,0,1) \\[4pt]
\Big\uparrow{\scriptstyle \chi} & & \Big\downarrow{\scriptstyle \chi} \\[4pt]
(0,1,0,0,0,0,1,0) & \xrightarrow{\;[\mathbf{Nor}_{Q_2^3}, \pi]\;} & (0,0,0,1,1,0,0,0)
\end{array}
$$

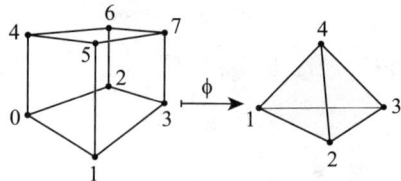

Fig. 4.16. A graph morphism from Q_2^3 to K_4.

Here is the key observation: We can compute the system state transition $(0, 1, 0, 0, 0, 0, 1, 0) \mapsto (0, 0, 0, 1, 1, 0, 0, 0)$ under $[\mathbf{Nor}_{Q_2^3}, \pi]$ using $[\mathbf{Nor}_{K_4}, \sigma]$. Therefore, we can obtain information about the phase space of $(Q_2^3, \mathbf{Nor}_{Q_2^3}, \pi)$ from the simpler and smaller SDS phase space of $(K_4, \mathbf{Nor}_{K_4}, \sigma)$.

We invite you to verify that χ induces a morphism of phase spaces $\Phi \colon \Gamma(Z, \mathbf{F}_Z, \sigma) \longrightarrow \Gamma(Y, \mathbf{F}_Y, \pi)$. Accordingly, (φ, η, Φ) is an SDS morphism.

$\diamond$

In Section 4.4.3 we will give a more general answer to the question about existence of SDS morphisms. We will show that any covering map [Eq. (3.5)] induces an SDS morphism in a natural way.

4.4.1 Covering Maps

In this section we consider covering maps $\varphi \colon Y \longrightarrow Z$, that is, for all $v \in \mathrm{v}[Y]$ the restriction map

$$\varphi|_{\mathsf{Star}_Y(v)} \colon \mathsf{Star}_Y(v) \longrightarrow \mathsf{Star}_Z(\varphi(v)) \tag{4.68}$$

is a graph isomorphism. The graph $\mathsf{Star}_Y(v)$ is the subgraph of Y given by $\mathrm{e}[\mathsf{Star}_Y(v)] = \{e \in \mathrm{e}[Y] \mid \omega(e) = v \text{ or } \tau(e) = v\}$ and $\mathrm{v}[\mathsf{Star}_Y(v)] = \{v' \in \mathrm{v}[Y] \mid v' = \omega(e) \vee v' = \tau(e), \ e \in \mathrm{e}[\mathsf{Star}_Y(v)]\}$.

4.4.2 Properties of Covering Maps

In later proofs we will need the following lemma, which can be viewed as the graph equivalent of a basic property of covering maps over topological spaces; see [103, 104].

Lemma 4.43. *Let Y and Z be non-empty, undirected, connected graphs and let $\varphi \colon Y \longrightarrow Z$ be a covering map. Then we have*

$$\forall \, x, y \in \mathrm{v}[Z] \; : \; |\varphi^{-1}(x)| = |\varphi^{-1}(y)| \, . \tag{4.69}$$

Proof. Let x and y be two Z-vertices and assume $|\varphi^{-1}(x)| > |\varphi^{-1}(y)|$. Since Z is connected, we can without loss of generality assume that there exists an edge e in Z such that $\omega(e) = x$ and $\tau(e) = y$. For any $\xi \in \varphi^{-1}(x)$ local bijectivity guarantees the existence of a Y-edge e' such that $\omega(e') = \xi$ and $\tau(e') = \eta$ with $\eta \in \varphi^{-1}(y)$. But this is impossible in view of $|\varphi^{-1}(x)| > |\varphi^{-1}(y)|$ and the lemma follows by contradiction. $\square$

In the context of covering maps the set $\varphi^{-1}(x)$ is usually called the *fiber* over x. Since all fibers have the same cardinality, we conclude that the order of $\varphi(Y)$ divides the order of Y.

The following is another useful fact that is needed later. For the statement of the result we need the notion of distance of vertices in an undirected graph. Let $v, v' \in \mathrm{v}[Y]$. The distance between v and v' in Y is the length of a shortest path connecting v and v', or ∞ if no such path exists. We write the distance between v and v' in Y as $d_Y(v, v')$. It satisfies the usual properties of a metric, which you can easily verify.

Proposition 4.44. *Let* Y, Z *be undirected graphs and* $\varphi \colon Y \longrightarrow Z$ *a covering map. Then for any* $u \in \mathrm{v}[Z]$ *and* $v, v' \in \varphi^{-1}(u)$ *with* $v \neq v'$ *we have* $d_Y(v, v') \geq 3$. *In particular, the fiber over* $\varphi(v)$ *is an independent set for any* $v \in \mathrm{v}[Y]$.

Proof. Let $v, v' \in \varphi^{-1}(u)$ with $v \neq v'$, and suppose $d_Y(v, v') = 1$. Then $\varphi|_{\mathsf{Star}_Y(v)} \colon \mathsf{Star}_Y(v) \longrightarrow \mathsf{Star}_Z(u)$ cannot be a bijection. If $d_Y(v, v') = 2$, then let $v'' \in \mathrm{v}[Y]$ be a vertex with $d_Y(v, v'') = 1$ and $d_Y(v', v'') = 1$. Since both v and v' are mapped to u, the restriction map $\varphi|_{\mathsf{Star}_Y(v'')} \colon \mathsf{Star}_Y(v'') \longrightarrow \mathsf{Star}_Z(\varphi(v''))$ cannot be a graph isomorphism. The last statement is clear. $\square$

Building on the proof of Lemma 4.43 we also have the following result, which is a special case of a more general result from [105]. It can be considered as the graph equivalent of the unique path-lifting property of covering maps of topological spaces.

Lemma 4.45. *Let* $\varphi \colon Y \longrightarrow Z$ *be a covering map and let* $v \in \mathrm{v}[Y]$. *Then any subtree* T' *of* Z *containing* $\varphi(v)$ *lifts back to a unique subtree* T *of* Y *containing* v.

This only holds when T' is a subtree of Z but fails to hold for general subgraphs Z' of Z containing $\varphi(v)$. Why?

4.4.3 Reduction of SDS

In this section we prove that a covering map $\varphi \colon Y \longrightarrow Z$ induces an SDS-morphism in a natural way. Without loss of generality we may assume that Z is connected. We can then conclude using Lemma 4.45 that φ is surjective. In the following we set $n = |\mathrm{v}[Y]|$ and $m = |\mathrm{v}[Z]|$.

Constructing the update order map η_φ. Let $\pi \in S_Z$. We define $s(\pi_k)$ to be the sequence of elements from the fiber $\varphi^{-1}(\pi_k)$ ordered by some total order on $\mathrm{v}[Y]$. As the image of π under η_φ we take the concatenation of the sequences $s(\pi_1)$ through $s(\pi_m)$, that is,

$$\eta_\varphi(\pi) = (s(\pi_1)|s(\pi_2)|\dots|s(\pi_m)) . \tag{4.70}$$

The map η_φ naturally induces a map $\hat{\eta}_\varphi \colon \mathsf{Acyc}(Z) \longrightarrow \mathsf{Acyc}(Y)$ via the bijection f_Y such that the diagram

$$\begin{array}{ccc}
S_Z & \xrightarrow{\;\;\eta_\varphi\;\;} & S_Y \\
\downarrow{\scriptstyle f'_Y} & & \downarrow{\scriptstyle f'_Y} \\
\mathsf{Acyc}(Z) & \xrightarrow{\;\;\hat{\eta}_\varphi\;\;} & \mathsf{Acyc}(Y)
\end{array}$$

where $\sigma \mapsto f'_Y(\sigma) = \mathcal{O}(\sigma)$, is commutative.

4.30. Verify the commutative diagram above. [2-]

Example 4.46. We revisit Example 4.42 and consider the covering map $\varphi\colon Q_2^3 \longrightarrow K_4$. We observe that $\sigma = (1,3,2,4) \in S_Z$ is mapped to $\eta_\varphi(\sigma) = \pi = (0,7,2,5,1,6,3,4) \in S_Y$ and the acyclic orientation $\mathcal{O}_Z(\sigma)$ is mapped to the acyclic orientation $\mathcal{O}_Y(\pi)$. ◇

We are now ready to complete the construction by providing the digraph morphism Φ_φ.

Theorem 4.47. *Let $\varphi\colon Y \longrightarrow Z$ be a covering map of undirected, connected graphs Y and Z, and let $\chi\colon K^m \longrightarrow K^n$ be the map $(\chi(x))_{v'} = x_{\varphi(v')}$. Suppose all vertex functions over Y and Z are induced by the sequence $(g_k)_k$ of symmetric functions. Then the map*

$$\Phi_\varphi\colon \Gamma(Z, \mathbf{F}_Z, \pi) \longrightarrow \Gamma(Y, \mathbf{F}_Y, \eta_\varphi(\pi))$$

induced by χ is a morphism of directed graphs and

$$(\varphi, \eta_\varphi, \Phi_\varphi)\colon (Y, \mathbf{F}_Y, \eta_\varphi(\pi)) \longrightarrow (Z, \mathbf{F}_Z, \pi) \tag{4.71}$$

is an SDS *morphism.*

Proof. We already have our candidates for the two first components φ and η of the SDS morphism. It remains to prove that the map Φ_φ induced by χ is a morphism of (directed) graphs.

According to Lemma 4.45, φ is surjective and Proposition 4.44 guarantees that $\varphi^{-1}(v)$ is an independent set of Y for all $v \in \mathrm{v}[Z]$. Therefore, for any $v \in \mathrm{v}[Z]$ the (composition) product of local maps

$$\prod_{v' \in \varphi^{-1}(v)} F_{v'}$$

is independent of composition order and is accordingly well-defined. Moreover, since φ is a covering map, and since the maps g_k are symmetric, the vertex functions f_v satisfy

$$f_v(x[v; Z]) = f_{v'}((\chi(x))[v'; Y]) \tag{4.72}$$

for any $v \in \mathrm{v}[Z]$ and $v' \in \mathrm{v}[Y]$ such that $\varphi(v') = v$.

We claim that the diagram

$$
\begin{array}{ccc}
K^{|\mathbf{v}[Z]|} & \xrightarrow{\ \chi\ } & K^{|\mathbf{v}[Y]|} \\
\downarrow{\scriptstyle F_{v,Z}} & & \downarrow{\scriptstyle \prod\limits_{v'\in\varphi^{-1}(v)} F_{v',Y}} \\
K^{|\mathbf{v}[Z]|} & \xrightarrow{\ \chi\ } & K^{|\mathbf{v}[Y]|}
\end{array}
\tag{4.73}
$$

commutes, that is,

$$
\chi \circ F_{v,Z} = \prod_{v'\in\varphi^{-1}(v)} F_{v',Y} \circ \chi \,.
\tag{4.74}
$$

Let us first analyze $\prod_{v'\in\varphi^{-1}(v)} F_{v',Y} \circ \chi$. The local map $F_{v',Y}(\chi(x))$ updates the state of v' via the vertex function $f_{v'}$ as $f_{v'}((\chi(x))[v';Y])$. By definition, we have $(\chi(x))_{v'} = x_{\varphi(v')}$, and since $\varphi(B_Y(v')) = B_Z(v)$ we can conclude

$$
f_{v'}((\chi(x))[v';Y]) = f_{v'}(x[v;Z]) = f_v(x[v;Z]) \,.
$$

Therefore, $\prod_{v'\in\varphi^{-1}(v)} F_{v',Y}$ is a well-defined product of Y-local maps that updates the vertices $v' \in \varphi^{-1}(v)$ of Y based on the family of states $(x_{\varphi(v_j')} \mid \varphi(v_j') \in B_Z(v))$ to the state $(F_{v',Y}(\chi(x)))_{v'}$.

We next compute $\chi \circ F_{v,Z}(x)$. By definition, $F_{v,Z}(x)$ updates the state of the vertex v of Z using the vertex function f_v as $f_v(x[v;Z])$. In view of $(\chi(x))_{v'} = x_{\varphi(v')}$, we obtain for any Y-vertex v'

$$
(\chi \circ F_{v,Z}(x))_{v'} = (F_{v,Z}(x))_v \,.
$$

That is, $\chi \circ F_{v,Z}(x)$ updates the states of the vertices $v' \in \varphi^{-1}(v)$ in Y to the state $(F_{v,Z}(x))_v$. Since $f_v(x[v;Z]) = f_{v'}((\chi(x))[v';Y])$, we derive

$$
\forall\, v' \in \varphi^{-1}(v), \quad (F_{v,Z}(x))_v = (F_{v',Y}(\chi(x)))_{v'},
$$

from which we conclude

$$
\chi \circ F_{v,Z} = \prod_{v'\in\varphi^{-1}(v)} F_{v',Y} \circ \chi \,.
$$

To prove that the diagram

$$
\begin{array}{ccc}
K^{|\mathbf{v}[Z]|} & \xrightarrow{\ \chi\ } & K^{|\mathbf{v}[Y]|} \\
\downarrow{\scriptstyle [\mathbf{F}_Z,\pi]} & & \downarrow{\scriptstyle [\mathbf{F}_Y,\eta_\varphi(\pi)]} \\
K^{|\mathbf{v}[Z]|} & \xrightarrow{\ \chi\ } & K^{|\mathbf{v}[Y]|}
\end{array}
$$

is commutative, we observe that for $\pi = (\pi_1, \ldots, \pi_m)$

$$
[\eta_\varphi(\pi)]_Y = [(\varphi^{-1}(\pi_1), \ldots, \varphi^{-1}(\pi_m))]_Y
$$

holds, where $[\ ]_Y$ denotes the equivalence class with respect to $\sim_Y$ [Section 3.1.4, Eq. (3.13)]. This implies that

$$[\mathbf{F}_Y, \eta_\varphi(\pi)] = \prod_{v=\pi_1}^{\pi_m} \left[\prod_{v' \in \varphi^{-1}(v)} F_{v',Y} \right].$$

We inductively apply $\prod_{v' \in \varphi^{-1}(v)} F_{v,Y} \circ \chi = \chi \circ F_{v,Z}$ and conclude

$$\prod_{v=\pi_1}^{\pi_m} \left[\prod_{v' \in \varphi^{-1}(v)} F_{v',Y} \right] \circ \chi = \chi \circ \prod_{v=\pi_1}^{\pi_m} F_{v,Z} ,$$

or

$$[\mathbf{F}_Y, \eta_\varphi(\pi)] \circ \chi = \chi \circ [\mathbf{F}_Z, \pi] . \tag{4.75}$$

Hence, the χ-induced map Φ_φ is a morphism of (directed) graphs, and the proof of the theorem is complete. □

From Eq. (4.75) we see that the phase space of the SDS over Z is embedded in the phase space of the SDS over Y via Φ_φ.

Since the graph Z generally has fewer vertices than Y, it is clear that the Z phase space is smaller than the Y phase space, hence the term *reduction*. How much smaller is the Z phase space? If we assume, for instance, binary states and that φ is a double covering, that is, $m = n/2$ and each fiber has size 2, the number of states is $2^{n/2}$ and 2^n, respectively.

Example 4.48. Here we extend Example 4.42. For reference, the covering map $\varphi: Q_2^3 \longrightarrow K_4$ is given by $\varphi(0) = \varphi(7) = 1$, $\varphi(1) = \varphi(6) = 2$, $\varphi(2) = \varphi(5) = 3$, and $\varphi(3) = \varphi(4) = 4$, and it is illustrated in Figure 4.16.

Here Φ_φ maps $x = (x_1, x_2, x_3, x_4)$ into $(x_1, x_2, x_3, x_4, x_4, x_3, x_2, x_1)$. Further let $\sigma = (1, 2, 3, 4) \in S_Z$. The corresponding update order over Y is $\pi = \eta_\varphi(\sigma) = (0, 7, 1, 6, 2, 5, 3, 4)$.

Theorem 4.47 now gives us an embedding of the phase space of the SDS $(K_4, \mathbf{Minority}_{K_4}, \sigma)$ into the phase space of $(Q_2^3, \mathbf{Minority}_{Q_2^3}, \pi)$. As you can easily verify, $(K_4, \mathbf{Minority}_{K_4}, \sigma)$ has precisely two periodic orbits of length five and no fixed points. The two 5-orbits are shown in the left column of Figure 4.17. Note that for representational purposes we have encoded binary tuples as decimal numbers using (4.17), e.g., $(1, 1, 0, 0)$ is represented as the decimal number 3. Figure 4.17 shows that $\Gamma(K_4, \mathbf{Minority}_{K_4}, \sigma)$ is embedded in $\Gamma(Q_2^3, \mathbf{Minority}_{Q_2^3}, \pi)$. ◇

Example 4.49. As another illustration of Theorem 4.47, we consider SDS with vertex functions induced by nor₃ and nor₄ over the graphs Y and Z shown in Figure 4.18(a) on the top and bottom, respectively. Note that in this case the graphs are not regular. The map φ that identifies the vertices v and v' for $v = a, b, c, d, e$ is clearly a covering map and by Theorem 4.47 we have an SDS morphism where the two other maps are η_φ and

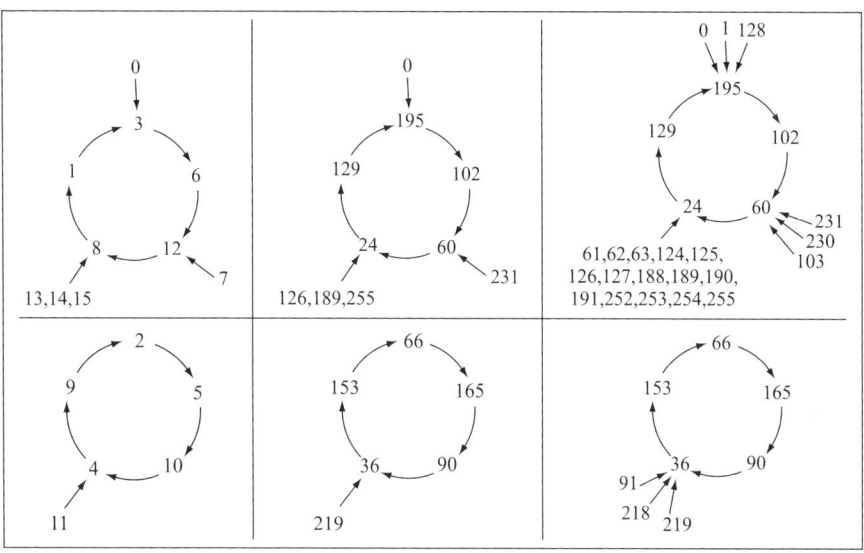

Fig. 4.17. Example 4.48: The left column shows the phase space of $(K_4, \mathbf{Minority}_{K_4}, \sigma)$. The middle column shows the image of $\Gamma(K_4, \mathbf{Minority}_{K_4}, \sigma)$ under the embedding map Φ_φ. The right column shows the components of $\Gamma(Q_2^3, \mathbf{Minority}_{Q_2^3}, \pi)$ that embed $\Gamma(K_4, \mathbf{Minority}_{K_4}, \sigma)$. Note that binary tuples are encoded as decimal numbers.

Φ_φ. Figure 4.18(b) illustrates the map η_φ and Figure 4.18(c) shows how the unique component containing a 3-cycle of $\Gamma(Z, \mathbf{F}_Z, (a, b, c, d, e))$ embeds into $\Gamma(Y, \mathbf{F}_Y, (a, a', b, b', c, c', d, d', e, e'))$. In fact, $\Gamma(Z, \mathbf{F}_Z, (a, b, c, d, e))$ contains four 2-cycles and one 3-cycle, while $\Gamma(Y, \mathbf{F}_Y, (a, a', b, b', c, c', d, d', e, e'))$ has fourteen 2-cycles, one 3-cycle, two 4-cycles, two 6-cycles and eight 8-cycles.

$\diamond$

4.31. What is the most general class of functions $(f_k)_k$ for which Theorem 4.47 still holds? Extend Theorem 4.47 to word-SDS.   [2-]

4.4.4 Dynamical Equivalence Revisited

In Proposition 4.30 we proved the conjugation formula

$$[\mathbf{F}_Y, \gamma\pi] = \gamma \circ [\mathbf{F}_Y, \pi] \circ \gamma^{-1} .$$

Using Theorem 4.47, we can derive the above conjugation formula directly since every graph automorphism is in particular a covering map. In fact, we can reframe the entire concept of equivalence of SDS using SDS morphisms.

Corollary 4.50. *Let* Y *be an undirected graph, let* $\gamma \in \mathsf{Aut}(Y)$, *and let* $\pi \in S_Y$. *For any sequence of symmetric functions* $(g_k)_k$ *with* $g_k \colon K^k \longrightarrow K$ *and*

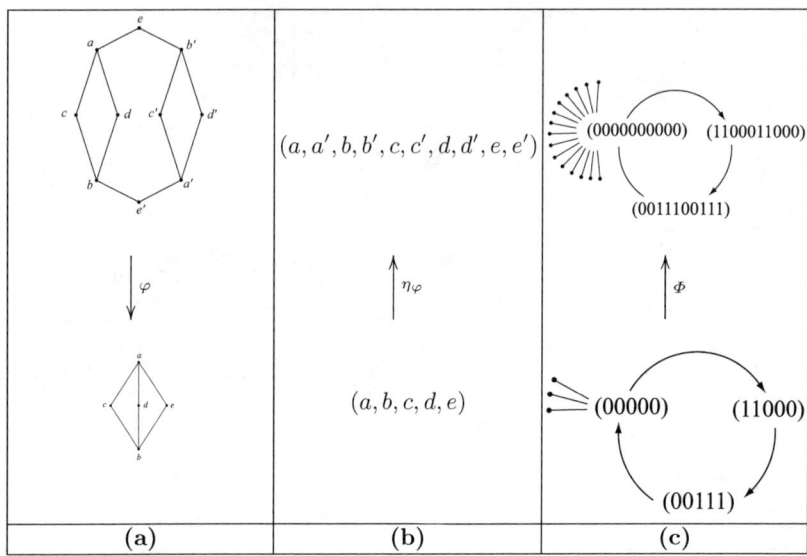

Fig. 4.18. An illustration of Theorem 4.47. The maps η_φ and Φ are shown for the covering map φ of Example 4.49.

any pair of induced SDS *of the form* $[\mathbf{F}_Y, \pi]$ *and* $[\mathbf{F}_Y, \eta_\gamma(\pi)]$, *we have an* SDS *isomorphism*

$$(\gamma, \eta_\gamma, \Phi) \colon [\mathbf{F}_Y, \eta_\gamma(\pi)] \longrightarrow [\mathbf{F}_Y, \pi] \,, \qquad (4.76)$$

where $\eta_\gamma(\pi) = \gamma^{-1}\pi$.

Proof. The proof is immediate since any graph automorphism is in particular a covering map. $\qquad\square$

4.4.5 Construction of Covering Maps

Theorem 4.47 shows that covering maps naturally induce SDS morphisms, and it thus motivates the study of covering maps over a given graph Y. This is similar, for instance, to group representation theory, where a given group is mapped into automorphism groups of vector fields. Here a given SDS is "represented" via its morphisms. To ask for all graphs that are covering images of a fixed undirected graph Y is a purely graph-theoretic question motivated by SDS and complements the research on graph covering maps which typically revolves around the problem of finding a common graph covering Y for a collection of graphs $\{Z_i\}$ as in [106].

In this section we will analyze covering maps from the generalized n-cube and the circle graph.

Cayley Graphs

Cayley graphs encode the structure of groups and play a central role in combinatorial and geometric group theory. There are more general definitions than the one we give below, but this will suffice here. We largely follow [107].

Definition 4.51. Let G be a group with generating set S. The Cayley graph $\mathrm{Cay}(G, S)$ is the directed graph with vertex set the elements of G and where (g, g') is an edge if and only if there exists $s \in S$ such that $g' = gs$.

If $g' = gs$, it is common to label the edge (g, g') with the element s.

Example 4.52. The group S_3 has generating set $\{(1,2), (1,2,3)\}$. Let $a = (1,2,3)$ and $b = (1,2)$. The Cayley graph $\mathrm{Cay}(S_3, \{a, b\})$ is shown in Figure 4.19. What is the group element $a^2 b a^2 b$? This is easy to answer using the

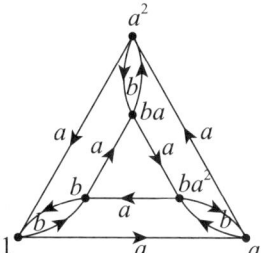

Fig. 4.19. The Cayley graph $\mathrm{Cay}(S_3, \{a = (1,2,3), b = (1,2)\})$.

Cayley graph. The directed walk starting at the identity element following the edges labeled a, a, b, a, a, and b in this order gives us the answer: 1.   ◇

Example 4.53. The cube Q_2^3 from the earlier examples is the Cayley graph of $\mathbb{F}_2^3$ viewed as an additive group G with generating set $S = \{e_1, e_2, e_3\}$ and the obvious relations, e.g., $2e_i = 0$ and $e_i + e_j = e_j + e_i$. The subgroup $H = \{(0,0,0), (1,1,1)\} < G$ acts on G by translation. This action naturally induces the orbit graph $H \backslash Q_2^3$ given by

$$\mathrm{v}[H \backslash Q_2^3] = \{H(0,0,0) := 0, H(1,0,0) := 1, H(0,1,0) := 2, H(0,0,1) := 3\},$$
$$\mathrm{e}[H \backslash Q_2^3] = \{\{0,1\}, \{0,2\}, \{0,3\}, \{1,2\}, \{1,3\}, \{2,3\}\},$$

that is, (a graph isomorphic to) the complete graph on four vertices. Accordingly, we have obtained the covering map from Example 4.42 as the projection map π_H induced by the subgroup H.   ◇

4.4.6 Covering Maps over Q_α^n

We now proceed to the general setting. Let F be the finite field with $|F| = \alpha = p^k$. Recall that the *generalized n-cube* is the combinatorial graph Q_α^n defined by

$$\mathrm{v}[Q_\alpha^n] = \{x = (x_1, \ldots, x_n) \in F^n\},$$
$$\mathrm{e}[Q_\alpha^n] = \{\{x, y\} \mid x, y \in F^n, \, d_H(x, y) = 1\}, \tag{4.77}$$

where $d_H(x, y)$ is the Hamming distance of $x, y \in F^n$, i.e., the number of coordinates in which x and y differ. The automorphism group of the generalized n-cube is isomorphic to the semidirect product of S_n and F^n, that is, $\mathsf{Aut}(Q_\alpha^n) \cong F^n \rtimes S_n$. The subgroup $\mathsf{Aut}_0(Q_\alpha^n) = \{\gamma \in \mathsf{Aut}(Q_\alpha^n) \mid \gamma(0) = 0\}$ of $\mathsf{Aut}(Q_\alpha^n)$ is isomorphic to S_n. Accordingly, any element $\gamma \in \mathsf{Aut}_0(Q_\alpha^n)$ is F-linear and we can consider $\mathsf{Aut}_0(Q_\alpha^n)$ as a subgroup of $\mathbf{GL}(F^n)$.

We can now generalize what we saw in Example 4.53. First, any sub-vectorspace $H < F^n$ can be considered as a subgroup of $\mathsf{Aut}(Q_\alpha^n)$, and we have the morphism

$$\pi_H \colon Q_\alpha^n \longrightarrow H \setminus Q_\alpha^n .$$

In Theorem 4.54 below we give conditions on the sub-vectorspace $H < F^n$ such that $\pi_H \colon Q_\alpha^n \longrightarrow H \setminus Q_\alpha^n$ is a covering map. In this construction a vertex v of Q_α^n is, of course, mapped to $v + H$ under π_H. We note that if the projection map is to be a covering map, then any vertex v and its neighbor vertices $v + ke_i$ in Q_α^n where $k \in F^\times$ and $i = 1, \ldots, n$ must be mapped to distinct vertices. Here $F^\times$ denotes the multiplicative group of F. Clearly, a necessary condition for $\pi_H \colon Q_\alpha^n \longrightarrow H \setminus Q_\alpha^n$ to be a covering map is

$$|F^n/H| \geq 1 + n|F^\times|, \tag{4.78}$$

since otherwise it would be impossible for π_H to be a local injection. By construction the projection map π_H is a local surjection, so if we can show that for all $k \in F^\times$ and for all $v, v' \in \{0, ke_i\}$ with $v \neq v'$: $(v + H) \cap (v' + H) = \varnothing$, then it would follow that π_H is also a local injection and thus a covering map. However, H may not satisfy this condition but may still satisfy (4.78). If this is the case, then Theorem 4.54 ensures that we can find a subspace H' isomorphic to H such that $\pi_{H'}$ is a covering map. Even though this is an existence theorem, the proof also gives an algorithm for constructing the covering maps. We outline the algorithm after the proof.

Theorem 4.54. *Let $G < F^n$ be a sub-vectorspace of F^n that satisfies*

$$|F^n/G| \geq 1 + n\,|F^\times| .$$

Then there exists a vectorspace H isomorphic to G such that the projection map $\pi_H \colon Q_\alpha^n \longrightarrow H \setminus Q_\alpha^n$ is a covering.

The proof of Theorem 4.54 will follow from Lemmas 4.55 and 4.56 below. Let us begin by introducing some notation. For a subspace $H < F^n$ we define the property (#) by

$$(\#) \quad \forall k \in F^\times \, \forall x \neq y, \, x, y \in \{0, ke_1, \ldots, ke_n\} \; : \; (x + H) \cap (y + H) = \varnothing .$$
$$\tag{4.79}$$

Clearly, this is the condition a sub-vectorspace H needs to satisfy in order for π_H to be a local injection.

Lemma 4.55. *For any subspace $G' < F^n$ we have*

$$|F^n/G'| \geq 1 + n\,|F^\times| \iff \exists\, G,\; G \cong G';\; G \text{ has property } (\#). \qquad (4.80)$$

Proof. Assume $|F^n/G'| \geq 1 + n\,|F^\times|$. We claim that the vectorspace F^n/G' contains $n|F^\times|$ distinct elements of the form $k\varphi_i + G'$, $i = 1, \ldots, n$, where $\{\varphi_1, \ldots, \varphi_n\}$ is a basis for F^n and $k \in F^\times$.

To prove this we take an arbitrary basis $\{v_1, \ldots, v_s\}$ of G' and extend it to a basis $\{v_1, \ldots, v_s, v_{s+1}, \ldots, v_n\}$ of F^n. Since $|F^n/G'| \geq 1 + n\,|F^\times|$, we have

$$|F^n/G' \setminus \{\, k\,v_i + G' \mid i = s+1, \ldots, n,\; k \in F \,\}| \geq s\,|F^\times|\,. \qquad (4.81)$$

The (Abelian) group $F^\times$ acts on F^n/G' via the restriction of scalar multiplication; hence,

$$F^n/G' = \{0 + G'\} \cup \bigcup_{j=s+1}^{n} {}^{\textstyle\cdot}\, F^\times(v_j + G') \cup \bigcup_{j=1}^{t} {}^{\textstyle\cdot}\, F^\times(w_j + G'),$$

and (4.81) guarantees that $t \geq s$. From this we conclude

$$\exists\, t \geq s;\quad F^n/G' \setminus \{\, k\,v_i + G' \mid i = s+1, \ldots, n,\; k \in F \,\} = \bigcup_{j=1}^{t} {}^{\textstyle\cdot}\, F^\times(w_j + G')\,.$$

We next define the sequence $(\varphi_i)_{1 \leq i \leq n}$ as follows:

$$\varphi_i = v_i + w_i \quad \text{for } i = 1, \ldots, s,$$
$$\varphi_i = v_i \qquad\;\; \text{for } i = s+1, \ldots, n\,.$$

In view of $\sum_i \lambda_i \varphi_i = \sum_i \lambda_i v_i + \sum_{i=1}^{s} \lambda_i w_i$, any linear relation of the form $\sum_i \lambda_i \varphi_i = 0$ implies that for $i = 1, \ldots, s$ we have $\lambda_i = 0$, since $\sum_{i=1}^{s} \lambda_i w_i$ is generated by $\{v_{s+1}, \ldots, v_n\}$. Therefore, we obtain $\sum_{i=1}^{s} \lambda_i w_i = 0$ and consequently we have $\lambda_i = 0$ for $i = s+1, \ldots, n$. Accordingly, $\{\varphi_1, \ldots, \varphi_n\}$ forms a basis of F^n. Since $\{w_i + G' \mid i = 1, \ldots, s\}$ is a set of representatives for the group action of $F^\times$ on F^n/G', we get

$$|\{k\varphi_i + G' \mid k \in F^\times,\; i = 1, \ldots, n\,\}| = |F^\times|\,n\,,$$

and the claim follows.

Let f be the F^n-isomorphism defined by $f(\varphi_i) = e_i$, for $i = 1, \ldots, n$. Clearly, the set $\{ke_i + f(G') \mid k \in F^\times,\; i = 1, \ldots, n\,\}$ has the property

$$|\{ke_i + f(G') \mid k \in F^\times,\; i = 1, \ldots, n\,\}| = n\,|F^\times|$$

and the proof is complete. $\qquad\square$

Lemma 4.56. *For each sub-vectorspace $H < F^n$ with property $(\#)$ the graph $H \setminus Q_\alpha^n$ is connected, undirected, and loop-free, and the natural projection*

$$\pi_H \colon Q_\alpha^n \longrightarrow H \setminus Q_\alpha^n, \quad v \mapsto H(v) = v + H$$

is a covering map.

Proof. The projection map π_H is linear and is a local surjection by construction. Property (#) ensures that π_H is locally injective. It remains to prove that π_H is a graph morphism. Since $\mathsf{Aut}(Q_\alpha^n) \cong F^n \rtimes S_n$, H is a subgroup of $\mathsf{Aut}(Q_\alpha^n)$ and acts on Q_α^n-edges; thus, π_H is a covering map $(\mathrm{e}[H \setminus Q_\alpha^n] = \{H(\{v, v + e_i\}) \mid i = 1, \ldots, n, \ v \in \mathrm{v}[Q_\alpha^n]\})$. Since π_H is locally injective, $H \setminus Q_\alpha^n$ is loop-free.   □

Here is an algorithm for computing the sub-vectorspace H in Theorem 4.54 and for deriving the covering map π_H.

Algorithm 4.57 (Construction of Q_α^n covering maps). Assume $G < F^n$ satisfies the conditions in Theorem 4.54. Using the same notation we can derive covering maps, and hence reduced dynamical systems, as follows:

1. Pick a basis $\{v_1, \ldots, v_s\}$ for G.
2. Extend this basis to a basis $\{v_1, \ldots, v_s, v_{s+1}, \ldots, v_n\}$ for F^n.
3. The action of $F^\times$ on F^n/G by scalar multiplication allows us to construct a collection of s vectors $(w_i)_1^s$ (orbit representatives) contained in $\mathrm{Span}(v_{s+1}, \ldots, v_n)$ that are not scalar multiples of each other or any of the vectors v_i for $s + 1 \leq i \leq n$. The set of s such vectors w_i can easily be "guessed," at least for small examples.
4. Define ϕ_i by

$$\phi_i = \begin{cases} v_i + w_i & \text{if } i = 1, \ldots, s, \\ v_i & \text{otherwise.} \end{cases}$$

5. Let f be the F^n-isomorphism given by $f(\phi_i) = e_i$ for $1 \leq i \leq n$.
6. The isomorphic vectorspace H is given by $H = f(G)$, and the covering map is given by $\pi_H \colon Q_\alpha^n \longrightarrow H \setminus Q_\alpha^n$.

The following examples illustrate the above algorithm.

Example 4.58. Consider the graph $Y = Q_3^4$. Let G be the two-dimensional subspace of $F^4 = \mathbb{F}_3^4$ spanned by $v_1 = (1, 0, 0, 0)$ and $v_2 = (0, 1, 0, 0)$. Clearly, G is not a distance-3 subspace. We have

$$|F^4/G| = 9 \geq 1 + 4 \cdot 2 = 1 + 4|F^\times|,$$

so by Theorem 4.54 there exists a subspace H isomorphic to G for which π_H is a covering map. By Proposition 4.44 we must have that H is a set with minimal Hamming distance 3. Attempting to construct the subspace H by trial and error may take some time and patience. However, with the help of the algorithm above it now becomes more or less mechanical. Here is how it can be done:

We extend the basis of G consisting of $v_1 = (1, 0, 0, 0)$ and $v_2 = (0, 1, 0, 0)$ to a basis for F^4 using the vectors $v_3 = (0, 0, 1, 0)$ and $v_4 = (0, 0, 0, 1)$. We need to find two vectors in $\mathrm{Span}\{v_3, v_4\}$ that are not scalar multiples of each other or of v_3 or v_4. Two such vectors are $w_1 = (0, 0, 1, 2)$ and $w_2 = (0, 0, 1, 1)$. By the algorithm we obtain

$$v_1 = (1,0,0,0) \qquad w_1 = (0,0,1,2) \qquad \phi_1 = (1,0,1,2)$$
$$v_2 = (0,1,0,0) \qquad w_2 = (0,0,1,1) \qquad \phi_2 = (0,1,1,1)$$
$$v_3 = (0,0,1,0) \qquad\qquad — \qquad\qquad \phi_3 = (0,0,1,0)$$
$$v_4 = (0,0,0,1) \qquad\qquad — \qquad\qquad \phi_4 = (0,0,0,1)$$

The F^4-isomorphism f satisfying $f(\phi_i) = e_i$ is straightforward to compute, and it has standard matrix representation

$$f_M = \begin{bmatrix} 1 & 0 & 0 & 0 \\ 0 & 1 & 0 & 0 \\ 2 & 2 & 1 & 0 \\ 1 & 2 & 0 & 1 \end{bmatrix} ,$$

which you should verify for yourself. The subspace H is now given as $H = f(G)$, and we get

$$H = \left\{ \begin{matrix} (0,0,0,0),\ (1,0,2,1),\ (2,0,1,2), \\ (0,1,2,2),\ (1,1,1,0),\ (2,1,0,1), \\ (0,2,1,1),\ (1,2,0,2),\ (2,2,2,0) \end{matrix} \right\} .$$

You should verify that $H = f(G)$ is a distance-3 set. What is the graph $H \setminus Q_3^4$? It is a combinatorial graph, it is connected, it is regular of degree 8, and it has size 9. It follows that $H \setminus Q_3^4$ equals K_9 (up to isomorphism).   ◇

When you compute the map f, it can be helpful to write the equations $f(\phi_i) = e_i$ in matrix form. If Φ denotes the matrix with the ϕ_i's as column vectors, we get

$$f\Phi = I_{n \times n} ,$$

and it is clear that f is the inverse of the matrix Φ.

Example 4.59. As another illustration of Theorem 4.54, we take the graph $Y = Q_3^3$ and ask if we can find a subspace $H < F^3 = \mathbb{F}_3^3$ with $\dim(F^3/H) = 2$ such that its induced orbit graphs $H \setminus Q_3^3$ are graphs of degree 6. If this is the case, then H must satisfy $d_{Q_3^3}(h, h') \geq 3$ for any $h, h' \in H$ with $h' \neq h$. Since a one-dimensional subspace G satisfies $F^3/G = 9 \geq 1 + 3 \cdot 2$, Theorem 4.54 guarantees that we can find such a subspace. In this case it is easy, and you can verify that $H = \{(000), (111), (222)\}$ is a distance-3 subset. Here H induces the covering map $\pi_H \colon Q_3^3 \longrightarrow K_{3,3,3}$, where $K_{3,3,3}$ is a complete 3-partite graph in which all vertex classes have cardinality 3.

We label the H-induced co-sets as follows:

$(0,0)$: $\{(0,0,0), (1,1,1), (2,2,2)\}$ $(1,0)$: $\{(1,0,0), (2,1,1), (0,2,2)\}$
$(1,2)$: $\{(0,1,2), (1,2,0), (2,0,1)\}$ $(0,1)$: $\{(0,1,0), (1,2,1), (2,0,2)\}$
$(2,1)$: $\{(0,2,1), (1,0,2), (2,1,0)\}$ $(2,2)$: $\{(0,0,1), (1,1,2), (2,2,0)\}$
$(2,0)$: $\{(2,0,0), (0,1,1), (1,2,2)\}$ $(0,2)$: $\{(0,2,0), (1,0,1), (2,1,2)\}$
$(1,1)$: $\{(0,0,2), (1,1,0), (2,2,1)\}$

Obviously, these labels correspond to Q_3^2-vertices, and it is straightforward to verify that $\{(0,0),(1,2),(2,1)\}$, $\{(1,0),(0,1),(2,2)\}$ and $\{(2,0),(0,2),(1,1)\}$ are exactly the vertex classes of $K_{3,3,3}$. Hence, $K_{3,3,3}$ contains Q_3^2 as a subgraph as it should according to Proposition 4.60 stated below.   ◇

The Orbit Graphs $H \setminus Q_\alpha^n$

In this section we study the orbit graphs $H \setminus Q_\alpha^n$.

Proposition 4.60. *Let H be an F^n-subspace and let $\pi_H \colon Q_\alpha^n \longrightarrow H \setminus Q_\alpha^n$ be the covering map induced by H with $\dim(F^n/H) = r$. Then*

$$Q_\alpha^r < H \setminus Q_\alpha^n , \qquad (4.82)$$

that is, $H \setminus Q_\alpha^n$ contains a subgraph isomorphic to Q_α^r.

Proof. Let $S = \{fe_i \mid f \in F^\times, \ i = 1,\ldots,n\}$. Then $Q_\alpha^n = (F^n, S)$, i.e., the Cayley graph over the group F^n with generating set S. The map π_H can then be written as

$$\pi_H \colon (F^n, S) \longrightarrow (F^n/H, S/H) .$$

Since S generates F^n, S/H generates F^n/H, and S/H contains a set of the form $S_0/H = \{fb \mid f \in F^\times, \ b \in \mathcal{B}\}$ where $\mathcal{B}$ is a basis of F^n/H. Clearly, we have an isomorphism $\eta \colon F^n/H \longrightarrow F^r$ and set $S' = \eta(S/H)$ and $S_0' = \eta(S_0/H)$. Without loss of generality we may assume that S_0' is of the form $S_0' = \{ke_i \mid k \in F^\times, \ i = 1,\ldots,r\}$ from which we immediately conclude $(F^r, S_0') \cong Q_\alpha^r$. In view of $S_0' \subset S'$, the embedding

$$(F^r, S_0') \longrightarrow (F^r, S') \quad (x_1,\ldots,x_r) \mapsto (x_1,\ldots,x_r) ,$$

is a graph morphism, and the proposition follows.   □

The following result shows that if H is a subspace of F^n with property (#) and if $H' = \eta(H)$ where $\eta \in \mathsf{Aut}_0(Q_\alpha^n)$, then the resulting orbit graphs are isomorphic.

Proposition 4.61. *Let $H < F^n$ be a (#)-sub-vectorspace. Then for any $\eta \in \mathsf{Aut}_0(Q_\alpha^n)$,*

$$\pi_{\eta(H)} \colon Q_\alpha^n \longrightarrow \eta(H) \setminus Q_\alpha^n$$

is a covering map and

$$H \setminus Q_\alpha^n \cong \eta(H) \setminus Q_\alpha^n . \qquad (4.83)$$

Furthermore, for two (#)-vectorspaces $H, H' < F^n$ with $H \setminus Q_\alpha^n \cong H' \setminus Q_\alpha^n$ there is in general no element $\eta \in \mathsf{Aut}_0(Q_\alpha^n)$ with the property $H' = \eta(H)$.

Proof. For any $\eta \in \mathsf{Aut}_0(Q_\alpha^n)$ the vectorspace $\eta(H)$ has property (#), so by Theorem 4.54 the map $\pi_{\eta(H)}$ is a covering map. Consider the map

$$\hat{\eta} : H \setminus Q_\alpha^n \longrightarrow \eta(H) \setminus Q_\alpha^n, \qquad \hat{\eta}(x + H) = \eta(x) + \eta(H) \;.$$

Since η is F-linear, we have $\hat{\eta}((x + h_1) + H_1) = \eta(x) + \eta(h_1) + \eta(H_1)$, proving that $\hat{\eta}$ is well-defined. It is clear that the map is injective, and the fact that it is a surjection is implied by η being surjective. It remains to show that $\hat{\eta}$ is a graph morphism. Let $\{x, y\} + H = \{\{x + h, y + h\} \mid h \in H\}$ be an edge in $H \setminus Q_\alpha^n$. We have

$$\hat{\eta}(\{x, y\} + H) = \{\eta(x), \eta(y)\} + \eta(H);$$

hence, $\hat{\eta}$ maps $H \setminus Q_\alpha^n$-edges into $\eta(H) \setminus Q_\alpha^n$-edges.

To prove the final statement, consider the two sub-vectorspaces $H = \langle(0,0,0),(1,2,2)\rangle$ and $H' = \langle(0,0,0),(1,1,1)\rangle$ of $F^3 = \mathbb{F}_3^3$. Since $\mathsf{Aut}_0(Q_\alpha^n) \cong S_n$ there exists no $\eta \in \mathsf{Aut}_0(Q_3^3)$ such that $H' = \eta(H)$, but it is straightforward to verify that

$$\langle(0,0,0),(1,2,2)\rangle \setminus Q_3^3 \cong \langle(0,0,0),(1,1,1)\rangle \setminus Q_3^3 \cong K_{3,3,3},$$

and the proposition follows. $\square$

Example 4.62. We will find all the covering maps of the form $\pi_H : Q_2^4 \longrightarrow H \setminus Q_2^4$. We first note that if H has dimension 2, then H has size 4. With $F = \mathbb{F}_2$ this leads to $|F^4/H| = 16/4 = 4 \not\geq 1 + n|F^\times| = 1 + 4 = 5$. In other words, if H has dimension 2, then we cannot get a covering map. If H has dimension 1, we have $|F^4/H| = 16/2 > 5$ and obtain covering maps.

There are five distance-3 subspaces. These are spanned by (1111), (0111), (1011), (1101), and (1110), respectively. Since the four last subspaces differ by an element of $\mathsf{Aut}_0(Q_2^4)$ (e.g., a permutation), the corresponding orbit graphs are all isomorphic by Proposition 4.61. Since the dimension of F^4/H is 3, it follows from Proposition 4.60 that $H \setminus Q_2^4$ contains Q_2^3 as a subgraph. We set

$$H_1 = \{0000, 1111\} \quad \text{and} \quad H_2 = \{0000, 1110\} \;.$$

We invite you to verify that the graph $H_1 \setminus Q_2^4$ is isomorphic to Q_2^3 with the four diagonal edges added. The graph $H_2 \setminus Q_2^4$ is isomorphic to Q_2^3 with four additional edges as shown on the right in Figure 4.20. Again, the significance of the map π_{H_1} is that it allows us to study dynamics over Q_2^4 in terms of dynamics over the smaller graph $H_1 \setminus Q_2^4$. However, we can only study those SDS over Q_2^4 that have an update order appearing as an image of $\eta_{\pi_{H_1}}$ and for which the vertex functions on $v \in \mathsf{v}[H_1 \setminus Q_2^4]$ and $v' \in \pi_{H_1}^{-1}(v)$ are identical. $\diamond$

4.32. Show that the orbit graphs $H_1 \setminus Q_2^4$ and $H_2 \setminus Q_2^4$ in Example 4.62 are not isomorphic.   [1]

4.33. Show that the two orbit graphs in Example 4.62 are the only covering images of Q_2^4.   [2C]

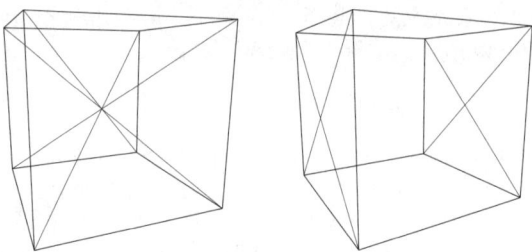

Fig. 4.20. The orbit graphs of Example 4.62.

Covering Maps into the Complete Graph

From the point of view of phase-space reductions, the best we can hope for is to have a covering map $\varphi : Q_\alpha^n \longrightarrow K_m$, where K_m is a complete graph over m vertices. (Why?) Note that $\mathsf{Aut}(K_m) \cong S_m$ and that in view of the group action $\gamma \bullet [\mathbf{F}_Y, \pi] = [\mathbf{F}_Y, \gamma\pi]$ (Section 4.3.3, Lemma 4.32) all SDS over K_m induced by symmetric functions are dynamically equivalent. As a special case of Theorem 4.54 we present a necessary and sufficient condition for the existence of covering maps $\varphi \colon Q_\alpha^n \longrightarrow K_m$.

Proposition 4.63. *There exists a covering map*

$$\varphi \colon Q_p^n \longrightarrow K_{1+(p-1)n} \tag{4.84}$$

if and only if $p^n \equiv 0 \mod 1 + (p-1)n$ holds.

Proof. Assume $\varphi \colon Q_p^n \longrightarrow K_{1+(p-1)n}$ is a covering map. Clearly, we have $|Q_p^n| = p^n$ and $|K_{1+(p-1)n}| = 1 + (p-1)n$, and Lemma 4.43 guarantees $p^n \equiv 0 \mod 1 + (p-1)n$.

Assume next that $p^n \equiv 0 \mod 1 + (p-1)n$. Corollary 4.64 below guarantees that there exists a subspace $G < \mathbb{F}_p^n$ with the property

$$\mathbb{F}_p^n = G(0) \;\dot\cup\; \dot{\bigcup}_{i=1}^n \dot{\bigcup}_{f \in \mathbb{F}_p^\times} G(fe_i) \,.$$

We observe that the mapping $\varphi \colon Q_p^n \longrightarrow G \setminus Q_p^n$ given by

$$\forall\, f \in \mathbb{F}_p^\times, \; i = 1, \ldots, n \;:\; \xi \in G(fe_i); \quad \varphi(\xi) = G(fe_i) \tag{4.85}$$

is a covering map. Clearly, $K_{1+(p-1)n} \cong G \setminus Q_p^n$ since by construction the graph $G \setminus Q_p^n$ is $(p-1)n$-regular and contains exactly $1 + (p-1)n$ vertices. $\quad\square$

The corollary below follows immediately from Lemma 4.55:

Corollary 4.64. *Let $n > 2$ be an integer and let p be a prime. Then we have $p^n \equiv 0 \mod 1 + (p-1)n$ if and only if there exists a subspace $G < \mathbb{F}_p^n$ with the property*

$$\mathbb{F}_p^n = G(0) \;\dot\cup\; \dot{\bigcup}_{i=1}^n \dot{\bigcup}_{f \in \mathbb{F}_p^\times} G(fe_i) \,.$$

Proof. Suppose we have $p^n \equiv 0 \mod 1 + (p-1)n$. Obviously, there exists a subspace $H < \mathbb{F}_p^n$ with $|\mathbb{F}_p^n/H| = 1 + (p-1)n$. The proof of Lemma 4.55 immediately shows that there exists some set of F^n/H-elements $\{f\varphi_i + H \mid i = 1, \ldots, n; \ f \in \mathbb{F}_p^\times\}$ such that $\{\varphi_i \mid i = 1, \ldots, n\}$ is an $\mathbb{F}_p^n$-basis. Let f' be the $\mathbb{F}_p$-morphism defined by $f'(\varphi_i) = e_i$ for $i = 1, \ldots, n$. Clearly, $G = f'(H)$ has the property

$$\mathbb{F}_p^n = G(0) \,\dot{\cup}\, \dot{\bigcup_{i=1}^{n}} \dot{\bigcup_{f \in \mathbb{F}_p^\times}} G(fe_i)\,,$$

and the corollary follows.   □

Example 4.65. We have already seen the example $\varphi \colon Q_2^3 \longrightarrow K_4$. Here 2^3 is congruent to 0 module $3 + 1 = 4$. Also, since 3^4 is congruent to 0 modulo $1 + 4 \cdot 2 = 9$, we have a covering map $\varphi' \colon Q_3^4 \longrightarrow K_9$; see Example 4.58.   ◇

4.34. Is there a covering map $\phi \colon Q_5^4 \longrightarrow K_{17}$? What is the smallest integer $n > 1$ such that there is a covering map of the form $\psi \colon Q_5^n \longrightarrow K_r$? What is r in this case?   [1]

There is a relation between covering maps $\varphi \colon Q_p^n \longrightarrow Z$ and algebraic codes. Any covering map $\varphi \colon Q_p^n \longrightarrow Z$ yields a 1-error-correcting code, and in particular, any perfect, 1-error-correcting code C in Q_p^n induces a covering map into $K_{1+(p-1)n}$, see [108]. We note that there are perfect, 1-error-correcting Hamming codes that are not groups as we ask you to show in Problem 4.35 below.

4.35. Let $\varphi \colon Q_\alpha^n \longrightarrow Z$ be a covering map. Show that $\varphi^{-1}(\varphi(0))$ is in general not a subspace of F^n.   [3]

4.4.7 Covering Maps over Circ_n

In this section we will study covering maps $\varphi \colon \mathsf{Circ}_n \longrightarrow Z$ where Z is connected. We will show that there exists a bijection between covering maps $\gamma \colon \mathsf{Circ}_n \longrightarrow Z$ where Z is connected, and subgroups $\langle \sigma^m \rangle < \mathsf{Aut}(\mathsf{Circ}_n)$, $m \geq 3$, $n \equiv 0 \mod m$ and $\sigma = (0, 1, \ldots, n-1)$. In fact, even more is true: If $\varphi \colon \mathsf{Circ}_n \longrightarrow Z$ is a covering map and Z is connected, then $Z \cong \langle \sigma^m \rangle \setminus \mathsf{Circ}_n$. Accordingly, covering maps over Circ_n are entirely determined by certain subgroups of $\mathsf{Aut}(Y)$.

Example 4.66. We have covering maps $\varphi \colon \mathsf{Circ}_{12} \longrightarrow \mathsf{Circ}_3$, $\varphi_1 \colon \mathsf{Circ}_{12} \longrightarrow \mathsf{Circ}_6$, and $\varphi_2 \colon \mathsf{Circ}_6 \longrightarrow \mathsf{Circ}_3$. Let $\sigma_{12} = (0, 1, 2, \ldots, 11)$ and $\sigma_6 = (0, 1, \ldots, 5)$ where we use cycle notation for permutations. The map φ is induced by σ_{12}^3 while φ_1 is induced by σ_{12}^6 and φ_2 is induced by σ_6^3. See Figure 4.21.   ◇

Elements of $\mathsf{Aut}(\mathsf{Circ}_n)$ are of the form $\tau\sigma^k$ where $\sigma = (0, 1, \ldots, n-1)$ and $\tau = \prod_{i=1}^{\lfloor n/2 \rfloor}(i, n-i)$. The covering maps from Circ_n are characterized by the following result:

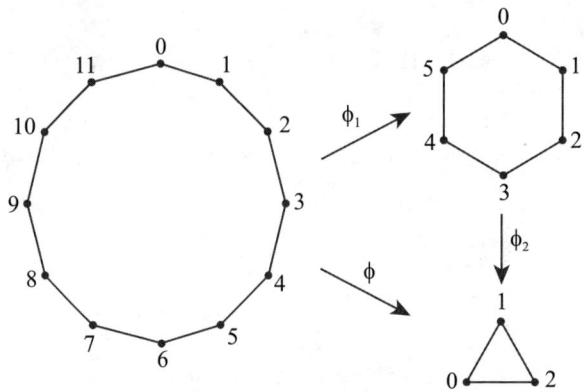

Fig. 4.21. Covering maps from Circ_{12} and Circ_6.

Proposition 4.67. *If* $\gamma\colon \mathsf{Circ}_n \longrightarrow Z$ *is a covering map, where* Z *is connected, then* $Z \cong \mathsf{Circ}_m$ *where* $n \equiv 0 \mod m$. *Accordingly, for any* γ *there is a subgroup* $H < \mathsf{Aut}(\mathsf{Circ}_n)$ *such that*

$$\mathsf{Circ}_n \overset{\gamma}{\mapsto} H \setminus \mathsf{Circ}_n \cong Z \tag{4.86}$$

holds. In particular, there are no nontrivial covering maps for $n < 6$.

Proof. Assume $\gamma\colon \mathsf{Circ}_n \longrightarrow Z$ is a covering map and that Z is connected. Then $\gamma\colon \mathsf{Circ}_n \longrightarrow Z$ is surjective. Since $\gamma\colon \mathsf{Circ}_n \longrightarrow Z$ is locally bijective, any vertex i in Z has degree 2. Thus, Z is a connected regular graph of degree 2, i.e., $Z \cong \mathsf{Circ}_m$. Lemma 4.43 implies $n \equiv 0 \mod m$ and $m \geq 3$. The subgroup $H = \langle \sigma^m \rangle$ satisfies $Z \cong H \setminus \mathsf{Circ}_n$ and gives us the desired covering map by $\gamma = \pi_H$,

$$\pi_{\langle \sigma^m \rangle}\colon \mathsf{Circ}_n \longrightarrow \langle \sigma^m \rangle \setminus \mathsf{Circ}_n \cong Z \ .$$

The last statement of the proposition follows from Lemma 4.43 and the fact that for every covering we have $d(i,j) \geq 3$ for any $i, j \in Y$ with $i \neq j$ and $i, j \in \gamma^{-1}(\gamma(i))$. □

There are various ways to construct covering maps from given covering maps. The following two problems illustrate the idea.

4.36. Let $\varphi_i\colon Y_i \longrightarrow Z_i$ for $i = 1, 2$ be covering maps. Show how to construct a covering map from $Y_1 \times Y_2$ to $Z_1 \times Z_2$ where $\times$ is the direct product of graphs. (Note that there are several types of possible graph products.)   [1+]

4.37. Let $\varphi\colon Y \longrightarrow Z$ be a covering map. Let Y' and Z' be the graphs obtained from Y and Z, respectively, by inserting a vertex on every edge. Show how to construct a covering map $\hat{\varphi}\colon Y' \longrightarrow Z'$. The process is illustrated in Figure 4.22.   [1+]

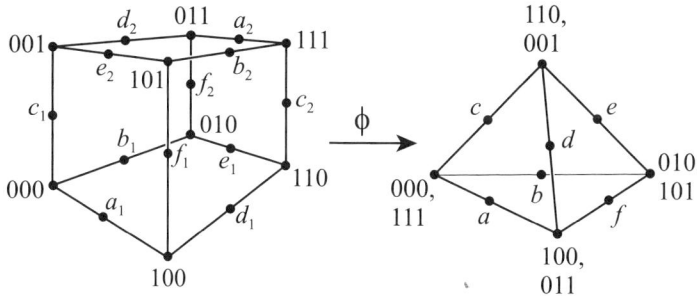

Fig. 4.22. An extension of the covering map $\varphi\colon Q_2^3 \longrightarrow K_4$.

Problems

4.38. In this problem we will consider covering maps of the form $\phi\colon Q_2^7 \longrightarrow$ $H \setminus Q_2^7$ where H is a sub-vectorspace of $F^7 = \mathbb{F}_2^7$. (i) Show that there are at most five non-isomorphic covering image graphs of the form $Z_H = H \setminus Q_2^7$ of order 64. (ii) Show that there exists a covering map $\phi'\colon Q_2^7 \longrightarrow K_8$ and give a four-dimensional, distance-3 sub-vectorspace H that induces the covering map ϕ'. [2]

Answers to Problems

4.1. $(3,1,2,0)$ has the representation $2 \cdot 4^2 + 1 \cdot 4^1 + 3 \cdot 4^0 = 32 + 4 + 3 = 39$. Since $1234 = 4^5 + 3 \cdot 4^3 + 4^2 + 2 \cdot 4^0$, we get $(2,0,1,3,0,1)$.

4.2. For Circ_6 we have $n[5] = (0,4,5)$ (where we have used the standard convention of ordering in the natural way).

The function Nor_5 is in this case given as

$$\mathrm{Nor}_5(x_0, x_1, x_2, x_3, x_4, x_5) = (x_0, x_1, x_2, x_3, x_4, \mathrm{nor}_3(x_4, x_5, x_0)) .$$

4.3. The phase space of $[\mathbf{Majority}_{\mathsf{Line}_3}, (2,3,1)]$ is shown in the figure below:

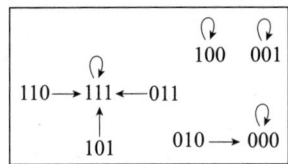

4.4. See, for example, R. A. Hernández Toledo's article *"Linear finite dynamical systems"* [33].

4.5. Proposition 4.11 does not hold for SDS with word update orders. For instance, in the somewhat pathological case where the word w equals the empty word, all states are fixed. Even if we restricted our attention to fair words, which are words where every vertex of the graph Y appears at least once, Proposition 4.11 does not hold. For instance, if a permutation-SDS with update order π has a period-2 orbit $\{x, y\}$, then the corresponding word-SDS with update order $w = (\pi|\pi)$ has x and y as fixed points.

4.6. The phase space is a union of cycles.

4.7. The solution follows by inspecting the function table. For the map f to induce invertible SDS over Circ_n, we must have that $a_7 = 1 + a_5$, $a_6 = 1 + a_4$, $a_3 = 1 + a_1$, and $a_2 = 1 + a_0$ (where additions are modulo 2). Thus, we can freely assign values to four of the a_i's, and thus there are 16 such maps.

If such a function is to be symmetric, it must have the same value for (001), (010), and (100). It must also have the same value on (011), (101), and (110). We see that this comes down to $a_6 = a_5 = a_3$ and $a_4 = a_2 = a_1$. If $a_0 = 0$, we get $a_1 = a_2 = a_4 = 1$. Furthermore, we have $a_6 = 1 + a_4$ and $a_3 = 1 + a_1$ so that $a_6 = a_5 = a_3 = 0$. Finally, $a_7 = 1 + a_5 = 1$. You can verify that the function we get is parity$_3$, which is rule $128 + 16 + 4 + 2 = 150$. If $a_0 = 1$, we get the function $1 + \text{parity}_3$ with rule number 105.

The rule numbers according to the Wolfram encoding of all the functions inducing invertible SDS are 51, 54, 57, 60, 99, 102, 105, 108, 147, 150, 153, 156, 195, 198, 201, and 204 .

You may have noticed that the functions come in pairs that add to 255. By flipping zeros and ones in the function table, we get rules with isomorphic phase-space digraphs. It is clear that if one function gives invertible SDS, then so must the "255 complement function."

4.8. For each degree d in the graph Y, the argument is virtually identical to the argument in Example 4.14.

4.9. $(p!)^{p^d}$.

4.10. NA

4.11. Consider the mapping ϑ of Eq. (4.35):

$$\vartheta \colon S_m \setminus Q_\kappa^m \longrightarrow P(K), \quad \vartheta(S_m(x)) = \{x_{v_{j_i}} \mid 1 \le i \le m\} \ .$$

We show that if ϑ contains two different elements $x_{v_{j_w}} \neq x_{v_{j_q}}$, then we have $N(x_{v_{j_w}}) \cap N(x_{v_{j_q}}) = \varnothing$ [Eq. (4.36)]. Suppose x contains $x_{v_{j_w}}$ and $x_{v_{j_q}}$ m_{j_w} and m_{j_q} times, respectively. Any element of $N(x_{v_{j_w}})$ contains $x_{v_{j_q}}$ at least m_{j_q} times and any element of $N(x_{v_{j_q}})$ contains $x_{v_{j_w}}$ at least m_{j_w} times. An element $\xi \in N(x_{v_{j_w}}) \cap N(x_{v_{j_q}})$ would therefore contain $x_{v_{j_w}}$ and $x_{v_{j_q}}$ at least m_{j_w} and m_{j_q} times, respectively. In addition, ξ is a neighbor of x, obtained by altering exactly *one* of the coordinates $x_{v_{j_w}}$ or $x_{v_{j_q}}$, which is impossible.

4.12. The graph $G = S_3 \setminus Q_3^3$ is shown in Figure 4.8. We have three choices for the "color" of the vertex [000] and two choices for the color of [001]. With these values set the remaining $\binom{5}{2} - 2 = 8$ vertex colors are fixed. Thus, there are six such vertex colorings and therefore six symmetric functions $f \colon \mathbb{F}_3^3 \longrightarrow \mathbb{F}_3$ that induce invertible local functions. Clearly, s_3 is the coloring that assigns 0 to [000] and 1 to [001].

4.13. The graph $G = S_3 \setminus Q_4^3$ is shown in Figure 4.23. (We have labeled the elements of the field 0, 1, 2, and 3.) The graph G has $\binom{3+4-1}{3} = 20$ vertices.

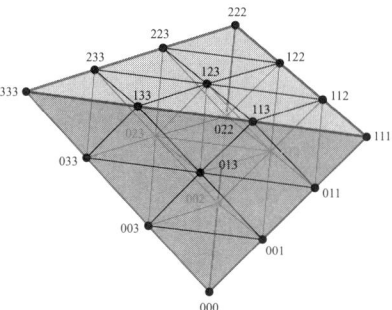

Fig. 4.23. The graph $G = S_3 \setminus Q_3^3$.

4.14. $\binom{\alpha+m-1}{\alpha-1}$.

4.15. $a(\mathsf{Wheel}_n) = 3^n - 3$. Pick $e = \{(0, (n-1)\}$. Observe that the graph Y_e'' is isomorphic to Wheel_{n-1}. Let W_n' be the graph obtained from Wheel_n by deleting e. Use the recursion relation for $a(Y)$ to find a recursion relation for $a(W_n')$ and find an explicit expression. Use this in the original recursion relation for $a(\mathsf{Wheel}_n)$.

4.16. NA

4.17. (i) $\widehat{\pi} = (0, 1, 2, 5, 3, 6, 4, 7, 8, 9)$. (ii) We need six computation cycles as there are six rank layers, and we have (iii) $a(E_n) = 3^n(2^n - 2)$.

4.18. We only need the functions to be "outer-symmetric" or symmetric in the "neighbor" arguments. A graph automorphism maps 1-neighborhoods to 1-neighborhoods and preserves the center vertex. The SDS does not need to be induced, but all functions $f_{v'}$ with $v' \in \mathsf{Aut}(Y)(v)$ must be the same. The proposition also holds for any pair of words w and w' that are "related" by a graph automorphism. We will get back to what "related" means in Chapter 7.

4.19. Let $\gamma, \eta \in \mathsf{Aut}(Y)$. We need to show that $(\eta\gamma)\mathcal{O} = \eta(\gamma\mathcal{O})$. To this end let e be an edge of Y. By definition we have

$$(\eta\gamma)\mathcal{O}(e) = (\eta\gamma)\mathcal{O}((\eta\gamma)^{-1}(e)) .$$

We also have

$$
\begin{aligned}
(\eta(\gamma\mathcal{O}))(e) &= \eta((\gamma\mathcal{O})(\eta^{-1}(e))) \\
&= \eta(\gamma(\mathcal{O}(\gamma^{-1}(\eta^{-1}(e))))) \\
&= (\eta\gamma)(\mathcal{O}((\eta\gamma)^{-1}(e))) .
\end{aligned}
$$

Clearly, $\mathrm{id}\,\mathcal{O} = \mathcal{O}$ for any acyclic orientation, and we have established that we have a group action.

It remains to show that $\gamma\mathcal{O}_Y(\pi) = \mathcal{O}_Y(\gamma\pi)$. Note that $\mathcal{O}_Y(\pi)$ is defined for combinatorial graphs Y. Let $\{v, v'\} \in \mathrm{e}[Y]$. We have

$$(\gamma\mathcal{O}_Y(\pi))(\{v, v'\}) = \gamma(\mathcal{O}_Y(\pi)(\gamma^{-1}\{v, v'\}))$$

$$= \begin{cases} (v, v') & \text{if } \gamma^{-1}(v) <_\pi \gamma^{-1}(v'), \\ (v', v) & \text{otherwise.} \end{cases}$$

Again by definition we have

$$\mathcal{O}_Y(\gamma\pi)(\{v, v'\}) = \begin{cases} (v, v') & \text{if } v <_{\gamma\pi} v', \\ (v', v) & \text{otherwise.} \end{cases}$$

But it is clear that

$$
\begin{aligned}
v <_{\gamma\pi} v' &\iff v = \gamma\pi(k),\ v' = \gamma\pi(k') \text{ with } k < k' \\
&\iff \gamma^{-1}(v) = \pi(k),\ \gamma^{-1}(v') = \pi(k') \text{ with } k < k' \\
&\iff \gamma^{-1}(v) <_\pi \gamma^{-1}(v')
\end{aligned}
$$

and equality follows.

4.20. In general the answer is no, but there are special cases/graphs where it does hold. We leave it to you to identify the conditions.

4.21. The bound $\Delta(Y)$ is the number of orbits in $S_n/\sim_Y$ under the action of $\mathsf{Aut}(Y)$. We have $\mathsf{Aut}(K_n) = S_n$ and we therefore have only one orbit, so $\Delta(K_n) = 1$.

4.22. NA

4.23. The bound is sharp. One way to see this is to pick representative update orders from the three $\mathsf{Aut}(\mathsf{Circ}_4)$-orbits and show that the three SDS induced by nor-functions have pairwise non-isomorphic phase spaces.

4.24. No, the bound is not sharp. If you do the math, you will find that $\Delta(\mathbf{Parity}_{\mathsf{Circ}_4}) = 2$.

4.25. There are 8! different permutation update orders, we can get 1862 functionally different permutation SDS since $a(Q_2^3) = 1862$, and we can get $\Delta(Q_2^3) = 54$ dynamically nonequivalent induced SDS; see [109]. The bound is sharp. To show this requires a lot of tedious comparisons of phase spaces, unless you find an approach that we are not aware of.

4.26. For $p > 2$ a prime the sum in (4.53) has only one term:

$$\Delta(\mathsf{Circ}_p) = \frac{1}{2p}\phi(1)(2^p - 2) = (2^{p-1} - 1)/p \ .$$

4.27. An element γ of $\mathsf{Aut}(\mathsf{Star}_{l,m})$ necessarily maps K_l vertices in $\mathsf{Star}_{l,m}$ into K_l vertices since automorphisms are degree-preserving. Since γ also preserves adjacency, the vertices of degree 1 attached to vertex i can only be permuted among themselves and moved such that they are adjacent to $\gamma(i)$. Thus, we see that $\mathsf{Aut}(\mathsf{Star}_{l,m}) = KH = HK$ where $H, K < S_{l(1+m)}$ are the groups

$$K = \left\{ \begin{pmatrix} 1 & 1_1 & \cdots & 1_m & | \cdots | & l_1 & \cdots & l_m \\ 1 & \sigma_{1,1} & & \sigma_{1,m} & | \cdots | & \sigma_{l,1} & & \sigma_{l,m} \end{pmatrix} \mid \sigma_i \in S(i_1, \ldots, i_m) \right\} \quad (4.87)$$

and

$$H = \{ \sigma \in S_{l(m+1)} \mid \sigma(i) = j \Rightarrow \forall k \in \mathbb{N}_m : \sigma(i_k) = j_k \} \ . \quad (4.88)$$

We must show that K is normal in G. Let $k \in K$ and $g = h \cdot k_1 \in \mathsf{Aut}(\mathsf{Star}_{l,m})$. Then we have

$$g \cdot k \cdot g^{-1} = h \cdot k_1 \cdot k \cdot k_1^{-1} \cdot h^{-1}$$
$$= h \cdot k_2 \cdot h^{-1},$$

where $k_2 = k_1 \cdot k \cdot k_1^{-1}$. In view of $h \cdot k_2 \cdot h^{-1} \in K$, we derive $K \triangleleft G$, and consequently $G = K \rtimes H$ follows. Since $K \cong S_m^l$ and $H \cong S_l$, we are done.

4.28. We will establish Eq. (4.67) by computing the sum in (3.31) directly. First, we know from Lemma 4.39 that $|\mathsf{Aut}(\mathsf{Star}_{l,m})| = l! \times m!^l$. We write automorphisms as $\gamma = (\sigma_l, \pi_1, \ldots, \pi_l)$, where σ_l is the permutation of the vertices of the K_l subgraph and π_i denotes the permutation of the vertices $i_1, \ldots, i_m$. We observe that $\gamma \in \mathrm{Aut}(\mathrm{Star}_{l,m})$ does only contribute to the sum in (3.31) when $\sigma_l = \mathrm{id}$ since the graph $\langle \gamma \rangle \setminus \mathrm{Star}_{l,m}$ would otherwise contain at least one loop and would thus not allow for any acyclic orientations. Now with $\sigma_l = \mathrm{id}$ it is clear that $\langle \gamma \rangle \setminus \mathrm{Star}_{l,m}$ will be the graph K_l with $c(\pi_i)$ vertices attached to vertex i of K_l. Here $c(\gamma)$ denotes the number of cycles in the the the cycle decomposition of γ where cycles of length 1 are included. Thus, the number of acyclic orientations of the reduced graph $\langle \gamma \rangle \setminus \mathrm{Star}_{l,m}$ in this case is $l! \times 2^{c(\gamma)}$. We now get

$$
\begin{aligned}
\Delta(\mathrm{Star}_{l,m}) &= \frac{1}{|\mathrm{Aut}(\mathrm{Star}_{l,m})|} \sum_{\gamma \in \mathrm{Aut}(\mathrm{Star}_{l,m})} a(\langle \gamma \rangle \setminus \mathrm{Star}_{l,m}) \\
&= \frac{1}{|\mathrm{Aut}(\mathrm{Star}_{l,m})|} \sum_{\gamma} a(\langle \gamma = (\mathrm{id}, \pi_1, \ldots, \pi_l) \rangle \setminus \mathrm{Star}_{l,m}) \\
&= \frac{1}{l! \times m!^l} \cdot l! \cdot \left(\sum_{\gamma \in S_m} 2^{\#(\gamma)} \right)^l \\
&= \left(\frac{\sum_{\gamma \in S_m} 2^{\#(\gamma)}}{m!} \right)^l \\
&= (m+1)^l,
\end{aligned}
$$

where the last equality follows by induction, and we are done.

4.29. Any interesting results here would probably make for a research paper.

4.31. As for dynamical equivalence the functions f_k need to be outer-symmetric. The extension to words is clear — all that needs to be done is to modify the map η_φ. If $w = (w_1, \ldots, w_k)$ is a word over $v[Z]$, then $\eta_\varphi(w) = (s(w_1) \mid \ldots s(w_k))$.

4.32. One way to see this is that the graph $H_2 \setminus Q_2^4$ contains triangles, which is not the case for $H_1 \setminus Q_2^4$.

4.33. NA

4.34. There is no covering map ϕ since, for example, 5^4 is not divisible by 17. A necessary and sufficient condition for the covering map ϕ to exist is that $r - 1 = 4n$ and that r divides 5^n. Thus, we have to have $4n + 1 | 5^n$, which happens for $n = 6$ in which case $r = 25$.

4.35. We show this by constructing a covering map $\varphi \colon Q_2^{15} \longrightarrow K_{16}$ where $\varphi^{-1}(\varphi(0))$ is not a subspace of $\mathbb{F}_2^{15}$.

According to Proposition 4.63, there exists a covering map $\pi_H \colon Q_2^7 \longrightarrow K_8$ for a $(\#)$-sub-vectorspace $H < \mathbb{F}_2^7$ such that $|H| = 2^4$ holds. Let $f \colon H \longrightarrow \mathbb{F}_2$ be defined by $f(0) = 0$ and $f(h) = 1$ otherwise. Using a well-known construction from coding theory (see, e.g., [108]), we introduce the set

$$H' = \left\{ \left(x, x + h, \sum_i x_i + f(h) \right) \mid x \in \mathbb{F}_2^7,\ h \in H \right\}.$$

We claim that $\mathbb{F}_2^{15} = H' \cup \left(\bigcup_{i=1}^{15} e_i + H' \right)$. To prove the claim we first show

$$\forall h_1', h_2' \in H' \ : \ d(h_1', h_2') \geq 3 . \tag{4.89}$$

Each H'-element is of the form $h_i' = (x_i, x_i + h_i, z_i)$ with $x_i \in \mathbb{F}_2^7$, $h_i \in H$ and $z_i \in \mathbb{F}_2$. Suppose now $h_1 = h_2$. Obviously, $z_1 = z_2$ implies $x_1 = \sigma(x_2)$ where $\sigma(x_2) = ((x_2)_{\sigma(1)}, \ldots, (x_2)_{\sigma(7)})$ and accordingly $d(x_1, x_2) = d(x_1 + h_1, x_2 + h_2) \geq 2$. For $z_1 \neq z_2$ we obtain $d(x_1, x_2) = d(x_1 + h_1, x_2 + h_2) \geq 1$; hence, $d(h_1', h_2') > 2$.

Assume next that $h_1 \neq h_2$ holds and observe that then $d(h_1 - h_2, 0) \leq d(h_1 - h_2, x_2 - x_1) + d(x_2 - x_1, 0)$. In view of $d(h_1 - h_2, 0) = d(h_1, h_2)$, $d(h_1 - h_2, x_2 - x_1) = d(h_1 + x_1, h_2 + x_2)$, and $d(x_2 - x_1, 0) = d(x_1, x_2)$, we have established (4.89). Clearly, (4.89) implies $(e_i + H') \cap (e_j + H') = \varnothing$ for $i \neq j$, $i, j = 1, \ldots, 15$, and since $|H'| = 2^{11}$ and the claim follows.

It remains to show that H' is not a group. We consider $h_1' = (x, x + h_1, z_1)$ and $h_2' = (x, x + h_2, z_2)$ with $h_1 \neq h_2$, $h_1 \neq 0$, and $h_2 \neq 0$. Then we have

$$h_1' + h_2' = (0, h_2 + h_1, f(h_1) + f(h_2)) \neq (0, h_1 + h_2, f(h_1 + h_2)) ,$$

i.e., the sum of h_1' and h_2' is not contained in H', which is therefore not a group. Accordingly, the map

$$\varphi \colon Q_2^{15} \longrightarrow K_{16}, \quad \varphi(x) = \begin{cases} 0 & \text{if and only if } x \in H', \\ i & \text{if and only if } x \in e_i + H', i = 1, \ldots, 15, \end{cases}$$

is a well-defined covering map for which $\varphi^{-1}(\varphi(0))$ is not a vectorspace.

4.36. NA

4.37. NA

4.38. (i) If Z_H has order 64, then the sub-vectorspace H must be one-dimensional. Additionally, H has to be a distance-3 set. Since sub-vectorspaces differing by a permutation give isomorphic covering images, we see that there are precisely five covering images of the form Z_H, and five representative sub-vectorspaces are $H_1 = \{0000000, 1110000\}$, $H_2 = \{0000000, 1111000\}$, $H_3 = \{0000000, 1111100\}$, $H_4 = \{0000000, 1111110\}$, and $H_5 = \{0000000, 1111111\}$.

(ii) For $p = 2$ we know there exists a covering map $\phi \colon Q_2^n \longrightarrow K_{1+n}$ if and only if 2^n is divisible by $n + 1$. Here $n + 1$ equals 8 so we have a covering map $\phi \colon Q_2^7 \longrightarrow K_8$. Here H must be a four-dimensional, distance-3 sub-vectorspace. The algorithm 4.57 leads us to choose a basis consisting of $v_i = e_i$ with $i = 1, \ldots, 4$ for H', and this basis is extended to a basis for

$\mathbb{F}_2^7$ by adding the vectors $v_i = e_i$ with $i = 5, 6, 7$. We now need to pick four vectors w_i in $\mathrm{Span}\{v_5, v_6, v_7\}$ that are not scalar multiples of each other nor scalar multiples of v_5, v_6, or v_7. We see that $w_1 = v_5 + v_6$, $w_2 = v_5 + v_7$, $w_3 = v_6 + v_7$, and $w_4 = v_5 + v_6 + v_7$ is one such choice. We set $\phi_i = v_i + w_i$ for $i = 1, \ldots, 4$ and $\phi_i = v_i$ for $i = 5, 6, 7$. The linear map f is the map given by $f(\phi_i) = e_i$ or, using matrix notation, $f\Phi = I$ where Φ is the matrix with the ϕ_i's as columns. We see that Φ is its own inverse so $f = \Phi$ and we get $H = f(H')$.

Explicitly, we have

$$H' = \left\{ \begin{array}{l} (0,0,0,0,0,0,0),\ (1,0,0,0,0,0,0),\ (0,1,0,0,0,0,0),\ (0,0,1,0,0,0,0), \\ (0,0,0,1,0,0,0),\ (1,1,0,0,0,0,0),\ (1,0,1,0,0,0,0),\ (1,0,0,1,0,0,0), \\ (0,1,1,0,0,0,0),\ (0,1,0,1,0,0,0),\ (0,0,1,1,0,0,0),\ (1,1,1,0,0,0,0), \\ (1,1,0,1,0,0,0),\ (1,0,1,1,0,0,0),\ (0,1,1,1,0,0,0),\ (1,1,1,1,0,0,0) \end{array} \right\} ,$$

$$f = \begin{bmatrix} 1 & 0 & 0 & 0 & 0 & 0 & 0 \\ 0 & 1 & 0 & 0 & 0 & 0 & 0 \\ 0 & 0 & 1 & 0 & 0 & 0 & 0 \\ 0 & 0 & 0 & 1 & 0 & 0 & 0 \\ 1 & 1 & 0 & 1 & 1 & 0 & 0 \\ 1 & 0 & 1 & 1 & 0 & 1 & 0 \\ 0 & 1 & 1 & 1 & 0 & 0 & 1 \end{bmatrix} ,$$

and $H = f(H')$ is given as

$$\left\{ \begin{array}{l} (0,0,0,0,0,0,0),\ (1,0,0,0,1,1,0),\ (0,1,0,0,1,0,1),\ (0,0,1,0,0,1,1), \\ (0,0,0,1,1,1,1),\ (1,1,0,0,0,1,1),\ (1,0,1,0,1,0,1),\ (1,0,0,1,0,0,1), \\ (0,1,1,0,1,1,0),\ (0,1,0,1,0,1,0),\ (0,0,1,1,1,0,0),\ (1,1,1,0,0,0,0), \\ (1,1,0,1,1,0,0),\ (1,0,1,1,0,1,0),\ (0,1,1,1,0,0,1),\ (1,1,1,1,1,1,1) \end{array} \right\} .$$

Phase-Space Structure of SDS and Special Systems

In this chapter we will study the phase spaces of special classes of SDS. The first part is concerned with computing the fixed-point structure of sequential dynamical systems and cellular automata over a subclass of the *circulant graphs* [83]. We then proceed to analyze SDS over special graph classes such as the complete graph, the line graph, and the circle graphs. We will also see that the periodic points of SDS induced by $(\mathrm{nor}_k)_k$ and $(\mathrm{nor}_k + \mathrm{nand}_k)_k$ do not depend on the choice of update order. This fact is needed in Chapter 6, where we will study groups associated to a certain class of SDS.

5.1 Fixed Points for SDS over Circ_n and $\mathsf{Circ}_{n,r}$

The fixed points of a dynamical system are usually easier to obtain than the periodic points of period $p \geq 2$. However, determining all the fixed points of an SDS is in general a computationally hard problem, and brute-force checking is the best approach. However, for certain graph classes we can characterize all fixed points efficiently. Here we will demonstrate this for $Y = \mathsf{Circ}_n$ and the more general class of graphs $\mathsf{Circ}_{n,r}$, $r \in \mathbb{N}$, defined below in the case of permutation-SDS. For similar constructions in the context of cellular automata, see, for example, [110]. The advantage of the approach here is that our construction extends directly to general graphs.

The permutation-SDS we consider here will be over the graph Circ_n, or more generally $\mathsf{Circ}_{n,r}$, and the functions f_v will all be induced by a common function ϕ. The graph $\mathsf{Circ}_{n,r}$, $r \in \mathbb{N}$, is given by

$$\mathrm{v}[\mathsf{Circ}_{n,r}] = \mathrm{v}[\mathsf{Circ}_n] = \{0, 1, 2, \ldots, n-1\}, \tag{5.1}$$
$$\mathrm{e}[\mathsf{Circ}_{n,r}] = \{\{i, j\} \mid -r \leq i - j \leq r\}.$$

The graph $\mathsf{Circ}_{6,2}$ in shown in Figure 5.1. In the case of Circ_n the function ϕ is of the form $\phi \colon \mathbb{F}_2^3 \longrightarrow \mathbb{F}_2$ and for $\mathsf{Circ}_{n,r}$ it is of the form $\phi \colon \mathbb{F}_2^{2r+1} \longrightarrow \mathbb{F}_2$. As for cellular automata we call r the *radius* of the rule ϕ. Here we assume

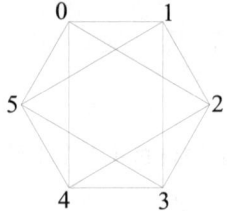

Fig. 5.1. The graph $\mathsf{Circ}_{6,2}$.

that $2r + 1 < n$ since there are only n vertex states. The state of each vertex i is updated as

$$x_i \mapsto \phi(x_{i-2r}, \ldots, x_{i-1}, x_i, x_{i+1}, \ldots, x_{i+2r}) ,$$

where all subscripts are taken modulo n. The idea in our approach works for any graph. Refer to Figure 5.2. We can construct a *local fixed point* at vertex 1 as a 5-tuple $(x_1, x_2, x_5, x_6, x_9)$ that satisfies $f_1(x_1, x_2, x_5, x_6, x_9) = x_1$. We can do the same for vertex 2, that is, we can find a 5-tuple $(x_1', x_2', x_3', x_4', x_5')$ such that $f_2(x_1', x_2', x_3', x_4', x_5') = x_2'$. The idea is to patch local fixed points together to create *global fixed points* for the full SDS. In order to patch together local fixed points, we need them to be compatible wherever they overlap. In the example this means that we must have $x_1 = x_1'$, $x_2 = x_2'$ and $x_5 = x_5'$. We formalize this idea as follows in the special case of the graph $\mathsf{Circ}_{n,r}$.

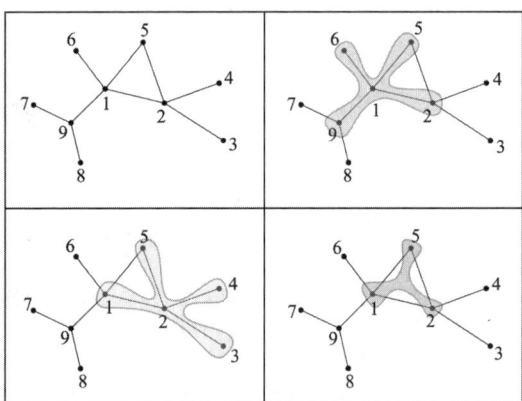

Fig. 5.2. The graph Y is shown in the upper left, $n[2] = (1, 2, 3, 4, 5)$ and $n[1] = (1, 2, 5, 6, 9)$ are highlighted in the lower left and upper right, respectively. In the lower right the vertices contained in both $n[1]$ and $n[2]$ are highlighted.

Definition 5.1 (Compatible, local fixed points for $\mathsf{Circ}_{n,r}$). Let r be a positive integer and let $x, x' \in \mathbb{F}_2^{2r+1}$. Then $x = (x_1, \ldots, x_{2r+1})$ is compatible with x' if $x_{i+1} = x'_i$, $1 \le i \le 2r$, which we write as $x \rhd x'$. A sequence $C = (x^i \in \mathbb{F}_2^{2r+1})_{i=0}^{n-1}$ is a compatible covering of $\mathsf{Circ}_{n,r}$ if

$$x^0 \rhd x^1 \rhd \cdots \rhd x^{n-1} \rhd x^0 .$$

Let $\phi\colon \mathbb{F}_2^{2r+1} \longrightarrow \mathbb{F}_2$. A compatible covering $C = (x^i)_{i=0}^{n-1}$ of $\mathsf{Circ}_{n,r}$ is a compatible fixed-point covering with respect to ϕ if $\phi(x^i) = x^i_{r+1}$ for $0 \le i \le n-1$. The set of all compatible fixed-point coverings of $\mathsf{Circ}_{n,r}$ with respect to ϕ is denoted $\mathcal{C}_\phi(n,r)$.

For $\mathsf{Circ}_{n,r}$ we can organize the local fixed points in a directed graph. Since each function ϕ gives such a graph we have a map $G\colon \mathsf{Map}(\mathbb{F}_2^{2r+1}, \mathbb{F}_2) \longrightarrow$ **Graph** that assigns to each map ϕ the directed graph $G = G(\phi)$ given by

$$\begin{align}
v[G] &= \{x \in \mathbb{F}_2^{2r+1} \mid \phi(x) = x_{r+1}\} , \tag{5.2}\\
e[G] &= \{(x, x') \mid x, x' \in v[G] : x \rhd x'\} .
\end{align}$$

Thus, G has vertices all local fixed points, and the directed edges encode compatibility.

Example 5.2. Let $r = 1$ and let $\phi = \mathrm{majority}_3\colon \mathbb{F}_2^3 \longrightarrow \mathbb{F}_2$. Recall that $\mathrm{majority}_3$ returns 1 if two or more of its arguments are 1 and returns 0 otherwise. We will compute the local fixed points of the form (x_{i-1}, x_i, x_{i+1}). For example, with $x_{i-1} = 0$, $x_i = 0$, and $x_{i+1} = 1$ we get $\mathrm{majority}_3(x_{i-1}, x_i, x_{i+1}) = 0 = x_i$ so that $(0,0,1)$ is a local fixed point. On the other hand, if $x_{i-1} = 0$, $x_i = 1$, and $x_{i+1} = 0$, we have $\mathrm{majority}_3(x_{i-1}, x_i, x_{i+1}) = 0 \ne x_i$ and we conclude that $(0,1,0)$ is not a local fixed point. You should verify that the local fixed points are as given in Table 5.1.

(x_{i-1}, x_i, x_{i+1})	$\mathrm{majority}_3$	Local fixed point?
$(0,0,0)$	0	Yes
$(0,0,1)$	0	Yes
$(0,1,0)$	0	No
$(0,1,1)$	1	Yes
$(1,0,0)$	0	Yes
$(1,0,1)$	1	No
$(1,1,0)$	1	Yes
$(1,1,1)$	1	Yes

Table 5.1. Local fixed points for SDS over Circ_n induced by $\mathrm{majority}_3$.

From the table it is clear that there are six local fixed points. Consider the local fixed point $(0,0,0)$. The local fixed points x such that $(0,0,0) \rhd x$ are $(0,0,0)$ and $(0,0,1)$ since the two last coordinates of $(0,0,0)$ must agree with the first two coordinates of x. Therefore, in the graph G there are a directed edge from $(0,0,0)$ to itself, and a directed edge from $(0,0,0)$ to $(0,0,1)$. You should check that the graph G is as shown in Figure 5.3. ◇

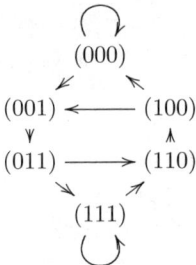

Fig. 5.3. The local fixed-point graph for majority$_3$.

The fixed-point graph G has at most 2^{2r+1} vertices. By definition, a cycle of length n in G corresponds to a compatible fixed-point covering of $\mathsf{Circ}_{n,r}$ for a given function ϕ. Each $C \in \mathcal{C}_\phi(n,r)$ corresponds uniquely to a fixed point of a corresponding permutation-SDS. To make this clear, we define the one-to-one map ψ by

$$\psi: \mathcal{C}_\phi(n,r) \longrightarrow \mathsf{Fix}[\mathbf{F}_{\mathsf{Circ}_{n,r}}, \pi] \,, \tag{5.3}$$
$$\psi(x^0, x^1, \ldots, x^{n-1}) = (x^0_{r+1}, x^1_{r+1}, \ldots, x^{n-1}_{r+1}) \,.$$

In other words, the map ψ extracts the center state of each local fixed point of a compatible fixed-point covering to create a fixed point for the SDS.

We can enumerate and characterize the fixed points of permutation-SDS induced by a function ϕ over $\mathsf{Circ}_{n,r}$ through the graph G.

Theorem 5.3. *Let $\phi: \mathbb{F}_2^{2r+1} \longrightarrow \mathbb{F}_2$, let L_n be the number of fixed points of a permutation-SDS over $\mathsf{Circ}_{n,r}$ induced by ϕ, and let A be the adjacency matrix of the graph $G(\phi)$. Then we have*

$$L_n = |\mathcal{C}_\phi(n,r)| = \mathsf{Tr}\, A^n \,. \tag{5.4}$$

Let $\chi_A(x) = \sum_{i=0}^k a_i x^{k-i}$ be the characteristic polynomial of A. The number of fixed points L_n satisfies the recursion relation

$$\sum_{i=0}^k a_i L_{n-i} = 0 \,. \tag{5.5}$$

Proof. The first equality in Eq. (5.4) follows since ψ is one-to-one. The second equality follows from Proposition 3.7 since $[A^n]_{ii}$ is the number of cycles of length n starting at vertex i. The last part of (5.4) can be rewritten as

$$L_n = \mathsf{Tr}\, A^n = \sum_{i=1}^k [A^n]_{ii} = \sum_{i=1}^k e_i A^n e_i^T \,,$$

where e_i is the ith unit vector. The left-hand side of (5.5) now becomes

$$\sum_{i=0}^{k} a_i L_{n-i} = \sum_{i=0}^{k} a_i \Big(\sum_{j=1}^{k} e_j A^{n-i} e_j^T \Big)$$

$$= \sum_{j=1}^{k} \Big(\sum_{i=0}^{k} e_j a_i A^{n-i} e_j^T \Big)$$

$$= \sum_{j=1}^{k} e_j (a_0 A^n + a_1 A^{n-1} + \cdots + a_k A^{n-k}) e_j^T$$

$$= \sum_{j=1}^{k} e_j \chi_A(A) A^{n-k} e_j^T$$

$$= 0,$$

where the last equality follows from the Hamilton–Cayley theorem (see Theorem 3.9, page 45).   □

Example 5.4. Let $r = 1$ and let $\phi = \mathrm{parity}_3 : \mathbb{F}_2^3 \longrightarrow \mathbb{F}_2$. Recall that

$$\mathrm{parity}_3(x_1, x_2, x_3) = x_1 + x_2 + x_3 \pmod 2 .$$

In this case it is actually easy to see what the fixed points are, so let us do that as a sanity check before we start up the machinery from Theorem 5.3. First the state x with all 0's is fixed, that is, $x = (0, 0, \ldots, 0)$. We also see that the state with all 1's is fixed. Otherwise, if we have a fixed point such that $x_i = 0$ that does not consist entirely of zeros, we must have $x_{i-1} = 1$ and $x_{i+1} = 1$. But if $x_{i+1} = 1$, then we must have $x_{i+2} = 0$ to have a fixed point. So we see that there are two other fixed-point candidates, namely the states with alternating 0's and 1's, but we need an even number of states to get these. Thus, we always have two fixed points, and when n is even we have two additional fixed points. Let's see what we get using the theorem.

You should first check that we get the local fixed points given in the table below:

(x_{i-1}, x_i, x_{i+1})	parity_3	Local fixed point?
$(0, 0, 0)$	0	Yes
$(0, 0, 1)$	1	No
$(0, 1, 0)$	1	Yes
$(0, 1, 1)$	0	No
$(1, 0, 0)$	1	No
$(1, 0, 1)$	0	Yes
$(1, 1, 0)$	0	No
$(1, 1, 1)$	1	Yes

From the local fixed points we construct the graph G shown in Figure 5.4. We see that the three components in G encode the fixed points we found

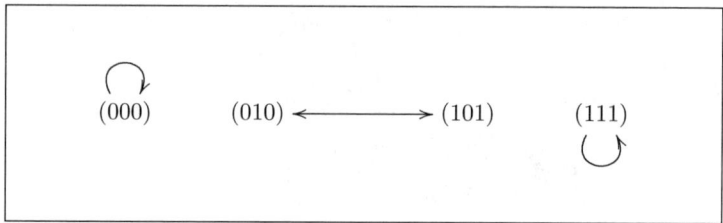

Fig. 5.4. The local fixed-point graph G for parity$_3$.

earlier. Now, we need the adjacency matrix of G, so let's index the four local fixed points as $1 : (0,0,0)$, $2 : (0,1,0)$, $3 : (1,0,1)$, and $4 : (1,1,1)$. The adjacency matrix A is then

$$\begin{bmatrix} 1 & 0 & 0 & 0 \\ 0 & 0 & 1 & 0 \\ 0 & 1 & 0 & 0 \\ 0 & 0 & 0 & 1 \end{bmatrix} ,$$

and the characteristic polynomial is $\chi_A(x) = \det(xI - A) = (x - 1)(x - 1)(x^2 - 1) = (x^2 - 2x + 1)(x^2 - 1) = x^4 - 2x^3 + 2x - 1$. We can write this as $\chi_A(x) \sum_{i=0}^{4} a_i x^{4-i}$ where $a_0 = 1$, $a_1 = -2$, $a_2 = 0$, $a_3 = 2$, and $a_4 = -1$. We therefore have the recursion relation $a_0 L_n + a_1 L_{n-1} + a_2 L_{n-2} + a_3 L_{n-3} + a_4 L_{n-4} = 0$, so that, after rearranging,

$$L_n = 2L_{n-1} - 2L_{n-3} + L_{n-4} . \tag{5.6}$$

As initial values for this recursion we have (from our initial discussion) $L_3 = 2$, $L_4 = 4$, $L_5 = 2$, and $L_6 = 4$. Note that we do not want to involve L_2 or L_1 since we want $n \geq 3$ in the circle graph. Based on this we can compute L_7 and L_8 as

$$L_7 = 2L_6 - 2L_4 + L_3 = 2 \cdot 4 - 2 \cdot 4 + 2 = 2$$

and

$$L_8 = 2L_7 - 2L_5 + L_4 = 2 \cdot 2 - 2 \cdot 2 + 4 = 4 ,$$

which is consistent with our above findings.   ◇

This was a pretty detailed example. In the next example we omit some of the details and consider the case with $r = 2$ and the function parity$_5$.

Example 5.5 (Parity). We want to enumerate the fixed points over $\mathsf{Circ}_{n,2}$ for SDS induced by parity$_5$. We proceed exactly as in the previous example, the only difference being that here we have to consider 5-tuples $(x_{i-2}, x_{i-1}, x_i, x_{i+1}, x_{i+2})$ for the local fixed points. You should verify that we get the local fixed-point graph G shown in Figure 5.5. By inspection you will now find that an SDS induced by parity$_5$ over $\mathsf{Circ}_{n,2}$ has 16 fixed points when $n \equiv 0 \pmod 6$, 8 fixed points if $n \equiv 0 \pmod 3$ and $n \not\equiv 0 \pmod 2$, 4 fixed points if $n \equiv 0 \pmod 2$ and $n \not\equiv 0 \pmod 3$, and 2 fixed points otherwise.

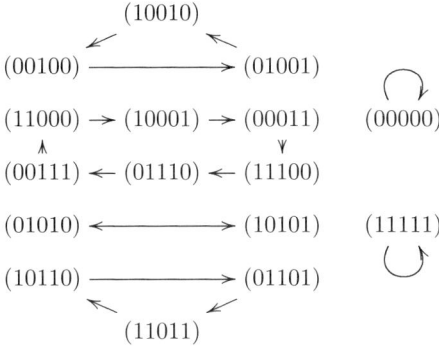

Fig. 5.5. The graph $G(\mathrm{parity}_5)$.

The adjacency matrix is a 16×16 matrix so you may want to use some suitable software to compute the characteristic polynomial and derive the recursion relation.   ◇

The next example is more involved. In this case we have $r = 2$ and we use the function $\mathrm{majority}_5$.

Example 5.6 (Majority). For an SDS over $\mathsf{Circ}_{n,2}$ induced by $\mathrm{majority}_5$ we get the following vertices for G_{majority_5}, which the reader should verify. (We have grouped the local fixed points by H-class. The elements of H-class k are all the tuples with exactly k entries that are 1.)

H-class	Vertices
0	(00000)
1	(00001), (00010), (01000), (10000)
2	(11000), (10010), (10001), (01010), (01001), (00011)
3	(11100), (10101), (00111), (01110), (01101), (10110)
4	(11110), (11101), (10111), (01111)
5	(11111)

The graph $G(\mathrm{majority}_5)$ is shown in Figure 5.6. By carefully inspecting the graph G, we see that the states (01000), (00010), (11101), (10111), (10010), (01001), (10110), and (01101) *cannot* be a part of a cycle in G. They are "absorbing" or "repelling." We can therefore omit these nodes from G for the purpose of counting cycles of length n. You can check that the graph G' obtained from G by deleting these vertices has adjacency matrix with characteristic polynomial $\chi(r) = r^{14} - 2r^{13} + 2r^{11} - r^{10} - r^8 + r^6$. Thus, the number of fixed points L_n of an SDS over $\mathsf{Circ}_{n,2}$ induced by $\mathrm{majority}_5$ satisfies the recursion relation

$$L_n = 2L_{n-1} - 2L_{n-3} + L_{n-4} + L_{n-6} - L_{n-8} \,,$$

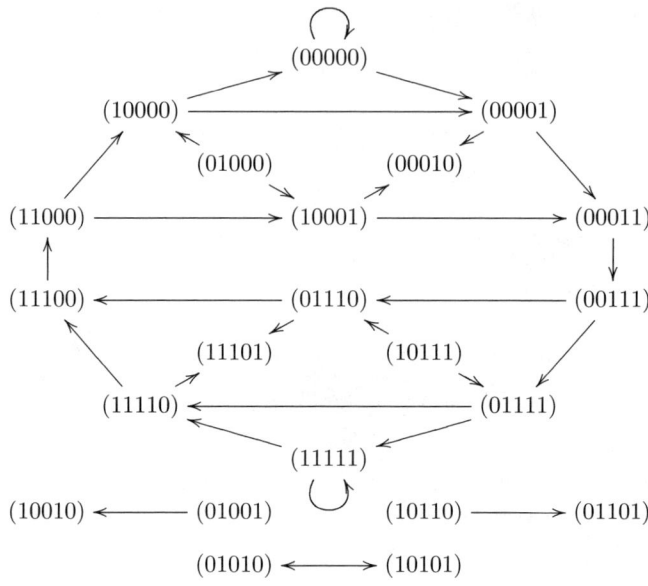

Fig. 5.6. The graph $G(\text{majority}_5)$.

and we have $L_5 = 2$, $L_6 = 10$, $L_7 = 16$, $L_8 = 28$, $L_9 = 38$, $L_{10} = 54$, $L_{11} = 68$, $L_{12} = 94$, and $L_{13} = 132$. Note that finding these initial values is probably best done looking at the right powers of the adjacency matrix and by using the Tr-formula in Eq. (5.4). ◇

5.1. Why did we only consider the graph $\text{Circ}_{n,r}$ and not arbitrary graphs? What goes wrong in the case of, for example, Wheel_n? [1+]

In Section 4.2.2 we have already seen that permutation-SDS induced by the nor or nand function never has fixed points. For $r = 1$ Theorem 5.3 reestablishes this for the special case:

Corollary 5.7. *Let $K = \mathbb{F}_2$, let $Y = \text{Circ}_n$, and let f be a symmetric function $f : K^3 \longrightarrow K$. If the permutation SDS over Y induced by f is fixed-point-free for any $n \geq 3$, then $f = \text{nor}_3$ or $f = \text{nand}_3$.*

Proof. Let $[\mathbf{F}_{\text{Circ}_n}, \pi]$ be an SDS induced by $f : K^3 \longrightarrow K$. From Theorem 5.3 we have that the non-existence of fixed points for any n is equivalent to G_f having no cycles or loops. Let a_i be the value of f on H-class i. Clearly, $a_0 = 1$ and $a_3 = 0$, since otherwise G_f would have loops. Now, $a_1 = 1$ implies $a_2 = 1$. Likewise, $a_1 = 0$ implies $a_2 = 0$. In the latter case we see that $f = \text{nor}_3$, and in the former case we have $f = \text{nand}_3$, and the proof is complete. □

We call a fixed point without any predecessors an *isolated fixed point*. From a computational point of view, such a fixed point is a "practically invisible" attractor in the sense that the probability of realizing such a particular state is $1/q^n$ for a graph on n vertices and with a state space of size q. Clearly, for the identity map all states are isolated fixed points. However, there are nontrivial examples of systems with such fixed points, as the following corollary shows.

Corollary 5.8. *Let $K = \mathbb{F}_2$. Then the permutation-SDS $[\mathbf{Majority}_{\mathsf{Circ}_n}, \pi]$ has isolated fixed points if and only if $n \equiv 0 \bmod 4$.*

Proof. From the graph $G(\text{majority}_3)$ of Circ_n we see that the fixed points of an SDS $[\mathbf{Majority}_{\mathsf{Circ}_n}, \pi]$ are all points without isolated zeros or ones, that is, if $x_i = 0$, then $x_{i-1} = 0$ or $x_{i+1} = 0$, and similarly for $x_i = 1$. If a fixed point $\overline{x}$ has three or more consecutive zeros, we can easily find a permutation such that there is a preimage of $\overline{x}$ different from $\overline{x}$ itself. To be explicit assume $x_{i-1} = x_i = x_{i+1} = 0$. Pick $\sigma \in S_n$ such that $i <_\sigma i - 1$ and $i <_\sigma i + 1$. Let $\hat{x}$ be the point obtained from $\overline{x}$ by setting x_i to 1. Clearly, $\hat{x}$ is a preimage of $\overline{x}$ under $[\mathbf{Majority}_{\mathsf{Circ}_n}, \sigma]$. The case with three or more consecutive states that are one is dealt with in the same way. Thus, the only candidates for isolated fixed points are points where two zeros are followed by two ones that again are followed by two zeros, and so on. These points clearly have no preimage apart from themselves. It is clear that $n \equiv 0 \bmod 4$ is necessary and sufficient for such points to exist, and the proof is complete. $\qquad\square$

From the proof it is also clear that there are precisely four isolated fixed points when $n \equiv 0 \bmod 4$.

5.2. (a) Derive a recursion relation for the number of fixed points L_n of $[\mathbf{Majority}_{\mathsf{Circ}_n}, \pi]$. (b) Give an asymptotic formula for L_n as a function of n. (c) Characterize the fixed points.   **[2+]**

5.2 Fixed-Point Computations for General Graphs

It is natural to ask to what extent the fixed-point characterization and enumeration for $\mathsf{Circ}_{n,r}$ can be generalized. The key features that we implicitly used were that (1) the graph is regular, (2) it has a Hamiltonian cycle, and (3) neighborhoods overlap in an identical, local manner along the Hamiltonian cycle. If any of these conditions fail to hold, then it is clear that we cannot construct the compact graph description $G(\phi)$ that we had for $\mathsf{Circ}_{n,r}$. A quick look at, for example, Q_2^3 or Wheel_n should clarify what goes wrong.

However, we can still consider local fixed points as well as compatible fixed-point coverings. Compatibility of two local fixed points x and x' still means that x and x' agree on the states that belong to the same vertex state in the graph Y. As before we can show that a compatible fixed-point covering corresponds to a fixed point. Although this may seem clear it takes a little bit of mathematical machinery to prove this rigorously. We will not do that here and will contend ourselves with the example below. The computationally inclined reader may not be that surprised to learn that computing all the fixed points of a finite dynamical system map $F\colon K^n \longrightarrow K^n$, even in the case of $K = \mathbb{F}_2$, is computationally intractable [20]. Note, however, that there are efficient algorithms for SDS if we restrict ourselves to special graph classes such as *tree-width bounded graphs* or to special function classes such as linear functions [20].

Example 5.9. We will compute all the fixed points for CA/SDS over the cube Q_2^3 induced by

$$\mathrm{xor}_4\colon \mathbb{F}_2^4 \longrightarrow \mathbb{F}_2, \qquad \mathrm{xor}_4(x) = \begin{cases} 1 & \text{if sum}(x_i)_i = 1, \\ 0 & \text{otherwise,} \end{cases} \tag{5.7}$$

by exhaustive enumeration. Here we have encoded the vertices of Q_2^3 in decimal such that, e.g., $(1,1,0) \leftrightarrow 3$. We set $V = \{0,4,5,7\} \subset \mathrm{v}[Q_2^3]$. Note that V is a dense subset of $\mathrm{v}[Y]$ in the sense that every vertex in Y is in V or is adjacent to a vertex in V. Since Q_2^3 is regular, and since we have the same local function for each vertex, the local fixed points are the same for each vertex. In the following we write the family of states of the vertices contained in $B_Y(v)$ such that the state of v is the first coordinate, for instance,

$$(1|000), \quad (0|000), \quad (0|011) \text{ and } (0|111).$$

The construction of fixed-point covers and the verification that the vertices in $\mathrm{v}[Y] \setminus V$ have fixed-point covers are given in Table 5.2. We get the table by starting at vertex 0, computing all of its local fixed points. For each such local fixed point we compute all possible local fixed points at vertex 4 that are compatible with the initial fixed point. We then branch to vertex 5 and vertex 7. Finally, we verify that the vertex state configurations around the vertices contained in $\mathrm{v}[Y] \setminus V$ are local fixed points. Note that by applying

$(x_0x_1x_2x_4)$	$(x_4x_0x_5x_6)$	$(x_5x_1x_4x_7)$	$(x_7x_3x_5x_6)$	1 2 3 6	Fixed point
(1000)	(0110)	(1000)	(0110)	y y y y	(10010100)
	(0101)	(0000)	(0101)	y y y y	(10010010)
	(0111)	(1000)	(0011)	y y y y	(10000110)
			(0111)	y y y y	(10010110)
(0000)	(0000)	(0000)	(0000)	y y y y	(00000000)
	(0011)	(1000)	(0110)	y y y y	(00010110)
(0111)	(1000)	(0111)	(1000)	y y y y	(01101001)
	(0110)	(0000)	(0000)	y y y y	(01101000)
(0011)	(1000)	(0011)	(1000)	y y y y	(00101001)
(0101)	(1000)	(0111)	(1000)	y y y y	(01001001)
(0110)	(0000)	(0101)	(1000)	y y y y	(01100001)

Table 5.2. The fixed-point computation for Q_2^3 with xor_4 as local functions.

the Q_2^3 automorphisms $\sigma = (0)(124)(365)(7)$ (cycle form) and σ^2 to the last fixed point we obtain the second-to-last and third-to-last fixed points. Note that in this case $\mathsf{Aut}(Y)$ acts on the set of fixed points.   ◇

5.3. Show that, under suitable conditions that you will need to identify, $\mathsf{Aut}(Y)$ acts on $\mathsf{Fix}[\mathbf{F}_Y, \pi]$.   [2]

Remark 5.10. In general, fixed points can be derived by considering a sheaf (of local fixed points) and computing its cohomology. This approach is based on category theory and generalized cohomology theories and is beyond the scope of this book.

5.3 Threshold SDS

Some SDS only have fixed points and no periodic points of period $p > 1$. While it is not the goal of this section to identify all such SDS, we will show that the class of *threshold* SDS has this property.

Definition 5.11. A function $f \colon \mathbb{F}_2^k \longrightarrow \mathbb{F}_2$ is a *threshold function* if it is symmetric and there exists $0 \leq m \leq k$ such that $f(x) = 0$ for all x in H-classes $H_i(x)$ with $0 \leq i \leq m$, and $f(x) = 1$ otherwise. An SDS is a *threshold* SDS if each function F_v is induced by a threshold function f.

An *inverted threshold function* is defined in exactly the same way but with the function values 0 and 1 interchanged. The two SDS $(Y, \mathbf{And}_Y, \pi)$ and $(Y, \mathbf{Majority}_Y, \pi)$ are examples of threshold SDS, while $(Y, \mathbf{Nor}_Y, \pi)$ is an example of an inverted threshold SDS.

Proposition 5.12. *A threshold* SDS *has no periodic points of period $p \geq 2$.*

The following lemma is a consequence of the inversion formula (4.25).

Lemma 5.13. *Let x be a periodic point of the permutation* SDS*-map $[\mathbf{F}_Y, \pi]$ over $\mathbb{F}_2^n$ with (prime) period $p > 1$. There is an index v that is maximal with respect to the ordering π for which*

$$([\mathbf{F}_Y, \pi](x))_{v'} = x_{v'}, \qquad v' >_\pi v,$$
$$([\mathbf{F}_Y, \pi](x))_v = 1 + x_v,$$
$$(F_{v,Y} \circ [\mathbf{F}_Y, \pi](x))_v = x_v.$$

Proof. By assumption x is periodic with period $p > 1$, so there is at least one vertex v' such that $[\mathbf{F}_Y, \pi](x)_{v'} \neq x_{v'}$, and thus there is a maximal (with respect to the order given by π) index v such that $([\mathbf{F}_Y, \pi])(x)_u = x_u$ for $u >_\pi v$. The last statement follows from the fact that restricted to an orbit $[\mathbf{F}_Y, \pi]$ is invertible with inverse $[\mathbf{F}_Y, \pi^*]$ and by using $[\mathbf{F}_Y, \pi^*] \circ [\mathbf{F}_Y, \pi] = \mathsf{id}$ on such an orbit.   □

Proof (Proposition 5.12). Let $[\mathbf{F}_Y, \pi]$ be a threshold SDS-map, and assume that x is a periodic point of period $p > 1$. Clearly, property 3 of Lemma 5.13 cannot hold for threshold systems and a contradiction results.   □

5.4. In Proposition 5.12 we can do better than threshold SDS: Let $A = \{a_1, \ldots, a_m\}$ be linearly ordered by $a_1 < a_2 < \cdots < a_m$. This gives us a partial order $\preceq$ on A^n by $x \preceq y$ if $x_i \leq y_i$ for $i = 1, \ldots, n$. For example, $(0, 1, 0, 0, 0) \preceq (0, 1, 1, 0, 0)$, but $(0, 1, 0, 0, 0)$ and $(1, 0, 0, 0, 0)$ are not comparable. A function $f \colon A^n \longrightarrow A$ is *monotone* if $x \preceq y$ implies $f(x) \leq f(y)$.
(a) Show that threshold functions are monotone. (b) Prove that permutation SDS where each vertex function f_v is monotone has no periodic points of period $p \geq 2$. (c) Does the statement in (b) hold for word-SDS?   **[1+]**

We next show that threshold systems can have long transient orbits:

Proposition 5.14. *For a given integer $m > 0$ there is a graph Y with $|v[Y]| \leq 2m$ and a permutation ordering π such that $[\mathbf{Majority}_Y, \pi]$ has points with transient length m.*

Proof. Let Y be the combinatorial graph with vertex set $v[Y] = \{1, 2, \ldots, 2m\}$ and edge set $e[Y] = \{\{1, 2\}, \ldots \{m, m+1\}, \{2, 2+m\}, \ldots, \{m, 2m\}\}$. Let x be the initial state with $x_i = 0$ for $1 \leq i \leq m$ and $x_i = 1$ for $m + 1 \leq i \leq 2m$ and let $\pi = (1, 2, \ldots, 2m)$. By direct calculation, it is clear that $[\mathbf{Majority}_Y, \pi]^l(x) \neq (1, 1, \ldots, 1)$ for $1 \leq l < m$ and $[\mathbf{Majority}_Y, \pi]^m(x) = (1, 1, \ldots, 1)$.   □

For further information on transient lengths of threshold systems see, for example, [14].

The following problem shows how the construction of *potential functions* can be used to conclude that certain threshold SDS only have fixed points.

5.5. Let $\mathrm{sign} \colon \mathbb{R} \longrightarrow \mathbb{R}$ be the function defined by $\mathrm{sign}(x) = 1$ if $x \geq 0$ and $\mathrm{sign}(x) = -1$ otherwise. In this problem we use the state space $K = \{-1, 1\}$ and vertex functions given by $f_v(x) = \mathrm{sign}(\sum_{v' \in B_1'(v)} x_{v'})$. Let Y be a combinatorial graph and let $\pi \in S_Y$. We define a potential function (or energy function) $E \colon K^n \longrightarrow \mathbb{R}$ by

$$E = - \sum_{\{u,v\} \in e[Y]} x_u x_v . \tag{5.8}$$

(i) Show that whenever the application of a Y-local map F_v leads to a change in the system state then the potential either stays the same or decreases.
(ii) Based on (i) show that this SDS has no periodic points of period $p > 1$.
    [2]

5.4 SDS over Special Graph Classes

We have seen many examples of SDS over Circ_n and the binary n-cubes. In this section we present a more systematic collection of results on the structure of SDS over special graphs classes. We start with the complete graph.

5.4.1 SDS over the Complete Graph

It is intuitively clear that for induced SDS the particular choice of permutation update order is not essential for dynamical equivalence when $Y = K_n$. This follows since we are free to relabel the vertices in any manner we like. More precisely, by the fact that $\mathsf{Aut}(K_n) = S_n$ and from Proposition 4.30 it follows that the induced SDS-maps $[\mathbf{F}_{K_n}, \sigma]$ and $[\mathbf{F}_{K_n}, \sigma']$ are dynamically equivalent for any choice of σ and σ'. To see this just choose $\gamma \in \mathsf{Aut}(K_n)$ such that $\sigma' = \gamma\sigma$ (we can always do this — why?) and conclude that

$$[\mathbf{F}_{K_n}, \sigma'] = [\mathbf{F}_{K_n}, \gamma\sigma] = \gamma \circ [\mathbf{F}_{K_n}, \sigma] \circ \gamma^{-1} .$$

In light of this, it is clearly enough to consider SDS with the identity update order $\mathrm{id} = (1, 2, 3, \ldots, n)$ in the case of $Y = K_n$. Again, note that this is generally only true for *induced* SDS. To start we make the following observation.

Lemma 5.15. *Let* $[\mathbf{F}_{K_n}, \mathrm{id}]$ *be the map of a permutation SDS induced by the symmetric Boolean function* $f_n \colon \mathbb{F}_2^n \longrightarrow \mathbb{F}_2$. *Let* $\mathcal{O}$ *be an orbit of* $[\mathbf{F}_{K_n}, \mathrm{id}]$ *and let* $\mathrm{res}_{\mathcal{O}} f_n$ *denote the restriction of* f_n *to* $\mathcal{O}$. *Suppose (a) that* $\mathrm{res}_{\mathcal{O}} f_n$ *satisfies the functional relation*

$$\phi(x_1, \ldots, x_{n-1}, \phi(x_1, \ldots, x_n)) = x_n \tag{5.9}$$

and (b) that we have the commutative diagram

$$(5.10)$$

where $\hat{\mathbb{F}}_2^n = \{x \in \mathbb{F}_2^{n+1} \mid x_{n+1} = \iota_f(x_1, x_2, \ldots, x_n)\}$ *and*

$$\mathsf{proj}(x_1, \ldots, x_n, x_{n+1}) = (x_1, \ldots, x_n),$$
$$\iota_f(x_1, \ldots, x_n) = (x_1, \ldots, x_n, f(x_1, \ldots, x_n)),$$
$$\sigma_{n+1}(x_1, x_2, \ldots, x_{n+1}) = (x_{n+1}, x_1, \ldots, x_n) .$$

Then we have $n + 1 \equiv 0 \pmod{|\mathcal{O}|}$.

Proof. Clearly, the commutative diagram implies $[\mathbf{F}_Y, \mathrm{id}]^\ell = (\mathsf{proj} \circ \sigma_{n+1} \circ \iota_f)^\ell$ and from the functional equation (5.9) we conclude $(\mathsf{proj} \circ \sigma_{n+1} \circ \iota_f)^2 = (\mathsf{proj} \circ \sigma_{n+1}^2 \circ \iota_f)$, and by induction

$$[\mathbf{F}_Y, \mathrm{id}]^\ell = \mathsf{proj} \circ \sigma_{n+1}^\ell \circ \iota_f.$$

In particular, for $\ell = n + 1$ we get $[\mathbf{F}_Y, \mathrm{id}]^{n+1} = \mathsf{proj} \circ \iota_f = \mathrm{id}$. $\square$

We also have

Lemma 5.16. *Let* $[\mathbf{F}_{K_n}, \mathrm{id}]$ *be the* SDS-*map induced by the symmetric function* f_n. *Let* $H_k = \{x = (x_1, \ldots, x_n) \mid \mathsf{sum}(x) = k\}$ *and let* $\mathcal{O}$ *be an orbit of the system. Suppose that for* $x \in \mathcal{O}$

$$\prod_{i=1}^{l} F_{i,K_n}(x) \in H_k \cup H_{k+1}, \quad 1 \le l \le n, \tag{5.11}$$

and that there exists at least one integer l_1 *with* $\prod_{i=1}^{l_1} F_{i,K_n}(x) \in H_k$ *and at least one integer* l_2 *with* $\prod_{i=1}^{l_2} F_{i,K_n}(x) \in H_{k+1}$. *Then* $n + 1 \equiv 0 \pmod{|\mathcal{O}|}$.

Proof. First, note that the conditions above imply that $f_n(x_1 \ldots, x_n) = 1$ for $x \in \mathcal{O} \cap A_k$ and $f_n(x_1 \ldots, x_n) = 0$ for $x \in \mathcal{O} \cap A_{k+1}$. The lemma now follows from the following two observations: First, for $x \in \mathcal{O}$ one has

$$f(x_1, \ldots, x_{n-1}, f(x_1, \ldots, x_n)) = x_n, \tag{5.12}$$

and second,

$$\forall\, 1 \le k \le n - 1 \quad ([\mathbf{F}_{K_n}, \pi](x))_{k+1} = (x)_k. \tag{5.13}$$

From this we conclude that (5.10) commutes, and the lemma follows.   □

5.6. Verify (5.12) and (5.13) in the proof of Lemma 5.16.   [1+]

In the following we describe the dynamics of SDS induced by the functions nor, parity, majority, and minority over K_n. We will use e_k to denote the kth unit vector, that is, the state $e_k \in \mathbb{F}_2^n$ with $(e_k)_k = 1$ and $(e_k)_j = 0$ for $k \ne j$. We set $\langle x, y \rangle = \sum x_i y_i$.

Proposition 5.17 (Nor). *Consider the* SDS-*map* $[\mathbf{Nor}_{K_n}, \mathrm{id}]$. *The states* x *for which* $\langle x, e_n \rangle = 1$ *are mapped to zero. If* $\langle x, e_n \rangle \ne 1$, *then* x *is mapped to* e_k *where* $k = 1 + \max\{i \mid x_i = 1\}$. *The set* $L = \{0, e_1, e_2, \ldots, e_n\}$ *is the unique periodic orbit of* $[\mathbf{Nor}_{K_n}, \mathrm{id}]$.

Proof. Clearly, all points are mapped into L. Also, 0 is mapped to e_1, e_k is mapped to e_{k+1} for $1 \le k \le n - 1$, and e_n is mapped to 0.   □

Proposition 5.18 (Parity). *For the* SDS-*map* $[\mathbf{Parity}_{K_n}, \mathrm{id}]$ *all states are contained in periodic orbits* $\mathcal{O}$ *and we have* $n + 1 \equiv 0 \pmod{|\mathcal{O}|}$.

Proof. By Problem 4.8, which is a straightforward corollary of Proposition 4.13, an SDS-map $[\mathbf{Parity}_Y, \pi]$ is bijective for any graph Y, and all states are periodic. It is clear that any orbit that contains at least two points satisfies the conditions in (5.11) in Lemma 5.16 for some odd integer k, and the last statement follows.   □

Proposition 5.19 (Minority). *For any periodic orbit* $\mathcal{O}$ *of the* SDS-*map* $[\mathbf{Minority}_{K_n}, \mathrm{id}]$ *we have* $n + 1 \equiv 0 \pmod{|\mathcal{O}|}$.

Proof. A periodic orbit for this system satisfies Eq. (5.11) for $k = \lfloor n/2 \rfloor$ and the proposition follows. $\square$

Proposition 5.20 (Majority). *For the* SDS*-map* $[\mathbf{Majority}_{K_n}, \mathrm{id}]$ *every state is fixed or eventually fixed. The only fixed points are* $(0, 0, \ldots, 0)$ *and* $(1, 1, \ldots, 1)$.

Proof. Obviously, $(0, 0, \ldots, 0)$ and $(1, 1, \ldots, 1)$ are fixed points. By definition, the application of majority$_n$ to a state x containing an equal number of vertex states that are 1 and 0 yields 1 as outcome, and hence such a state x is mapped to the fixed point $(1, 1, \ldots, 1)$. Clearly, any other point will be mapped to either $(0, 0, \ldots, 0)$ or $(1, 1, \ldots, 1)$ by a single application of $[\mathbf{Majority}_{K_n}, \mathrm{id}]$.
                                                                              $\square$

The following result is not over the complete graph, but on the *complete bipartite graph* of order (m, n) written $K_{m,n}$. This graph is the *graph union* of E_m and E_n, the empty graphs on m and n vertices, respectively.

Proposition 5.21. *For* $[\mathbf{Majority}_{K_{m,n}}, \pi]$ *all states are fixed or eventually fixed. There are* $\binom{m}{\lfloor m/2 \rfloor}\binom{n}{\lfloor n/2 \rfloor} + 2$ *fixed points.*

Proof. Recall that the function majority$_n$ yields 1 when applied to a state x containing an equal number of 0's and 1's. Let the vertex classes of $K_{m,n}$ be V_m and V_n. Call a state x *balanced* if the states contained in V_m have exactly $\lceil m/2 \rceil$ zeros and the states contained in V_n has exactly $\lceil n/2 \rceil$ zeros. Clearly, all balanced states are fixed and all other points eventually map to either $(0, 0, \ldots, 0)$ or $(1, 1, \ldots, 1)$. $\square$

Obviously a balanced state has no preimage apart from itself. The dynamics of this system is thus fully understood.

Remark 5.22. Note that for a majority-SDS over $K_{m,n}$ with $n = 2$ one has states with a minority of zeros that are mapped to $(0, 0, \ldots, 0)$ for some update orders and that are mapped to $(1, 1, \ldots, 1)$ for other update orders. In the context of a voting game with opportunistic voters we thus see that the right update order can completely change the outcome of the election based on the initial inclination of a small set of voters. (An opportunistic voter is a voter who votes the same as the majority of his contacts have voted already or are planning to vote.)

5.4.2 SDS over the Circle Graph

The circle graph has helped us illustrate many concepts so far. As we have seen in Chapter 2, this is also the graph that is frequently used in the studies of one-dimensional cellular automata in the case of periodic boundary conditions. Here we will give results on invertible dynamics on the circle graph. After the next section where we consider line graphs, we conclude with a problem that points to one of the central questions in analysis of graph dynamical systems: How can we relate the dynamics over two graphs that only differ by one edge?

Proposition 5.23. *The SDS-map* $[\mathbf{Parity}_{\mathrm{Circ}_n}, \mathrm{id}]\colon \mathbb{F}_2^n \longrightarrow \mathbb{F}_2^n$ *is conjugate to a right-shift of length* $n-2$ *on a subset of* $\mathbb{F}_2^{2n-2}$. *In particular,*

$$|\mathsf{Per}(x)| \equiv \begin{cases} 0 \mod n-1, & n \text{ even}, \\ 0 \mod 2n-2, & n \text{ odd}, \end{cases} \tag{5.14}$$

for all $x \in \mathbb{F}_2^n$. *The same statement holds for the corresponding* SDS *induced by* $(1 + \mathrm{parity}_3)$.

Proof. Define the embedding $\iota\colon \mathbb{F}_2^n \longrightarrow \mathbb{F}_2^{2n-2}$ by

$$\iota(x_0, \ldots, x_{n-1})$$
$$= (x_0, \ldots, x_{n-1}, x_{n-1} + x_0 + x_1, x_{n-1} + x_0 + x_2, \ldots, x_{n-1} + x_0 + x_{n-2}),$$

and set $\widehat{\mathbb{F}}_2^{2n-2} = \iota(\mathbb{F}_2^n)$. A direct calculation shows that the diagram

$$\begin{array}{ccc} \mathbb{F}_2^n & \xrightarrow{[\mathbf{Parity}_{\mathrm{Circ}_n}, \mathrm{id}]} & \mathbb{F}_2^n \\ \iota \downarrow & & \downarrow \iota \\ \widehat{\mathbb{F}}_2^{2n-2} & \xrightarrow{\sigma_{n-2}} & \widehat{\mathbb{F}}_2^{2n-2} \end{array} \tag{5.15}$$

commutes. Here $\sigma_{n-2}\colon \widehat{\mathbb{F}}_2^{2n-2} \longrightarrow \widehat{\mathbb{F}}_2^{2n-2}$ is defined by $\sigma_{n-2}(x_0, \ldots, x_{2n-3}) = (x_n, \ldots, x_{2n-3}, x_0, \ldots, x_{n-1})$. It is well-defined. Note that $\iota\colon \mathbb{F}_2^n \longrightarrow \widehat{\mathbb{F}}_2^{2n-2}$ is a bijection. Thus, the map σ and $[\mathbf{Parity}_{\mathrm{Circ}_n}, \mathrm{id}]$ are topologically conjugate (discrete topology) under ι.

Explicitly, we have

$$[\mathbf{Parity}_{\mathrm{Circ}_n}, \mathrm{id}](x_0, x_1, \ldots, x_{n-1})$$
$$= (x_{n-1} + x_0 + x_1, x_{n-1} + x_0 + x_2, \ldots, x_{n-1} + x_0 + x_{n-2}, x_0, x_1),$$

and then

$$\iota(x_{n-1} + x_0 + x_1, x_{n-1} + x_0 + x_2, \ldots, x_{n-1} + x_0 + x_{n-2}, x_0, x_1)$$
$$= (x_{n-1} + x_0 + x_1, x_{n-1} + x_0 + x_2, \ldots, x_{n-1}$$
$$+ x_0 + x_{n-2}, x_0, x_1, x_2, \ldots, x_{n-1}).$$

On the other hand, this also equals $(\sigma_{n-2} \circ \iota)(x_0, \ldots, x_{n-1})$, verifying the commutative diagram.

From the conjugation relation it is clear that the size of a periodic orbit under $[\mathbf{Parity}_{\mathrm{Circ}_n}, \mathrm{id}]$ must be a divisor of $(2n-2)/\gcd(n-2, 2n-2)$. The statement of the proposition follows from the fact that

$$\gcd(n-2, 2n-2) = \begin{cases} 1, & n \equiv 0 \mod 2, \\ 2, & \text{else}. \end{cases}$$

The proof for $[(1 + \mathbf{Parity})_{\mathrm{Circ}_n}, \mathrm{id}]\colon \mathbb{F}_2^n \longrightarrow \mathbb{F}_2^n$ and the details are left for the reader. $\square$

In analogy with the case of K_n we obtain that the phase space of $[\mathbf{Parity}_{\mathsf{Circ}_n}, \mathsf{id}]$ can be embedded in the phase space of the $(n-2)$th power of the elementary cellular automaton $\Phi \colon \mathbb{F}_2^{2n-2} \longrightarrow \mathbb{F}_2^{2n-2}$ induced by $\phi \colon \mathbb{F}_2^3 \longrightarrow \mathbb{F}_2$, $\phi(x_{i-1}, x_i, x_{i+1}) = x_{i-1}$ (rule 240), i.e.,

$$\Gamma[\mathbf{Parity}_{K_n}, \mathsf{id}] \hookrightarrow \Gamma(\Phi_{240,(2n-2)}^{n-2}). \tag{5.16}$$

For a followup to Proposition 5.23 see Problem 5.8.

5.4.3 SDS over the Line Graph

The graph Line_n differs from Circ_n by one edge, but as you may have expected the dynamics of SDS over these two graphs can be significantly different.

Proposition 5.24. *The SDS-map* $[\mathbf{Parity}_{\mathsf{Line}_n}, \mathsf{id}] \colon \mathbb{F}_2^n \longrightarrow \mathbb{F}_2^n$ *is conjugate to the composition* $\tau \circ \sigma_{-1}$ *where* $\tau \colon \mathbb{F}_2^{n+1} \longrightarrow \mathbb{F}_2^{n+1}$ *is given by*

$$\tau(x = (x_1, \ldots, x_{n+1})) = (x_{n+1} + x_i)_i$$

and $\sigma_{-1} \colon \mathbb{F}_2^{n+1} \longrightarrow \mathbb{F}_2^{n+1}$ *is given by*

$$\sigma_{-1}(x_1, \ldots, x_{n+1}) = (x_2, x_3, \ldots, x_{n+1}, x_1) .$$

In particular, $|\mathsf{Per}(x)| \equiv 0 \mod (n+1)$ *for all* $x \in \mathbb{F}_2^n$. *The same statement holds for the corresponding* SDS *induced by* $(1 + \mathrm{parity}_3)$.

Proof. We have the embedding $\iota \colon \mathbb{F}_2^n \longrightarrow \mathbb{F}_2^{n+1}$ given by

$$\iota(x_1, \ldots, x_n) = (x_1, \ldots, x_n, 0) .$$

A direct computation gives

$$\begin{aligned}
(\iota \circ [\mathbf{Parity}_{\mathsf{Line}_n}, \mathsf{id}])(x_1, \ldots, x_n) &= \iota(x_1 + x_2, \ldots, x_1 + x_n, x_1) \\
&= (x_1 + x_2, \ldots, x_1 + x_n, x_1, 0)
\end{aligned}$$

and

$$\begin{aligned}
\tau \circ \sigma_{-1} \circ \iota(x_1, \ldots, x_n) &= \tau \circ \sigma_{-1}(x_1, \ldots, x_n, 0) \\
&= \tau(x_2, x_3, \ldots, x_n, 0, x_1) \\
&= (x_1 + x_2, x_1 + x_3, \ldots, x_1 + x_n, x_1, 0) .
\end{aligned}$$

The rest is now clear. The proof of the last statement is left to the reader. $\square$

5.7. Investigate the dynamics of $[\mathbf{Nor}_{\mathsf{Line}_n}, \mathsf{id}]$.   [2+]

5.8. Let Y and Y' be combinatorial graphs that differ by exactly one edge e. Clearly, SDS over Y and Y' cannot have the same vertex functions $(f_v)_v$ since there are two vertices where the degrees do not match. However, we may consider induced SDS. For a fixed set of functions it would be very desirable to relate the dynamics of the two SDS. The addition or deletion of an edge is a key operation, and it would allow us to relate systems over different graphs by successive edge removals and additions. Using Propositions 5.23 and 5.24, what can be said about this problem in the particular case of SDS induced by parity functions over Circ_n and Line_n? [3]

5.9. What can be said about Problem 5.8 in the general case or in interesting special cases? That is, relate induced SDS over graphs that differ by precisely one edge. [5]

5.4.4 SDS over the Star Graph

We have already considered SDS over Star_n induced by nor functions when we showed that the bound $\Delta(\mathsf{Star}_n)$ is sharp. The graph Star_n often provides interesting examples since it has a large automorphism group. Here we will consider SDS induced by parity functions.

Proposition 5.25. *Let* $Y = \mathsf{Star}_n$, *let* $\pi \in S_Y$, *and set* $\phi = [\mathbf{Parity}_Y, \pi]$. *Then for all* $x \in \mathbb{F}_2^n$ *we have* $|\mathsf{Per}(x)| \equiv 0 \mod 3$ *for* n *even and* $|\mathsf{Per}(x)| \equiv 0 \mod 4$ *for* n *odd.*

Proof. Since $\mathsf{Aut}(\mathsf{Star}_n) \cong S_n$, each orbit in $U(Y)/\sim_Y$ under $\mathsf{Aut}(\mathsf{Star}_n)$ is fully characterized by the position of the center vertex 0 in the underlying permutations. It is now straightforward to verify that in all the $n + 1$ cases the statement of the proposition holds. We leave the details to the reader. $\square$

5.10. Characterize the dynamics of $[\mathbf{Minority}_{\mathsf{Star}_n}, \pi]$ for $\pi \in S_{\mathsf{Star}_n}$ up to dynamical equivalence. [2]

5.11. Determine the fixed points of permutation SDS over Star_n induced by the 2-threshold functions. Show that the number of fixed points is exponential in n. Let $x \in \mathbb{F}_2^{n+1}$, and let $\omega(x)$ denote the set of fixed points that can be reached from x for all fixed choices of permutation update order. Show that there exists a state x' such that $\omega(x')$ has size that is exponential in n. (See also [111].) [3]

5.5 SDS Induced by Special Function Classes

In this section we study systematically several SDS that we encountered before. For instance, we analyze SDS induced by nor functions, which proved to be helpful in establishing that the bound $|\mathsf{Acyc}(Y)|$ in Section 4.3.1 is sharp.

Here we will study the phase-space structure of SDS induced by nor and enumerate some of these configurations. Note that some of the results are valid only for permutation update orders and that some results are valid in the more general context of word update orders.

5.5.1 SDS Induced by $(\mathrm{nor}_k)_k$ and $(\mathrm{nand}_k)_k$

Here we will characterize properties of SDS induced by $(\mathrm{nor}_k)_k$ or $(\mathrm{nand}_k)_k$ more systematically. Our description will start at a general level and finish with some properties that apply for these systems for special graph classes such as Circ_n. To begin, recall the following fact:

Proposition 5.26. *Let Y be an combinatorial graph, let w be a word over $\mathrm{v}[Y]$, and let $K = \mathbb{F}_2$. Then*

$$[\mathbf{Nand}_Y, w] \circ \mathrm{inv} = \mathrm{inv} \circ [\mathbf{Nor}_Y, w] , \tag{5.17}$$

where the function inv *is the inversion map (4.22).*

Thus, whatever we can derive for SDS induced by nor functions applies to SDS induced by nand functions up to dynamical equivalence. For this reason we will omit the obvious statements for SDS induced by nand functions in the following.

Fixed Points and Periodic Points

As you have seen in the examples so far, permutation-SDS induced by nor functions never have any fixed points.

Proposition 5.27. *Let Y be a combinatorial graph. A permutation-SDS over Y induced by $(\mathrm{nor}_k)_k$ has no fixed points.*

The proof of this is straightforward and is left as an exercise.

5.12. Give the proof of Proposition 5.27.   [1]

5.13. Proposition 5.27 does not hold for word update orders. Why?   [2-]

We next establish what the periodic points are for Nor-SDS. It turns out that the periodic points only depend on the graph structure and not on the update order, a property we will need later when we study certain groups that describe the actual dynamics on the set of periodic points in Chapter 6. Moreover, this characterization of periodic points is also valid for *fair words*. Recall that a fair word over $\mathrm{v}[Y]$ is a word that contains each element of $\mathrm{v}[Y]$ at least once.

Theorem 5.28. *Let Y be a combinatorial graph on n vertices, let w be a fair word over $v[Y]$, and let $K = \mathbb{F}_2$. Then the set of periodic points of $[\mathbf{Nor}_Y, w]$ is*

$$\mathrm{Per}[\mathbf{Nor}_Y, w] = \{x \in \mathbb{F}_2^n \mid \forall v \ : \ x_v = 1 \Rightarrow \forall v' \in B_1'(v) \ : \ x_{v'} = 0\} \ . \quad (5.18)$$

In particular, $\mathrm{Per}[\mathbf{Nor}_Y, w]$ is independent of w and is in a bijective correspondence with $\mathcal{I}(Y)$, the set of independent sets of Y.

Proof. Let $w = (w_1, \ldots, w_k)$ be fair word over $v[Y]$ and introduce the set

$$P(Y) = \{(x_{v_1}, \ldots, x_{v_n}) \in \mathbb{F}_2^n \mid \forall v \ : \ x_v = 1 \Rightarrow \forall v' \in B_1'(v) \ : \ x_{v'} = 0\} \ .$$

We will execute the proof in three steps. The first step is to show that $\mathrm{Per}[\mathbf{Nor}_Y, w] \subset P(Y)$. Let $x \in \mathbb{F}_2^n$. We observe that the only circumstance in which x_v is mapped to 1 by Nor_v is when the state of all vertices in $B_Y(v)$ is 0. Since w is a fair word, it is clear that the image of x under $[\mathbf{Nor}_Y, w]$ is contained in $P(Y)$, and therefore that $\mathrm{Per}[\mathbf{Nor}_Y, w] \subset P(Y)$.

The next step is to show that the maps $\mathrm{Nor}_v \colon P(Y) \longrightarrow P(Y)$ are well-defined and invertible. Let $x \in P(Y)$. There are three cases to consider. Assume $x_v = 1$. Then by construction all states $x_{v'}$ with $v' \in B_1'(v)$ satisfy $x_{v'} = 0$. Thus, $\mathrm{Nor}_v(x)_v = 0$ and consequently $(\mathrm{Nor}_v^2)(x)_v = 1$. If $x_v = 0$, there are two cases to consider. In the first case all $x_{v'}$ with $v' \in B_1'(v)$ are zero. Clearly, in this case x_v is mapped to 1 under Nor_v, which is then mapped back to 0 by a subsequent application of Nor_v. The final case with $x_v = 0$ and where one or more neighbor vertex v' has $x_{v'} = 1$ is clear. There are two things to be learned from this. First, the map Nor_v maps P into P and is thus well-defined. Second, we have seen that for all $v \in v[Y]$

$$(\mathrm{Nor}_v)^2 \colon P(Y) \longrightarrow P(Y) = \mathrm{id} \colon P(Y) \longrightarrow P(Y) \ . \quad (5.19)$$

We next show that $P(Y) \subset \mathrm{Per}[\mathbf{Nor}_Y, w]$. By definition, $\mathrm{Per}[\mathbf{Nor}_Y, w]$ is the maximal subset of $\mathbb{F}_2^n$ over which $[\mathbf{Nor}_Y, w]$ is invertible. By our previous argument, each map Nor_v is invertible over $P(Y)$, and consequently all SDS

$$[\mathbf{Nor}_Y, w] = \prod_{j=1}^{k} \mathrm{Nor}_{w(j)} \colon P(Y) \longrightarrow P(Y)$$

are invertible maps. We therefore conclude that

$$P(Y) \subset \mathrm{Per}[\mathbf{Nor}_Y, w]$$

and hence that $P(Y) = \mathrm{Per}[\mathbf{Nor}_Y, w]$.

It only remains to verify that we have a bijective correspondence between $\mathrm{Per}[\mathbf{Nor}_Y, w]$ and $\mathcal{I}$. To this end define $\beta \colon \mathrm{Per}[\mathbf{Nor}_Y, w] \longrightarrow \mathcal{I}$ by

$$\beta(x_{v_1}, \ldots x_{v_n}) = \{v_k \mid x_{v_k} = 1\} \ . \quad (5.20)$$

The map is clearly well-defined, and it is clear that β is a bijection. $\square$

As a part of the proof of Theorem 5.28 we saw that

$$\mathrm{Per}[\mathbf{Nor}_Y, w] = [\mathbf{Nor}_Y, w](\mathbb{F}_2^n) .$$

This fact translates into the following corollary for transients states of Nor-SDS:

Corollary 5.29. *Let Y be a combinatorial graph and let w be a fair word over Y. The maximal transient length of any state under $[\mathbf{Nor}_Y, w]$ is 1.*

Example 5.30. Let $\phi = [\mathbf{Nor}_{\mathrm{Circ}_4}, w]$. In accord with Theorem 5.28 we have the following seven order-independent periodic points:

$$\mathrm{Per}[\mathbf{Nor}_{\mathrm{Circ}_4}, w] = \{(0,0,0,0), (0,0,0,1), (0,0,1,0), (0,1,0,0),$$
$$(1,0,0,0), (1,0,1,0), (0,1,0,1)\} ,$$

where w is a fair word. Clearly, $|\mathrm{Per}[\mathbf{Nor}_{\mathrm{Circ}_n}, w]| \equiv 0 \mod 7$. Later in Chapter 6 we will see that for any configuration of these seven points into cycles we can find a word w such that the corresponding Nor-SDS has exactly this cycle configuration as its periodic orbits. In particular, this means that we can find a word $w' \in W_Y$ such that the $[\mathbf{Nor}_{\mathrm{Circ}_4}, w']$ has exactly one periodic orbit of length 7 and another word w'' such that $[\mathbf{Nor}_{\mathrm{Circ}_4}, w'']$ has exactly seven fixed points. For example, a straightforward computation shows that the SDS

$$[\mathbf{Nor}_{\mathrm{Circ}_4}, (0, 1, 2, 3)]$$

has exactly one periodic orbit of length 7.   ◇

Enumeration of Periodic Points

Here we illustrate how to obtain information about $P(Y) = |\mathrm{Per}[\mathbf{Nor}_Y, w]|$ for $w \in W_Y'$ in the special case of $Y = \mathrm{Circ}_n$ through a recursion relation. Here W_Y' denotes the fair words over $v[Y]$. We will later find an explicit expression for P_n.

Proposition 5.31. *Let $n \geq 3$. Then we have the Fibonacci recursion*

$$P_{n+1} = P_n + P_{n-1} . \tag{5.21}$$

Proof. Set $\phi_n = [\mathbf{Nor}_{\mathrm{Circ}_n}, w]$ with $w \in W_Y'$. Since any periodic point x of ϕ_n can be extended to a periodic point of ϕ_{n+1} by $x \mapsto (x,0)$, we have a well-defined injection

$$a \colon \mathrm{Per}(\phi_n) \longrightarrow \mathrm{Per}(\phi_{n+1}), \quad x \mapsto (x,0) .$$

Moreover, we see that an element $x \in \mathrm{Per}(\phi_n)$ can be extended to two periodic points $(x,0)$ and $(x,1)$ of ϕ_{n+1} if and only if we have $x_0 = x_{n-1} = 0$. Let

$$p(a, b, c) = |\{x \in \mathsf{Per}(\phi_{n+1}) \mid x_{n-1} = a, \, x_n = b, \, x_0 = c\}| \, .$$

We then have

$$P_{n+1} = p(0, 1, 0) + p(1, 0, 0) + p(0, 0, 1) + p(0, 0, 0) + p(1, 0, 1) \, . \qquad (5.22)$$

The three first terms on the right in (5.22) add up to P_n. To give an interpretation of the last two terms in (5.22), we see that the map

$$b \colon \mathsf{Per}(\phi_{n-1}) \longrightarrow \mathsf{Per}(\phi_{n+1}), \quad x \mapsto \begin{cases} (x, 1, 0) & \text{if } x_0 = 1 \, , \\ (x, 0, 0) & \text{if } x_0 = 0 \, , \end{cases} \qquad (5.23)$$

is a well-defined injection with image size $p(0, 0, 0) + p(1, 0, 1)$. Equation (5.22) therefore becomes

$$P_{n+1} = P_n + P_{n-1} \, ,$$

and the proposition follows. □

Example 5.32. The values of P_n for small n are given in the table below.

n	3	4	5	6	7	8	9	10	11	12	13	14	15	16
$P(\mathsf{Circ}_n)$	4	7	11	18	29	47	76	123	199	322	521	843	1364	2207

◇

Here is an alternative approach for computing the number of periodic points of a Nor-SDS over $Y = \mathsf{Circ}_n$. It also gives an explicit formula for P_n as well as L_n, which is the number of periodic points of $[\mathbf{Nor}_{\mathsf{Line}_n}, w]$.

Proposition 5.33. *The number of periodic points of an* SDS *induced by* nor *functions on* Line_n *is* $L_n = F_{n+1}$ *where* F_n *denotes the nth Fibonacci number* $(F_0 = 1, \, F_1 = 1, \, \text{and} \, F_n = F_{n-1} + F_{n-2}, \, n \geq 2)$. *The number of periodic points of an* SDS *induced by* nor *functions on* Circ_n *is* $P_n = r^n + (-1/r)^n$, *where* $r = (1 + \sqrt{5})/2$ *(the golden ratio).*

Proof. The case of Line_n follows from the observation that for the periodic points with $x_n = 1$ one must have $x_{n-1} = 0$. Clearly, for the remaining coordinates there are as many choices as there are periodic points for $[\mathbf{Nor}_{\mathsf{Line}_{n-2}}, \pi]$. Thus, the number of periodic points of $[\mathbf{Nor}_{\mathsf{Line}_n}, \pi]$ with $x_1 = 1$ is L_{n-2}. Similarly, we get that the number of periodic points of $[\mathbf{Nor}_{\mathsf{Line}_n}, \pi]$ with $x_n = 0$ equals L_{n-1}. Thus, we have $L_n = L_{n-1} + L_{n-2}$ for $n \geq 3$ where $L_1 = 2$, $L_2 = 3$, and thus $L_n = F_{n+1}$ as claimed.

For the case of Circ_n we see that the number of periodic points with $x_0 = 1$ equals L_{n-3} while the number of periodic points with $x_0 = 0$ equals L_{n-1}, and we conclude that $P_n = L_{n-1} + L_{n-3} = F_n + F_{n-2}$ for $n \geq 4$. Using the formulas for the nth Fibonacci number gives $P_n = r^n + (-1/r)^n$ for $n \geq 4$. The formula also holds for $n = 3$, so we are done. □

5.14. Derive the recursion relation (5.21) from Proposition 5.33. [1]

5.15. Derive a recursion relation for the number of periodic points of a Nor-SDS over Wheel_n. [1+]

Further Characterization of Phase Space

In the remainder of this section we include some more results on the structure of the phase space of Nor-SDS. The proofs here are somewhat more technical and were derived as a part of the research that investigated whether or not the bound $\Delta(Y)$ is sharp.

Proposition 5.34. *Let Y be a combinatorial graph, let $\pi \in S_Y$, and let $K = \mathbb{F}_2$. The state zero has maximal indegree in $\Gamma[\mathbf{Nor}_Y, \pi]$.*

The proof is a direct consequence of the following lemma:

Lemma 5.35. *Let Y be a combinatorial graph, let $\pi \in S_Y$, and let $K = \mathbb{F}_2$. For $x \neq 0$ let $M(x) = \{v \in \mathrm{v}[Y] \mid x_v = 1\}$, and for $S \subset M(x)$ let x^S be the state with $x_v^S = x_v$ for $v \notin S$ and $x_v^S = 0$ for $v \in S$. We then have*

$$\forall x \in \mathbb{F}_2^n \ \forall S \subset M(x) \ : \ |[\mathbf{Nor}_Y, \pi]^{-1}(x)| \leq |[\mathbf{Nor}_Y, \pi]^{-1}(x^S)| \, , \qquad (5.24)$$

and in particular $|[\mathbf{Nor}_Y, \pi]^{-1}(x)| \leq |[\mathbf{Nor}_Y, \pi]^{-1}(0)|$.

Proof. Assume $|\mathrm{v}[Y]| = n$ and let $x \in \mathbb{F}_2^n$. The inequality (5.24) clearly holds for any x with $[\mathbf{Nor}_Y, \pi]^{-1}(x) = \varnothing$, so without loss of generality we may assume that $[\mathbf{Nor}_Y, \sigma]^{-1}(x) \neq \varnothing$. Since we have a Nor-SDS, this assumption implies that x is a periodic point.

Let $x \in \mathbb{F}_2^n$ be a periodic point such that $x \neq 0$ with $x_v = 1$. Without loss of generality we may assume that v is maximal with respect to π such that $x_v = 1$. From the characterization of the periodic points in Theorem 5.28 we know that $x_u = 0$ for all $u \in B_1'(v)$. Moreover, any $y \in [\mathbf{Nor}_Y, \pi]^{-1}(x)$ satisfies $y_u = 0$ for all $v \neq u \in B_Y^{\geq \pi}(v)$. Let $\hat{x}$ be the state defined by $\hat{x}_v = 0$ and $\hat{x}_u = x_u$ for $u \neq v$. We can now define a map

$$r_v \colon [\mathbf{Nor}_Y, \pi]^{-1}(x) \longrightarrow [\mathbf{Nor}_Y, \pi]^{-1}(\hat{x}) \qquad (5.25)$$

by $(r_v(z))_v = 1$ and $(r_v(z))_u = z_u$ otherwise. Clearly, this map is well-defined. Moreover, it is an injection, which in turn implies

$$|[\mathbf{Nor}_Y, \pi]^{-1}(x)| \leq |[\mathbf{Nor}_Y, \pi]^{-1}(\hat{x})| \, .$$

Equation (5.24) now follows by induction on $|\{v \mid x_v = 1\}|$ by successively replacing coordinates for which $x_v = 1$ by 0 and by working in decreasing order as given by π. Clearly, (5.24) implies that $|[\mathbf{Nor}_Y, \pi]^{-1}(x)| \leq |[\mathbf{Nor}_Y, \pi]^{-1}(0)|$ as this corresponds to choosing $S = M(x)$. $\qquad \square$

The next result is a further characterization of phase spaces of Nor-SDS. It turns out that the image of the state zero under $[\mathbf{Nor}_Y, \pi]$ has zero as its unique predecessor. For some graph classes the state zero is the unique state of maximal indegree for which its successor has this property. It is convenient to introduce the set $M(Y, \pi)$ as

$$M(Y, \pi) = \{x \in \mathbb{F}_2^n \mid \mathrm{indegree}(x) \text{ is maximal in } \Gamma[\mathbf{Nor}_Y, \pi] \qquad (5.26)$$
$$\text{and } [\mathbf{Nor}_Y, \pi]^{-1}([\mathbf{Nor}_Y, \pi](x)) = \{x\}\} \, .$$

Proposition 5.36. *Let Y be a combinatorial graph, let $[\mathbf{Nor}_Y, \pi]$ be a permutation SDS, and let $M(Y, \pi)$ be as in (5.26). Then*

(i) for any connected graph Y we have $0 \in M(Y, \pi)$,

(ii) for $Y = \mathsf{Line}_n$ or $Y = \mathsf{Circ}_n$ we have $M(Y, \pi) = \{0\}$, and

(iii) there exist graphs Y such that $|M(Y, \pi)| > 1$.

Thus, if we have two phase spaces of Nor-SDS over Circ_n (or Line_n) where the preimage sizes of the zero states are different, then we are guaranteed that the phase spaces are nonisomorphic as directed graphs. Proposition 5.36, when applicable, gives us a local criterion for determining the nonequivalence of Nor-SDS.

Proof. It is clear from Lemma 5.35 that the state 0 has maximal in-degree in $\Gamma[\mathbf{Nor}_Y, \pi]$ for any $\pi \in S_Y$. Thus, to prove statement (i) we only need to show that $[\mathbf{Nor}_Y, \pi]^{-1}([\mathbf{Nor}_Y, \pi](0)) = \{0\}$. Let $z = [\mathbf{Nor}_Y, \pi](0)$ and assume there exists $y \neq 0$ such that $[\mathbf{Nor}_Y, \pi](y) = [\mathbf{Nor}_Y, \pi](0) = z$. Since $y \neq 0$ there exists some vertex v with $y_v = 1$ and hence $[\mathbf{Nor}_Y, \pi](0)_v = 0$. By assumption we have $[\mathbf{Nor}_Y, \pi^*] \circ [\mathbf{Nor}_Y, \pi](0) = 0$, and since $z_v = 0$ we are forced to conclude that there exists a vertex $v \neq v' \in B_Y^{<\pi}(v)$ such that $z_{v'} = 1$. But this is clearly impossible since $y_v = 1$ implies that $[\mathbf{Nor}_Y, \pi](y)_{v'}$, and thus $z_{v'}$ equals 0. Thus, there exists no $y \neq 0$ that maps to z, and statement (i) follows.

For the proof of statements (ii) and (iii) we first prove two auxiliary results. Assume there exists $x \in M = M(Y, \pi)$ with $x \neq 0$, and let v be a vertex such that $x_v \neq 0$. Without loss of generality we can assume that v is minimal with respect to the order $<_\pi$ such that $x_v = 1$.

Claim 1. *For all $v' \in B_Y'(v)$ we have $v' <_\pi v$.*

We prove this by contradiction. Suppose there exists $v' \in B_Y'(v)$ such that $v' >_\pi v$, and let x^v be the n-tuple defined by $x_v^v = 1$ and $x_u^v = 0$ otherwise. By Lemma 5.35 we conclude that $|[\mathbf{Nor}_Y, \pi]^{-1}(x^v)| = |[\mathbf{Nor}_Y, \pi]^{-1}(0)|$ since $x \in M(Y, \pi)$. Moreover, in this case the map r_v in (5.25) is a bijection, and therefore the preimages of 0 correspond uniquely to the preimages z' of x^v, which have the property $z_v' = 1$. Define $z = (z_u)_u$ by $z_v = 0$ and $z_u = 1$ otherwise. Since there exists $v' >_\pi v$, we derive $[\mathbf{Nor}_Y, \pi](z) = 0$. But since $z_v = 0$, we have created an additional preimage of zero, which contradicts Lemma 5.35 since $|[\mathbf{Nor}_Y, \pi]^{-1}(x^v)| = |[\mathbf{Nor}_Y, \pi]^{-1}(0)|$, and the claim follows.

Since Y is connected, it follows that there exists v' adjacent to v with $v' <_\pi v$. Moreover:

Claim 2. *If $\mathrm{degree}(v') > 1$, then there exists $k \in B_1'(j)$ with $k <_\pi v'$.*

Assume that for all $k \in B_1'(v')$ we have $v' <_\pi k$. Then we define $x' = (x_u')_u$ by

$$x_u' = \begin{cases} 1 & u = v', \\ x_u & u \neq v'. \end{cases} \tag{5.27}$$

Since $x_v = 1$, we have $x_{v'} = 0$, so clearly $x \neq x'$. By the assumption that for all $k \in B'_1(v')$ we have $v' <_\pi k$, we can conclude that $[\mathbf{Nor}_Y, \pi](x') = [\mathbf{Nor}_Y, \pi](x)$, which is impossible by the same argument as in Claim 1, and Claim 2 follows.

Since v is minimal with respect to the ordering $<_\pi$ with the property $x_v = 1$, we have $x_k = 0$, and thus there exists no $s <_\pi k$ with $x_s = 1$.

To prove the second statement of the proposition, assume that there exists $x \in M$ with $x \neq 0$. For $Y = \text{Line}_n$ or $Y = \text{Circ}_n$ we can conclude from $x_k = 0$ that for any $y \in [\mathbf{Nor}_Y, \pi]^{-1}(x)$ we have $y_{v'} = 1$. Again since $|[\mathbf{Nor}_Y, \pi]^{-1}(x)| = |[\mathbf{Nor}_Y, \sigma]^{-1}(0)|$, we can construct a bijection r' analogous to r_v in (5.25):

$$r' : [\mathbf{Nor}_Y, \pi]^{-1}(x) \longrightarrow [\mathbf{Nor}_Y, \pi]^{-1}(0) \tag{5.28}$$

with the property $r'(y)_{v'} = 0$. We now derive a contradiction by showing that there exists a preimage $y' = (y'_u)_u$ of 0 with the property $y'_{v'} = 0$. For this purpose we define y' by

$$y'_u = \begin{cases} 0 & u = v', \\ 1 & u \neq v'. \end{cases} \tag{5.29}$$

Clearly, we have $[\mathbf{Nor}_Y, \pi](y') = 0$ and (ii) follows.

For statement (iii) consider the graph Y and the orientation $\mathcal{O}_Y$ as shown below.

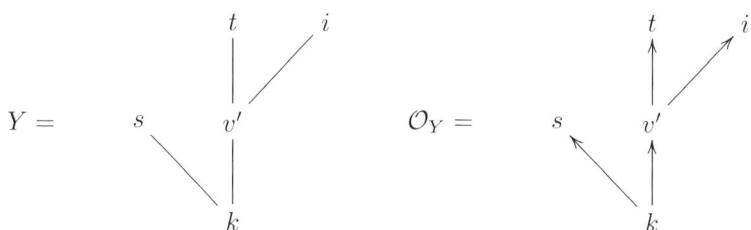

Let x be the 5-tuple $x = (x_k, x_s, x_{v'}, x_t, x_i) = (0, 0, 0, 0, 1)$, and let $\pi \in S_Y$ be an update order for which $\mathcal{O}_Y^\pi = \mathcal{O}_Y$. Then $z = [\mathbf{Nor}_Y, \pi](x) = (1, 0, 0, 1, 0)$. For any $y \in [\mathbf{Nor}_Y, \pi]^{-1}(x)$ we have $y_s = y_t = 1$ and $y_i = 0$ while y_k and $y_{v'}$ may take any value. For y' to be in the preimage of the state 0 we must have $y_s = y_t = y_i = 1$ while y_k and $y_{v'}$ are arbitrary. We see that we have a bijection

$$\rho : [\mathbf{Nor}_Y, \pi]^{-1}(x) \longrightarrow [\mathbf{Nor}_Y, \pi]^{-1}(0), \quad \rho(z)_h = \begin{cases} z_h & \text{for } h \neq i, \\ 1 & \text{for } h = i. \end{cases} \tag{5.30}$$

This is a particular instance of the map r_v in (5.25). Now let $\eta \in [\mathbf{Nor}_Y, \pi]^{-1}(x)$. Clearly, we have $\eta_k = \eta_t = 0$ and since $z_k = 1$ we have $\eta_r = \eta_{v'} = 0$. Finally, since $z_i = 0$, we must have $\eta_i = 1$. Thus, x is the only preimage of z under $[\mathbf{Nor}_Y, \pi]$, and we conclude that

$$[\mathbf{Nor}_Y, \pi]^{-1}([\mathbf{Nor}_Y, \pi](x)) = x \ ,$$

which proves statement (iii).   □

Again, the background of Proposition 5.36 is the analysis of the bound $\Delta(Y)$. The bound is conjectured to be sharp and to be realized if the vertex functions are induced by $(\text{nor}_k)_k$. The reader interested in pursuing this problem may want to refer to $[100, 109, 112]$.

5.16. Let $Y = \mathsf{Star}_5$ and let ϕ be the sequential dynamical system induced by nor functions with update order $(1, 2, 0, 3, 4)$. Construct the sets $\phi^{-1}(0, 0, 0, 0, 0)$ and $\phi^{-1}(0, 0, 0, 1, 0)$. We know that $\phi(0)$ has in-degree 1. What is the in-degree of $\phi(0, 0, 0, 1, 0)$? Based on this, what can you say about the set $M(Y, \pi)$? What is the bijection $r\colon \phi^{-1}(0, 0, 0, 1, 0) \longrightarrow \phi^{-1}(0, 0, 0, 0, 0)$ in this case?   [1+]

5.17. Research the dynamics of permutation SDS over Circ_n induced by nor functions. Use your analysis to decide if $\Delta(\mathbf{Nor}_{\mathsf{Circ}_n})$ equals $\Delta(\mathsf{Circ}_n)$.   [5-]

5.18. Describe the phase space of $[\mathbf{Nor}_{\mathsf{Line}_n}, \pi]$. Is $\Delta(\mathbf{Nor}_{\mathsf{Circ}_n}) = \Delta(\mathsf{Circ}_n)$?   [5-]

5.5.2 SDS Induced by $(\text{nor}_k + \text{nand}_k)_k$

We just saw that the SDS induced by $(\text{nor}_k)_k$ or $(\text{nand}_k)_k$ have periodic points that depend only on the graph Y. Perhaps somewhat surprisingly it turns out that the same holds for SDS induced by the sum of these functions, that is, SDS induced by $(\text{nor}_k + \text{nand}_k)_k$.

5.19. The previous statement may lead one to speculate if the function sequences that induce SDS with periodic points independent of the update order are a closed set under addition. This is, however, not the case. Give a counterexample proving this claim. *Hint.* You will find all you need using symmetric functions over Circ_4. We will return to this problem in Chapter 6.   [2]

Example 5.37. In Figure 5.7 we have shown the phase spaces of SDS over $Y = \mathsf{Circ}_4$, and $Y = \mathsf{Circ}_5$ using the update orders $(0, 1, 2, 3)$ and $(4, 3, 2, 1, 0)$ induced by the function $h_3 = \text{nor}_3 + \text{nand}_3$. Note that h_3 only returns 0 if its argument consists entirely of 0's or entirely of 1's.   ◇

It turns out that the periodic points of SDS induced by nor functions and the SDS induced by nor + nand functions essentially coincide. Again, the set W'_Y denotes the *fair words* over $\text{v}[Y]$.

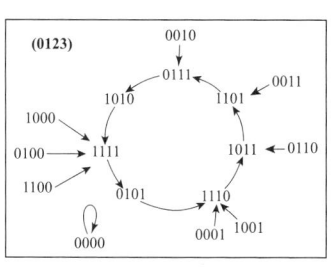
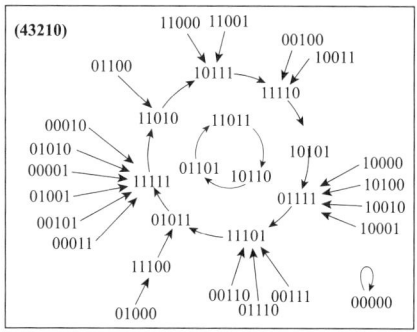

Fig. 5.7. The phase space $\Gamma[(\mathbf{Nor} + \mathbf{Nand})_{\mathsf{Circ}_4}, (0, 1, 2, 3)]$ (left) and the phase space $\Gamma[(\mathbf{Nor} + \mathbf{Nand})_{\mathsf{Circ}_5}, (4, 3, 2, 1, 0)]$ (right).

Proposition 5.38. *Let Y be a combinatorial graph and let $[\mathbf{F}_Y, w]$ be an* SDS *over Y induced by $(\mathrm{nor}_k + \mathrm{nand}_k)_k$ with a word $w \in W'_Y$. We then have*

$$\mathsf{Per}[\mathbf{F}_Y, w] = \{0\} \cup \{x \in \mathbb{F}_2^n \mid \forall v : x_v = 0 \Rightarrow \forall\, v' \in B'_1(v) : x_{v'} = 1\} \, . \quad (5.31)$$

We will show that the set M

$$M = \{0\} \cup \{x \in \mathbb{F}_2^n \mid \forall v : x_v = 0 \Rightarrow \forall v' \in B'_1(v) : x_{v'} = 1\} \quad (5.32)$$

is a maximal, invariant set for all the SDS ϕ induced by $(\mathrm{nor}_k + \mathrm{nand}_k)_k$ such that the restriction of ϕ to M is a bijection.

Proof. Let M be as in (5.32). We first show that $M \subset \mathsf{Per}[F_Y, w]$, and to prove this we verify that

$$F_{v,Y} : M \longrightarrow M$$

is a well-defined map. Clearly, $F_{v,Y}(0) = 0$. If $0 \neq x \in M$ has $x_v = 0$, then by definition we have $x_{v'} = 1$ for all $v' \in B'_1(v)$. Hence, we have

$$F_{v,Y}(x)_v = (\mathrm{nor}_k + \mathrm{nand}_k)(x[v]) = 1 \, , \quad (5.33)$$

where $k = d(v) + 1$, and thus $F_{v,Y}(x) \in M$. If $x_v = 1$, there are two cases two consider. If $x'_v = 1$ for all $v' \in B'_v(Y)$, then $F_{v,Y}(x)_v = 0$, and if there is (precisely) one $v' \in B'_1(Y)$ with $x_{v'} = 0$, then $F_{v,Y}(x)_v = 1$. In either case we see that $F_{v,Y}(x) \in M$ and in summary that $F_{v,Y} : M \longrightarrow M$ is well-defined.

We claim that the composed map $F_{v,Y}^2 : M \longrightarrow M$ satisfies

$$F_{v_i,Y}^2 = \mathsf{id} \, .$$

This follows by an identical three-case argument like the one we did in the proof for the periodic points of Nor-SDS in Theorem 5.28. We leave the verification of this to the reader. By a straightforward extension of Proposition 4.13 to words, we conclude that $[\mathbf{F}_Y, w] : M \longrightarrow M$ is invertible.

Since M is invariant under all SDS induced by $(\mathrm{nor}_k + \mathrm{nand}_k)_k$ and fair words, it is clear that $M \subset \mathrm{Per}[\mathbf{F}_Y, w]$. We next show that we have the inclusion $\mathrm{Per}[\mathbf{F}_Y, w] \subset M$ as well. Let $x \in \mathbb{F}_2^n$ with $x \neq (0, 0, \dots, 0)$. We see that

$$
F_{v,Y}(x)_v = \begin{cases} x_v & \begin{aligned} &\text{when } (\forall\, v' \in B_1'(v);\ x_{v'} = 0)\ \vee \\ &\quad (x_v = 1 \ \wedge\ \exists\, v' \in B_1'(v_i);\ x_{v'} = 0)\,, \end{aligned} \\[1em] 1 + x_v & \begin{aligned} &\text{when } (\forall\, v' \in B_1'(v);\ x_{v'} = 1)\ \vee \\ &\quad (x_v = 0 \ \wedge\ \exists\, v' \in B_1'(v);\ x_{v'} = 1)\,. \end{aligned} \end{cases}
$$

From this it follows that an x-coordinate with $x_v = 0$ is mapped to 1 if and only if at least one Y-neighbor has state 1, and an x-coordinate with $x_v = 1$ changes into 0 if and only if all its Y-neighbor states are 1. Since by assumption $x \neq (0, \dots, 0)$ and Y is connected, we conclude that there exists $h \in \mathbb{N}$ such that

$$
[\mathbf{F}_Y, w]^h(x) \in M .
$$

In particular this holds for any nonzero periodic point p of period, say, r. That is, there exists h such that $q = [\mathbf{F}_Y, w]^h(p) \in M$. Moreover, there exists $0 \leq t < r$ such that $[\mathbf{F}_Y, w]^t(q) = p$ since q and p are on the same orbit. Since M is an invariant set, it follows that $p \in M$ as well, and Proposition 5.38 follows. □

If we conjugate the function nor + nand (we omit the subscript k here) with the inversion map inv, we obtain the relation

$$
\mathrm{inv} \circ (\mathrm{nor} + \mathrm{nand}) \circ \mathrm{inv} = 1 + \mathrm{nor} + \mathrm{nand} = \mathrm{or} + \mathrm{nand} = \mathrm{nor} + \mathrm{and} , \quad (5.34)
$$

which leads to

Corollary 5.39. *Let Y be a combinatorial graph and let $w \in W_Y'$. Then we have*

$$
\mathrm{Per}[(1 + \mathbf{Nand} + \mathbf{Nor})_Y, w] = \mathrm{inv}(\mathrm{Per}[(\mathbf{Nand} + \mathbf{Nor})_Y, w]) . \quad (5.35)
$$

We can now also state precisely what we mentioned earlier about the relation to periodic points of Nor-SDS:

Corollary 5.40. *Let Y be a combinatorial graph and let $w \in W_Y'$. Then the periodic points of $[(1 + \mathbf{Nand} + \mathbf{Nor})_Y, w]$ of period $p > 1$ are precisely the periodic points of $[(\mathbf{Nor})_Y, w]$.*

Proof. From Corollary 5.39 it is clear that in addition to the fixed point $(1, 1, \dots, 1)$ the periodic points of $[(1 + \mathbf{Nand} + \mathbf{Nor})_Y, w]$ are all $x \in \mathbb{F}_2^n$ with the property that for all v we have $x_v = 1$ implies $x_{v'} = 0$ for all $v' \in B_Y'(v)$, but this is precisely the periodic points of $[(\mathbf{Nor})_Y, w]$. □

Even though we have the same set of periodic points, the transient structure of the two types of SDS are different. For example, (nor + nand)-SDS can have transients lengths exceeding 1, as illustrated in Figure 5.7.

Enumeration of Periodic Points

It is now straightforward to derive a recursion relation for $P_n'(Y) = |\text{Per}[(\textbf{Nor}+\textbf{Nand})_Y, w]|$ for $Y = \text{Circ}_n$.

Proposition 5.41. *Let $w \in W_Y'$. Then $P_n' = P_n'(\text{Circ}_n)$ satisfies the recursion*

$$P_n' = P_{n-1}' + P_{n-2}' - 1 \,. \tag{5.36}$$

Proof. From Proposition 5.38 it is clear that $P_n'(Y) = P_n(Y) + 1$ where $P_n(Y) = |\text{Per}[\textbf{Nor}_Y, w]|$. Specializing to the graph $Y = \text{Circ}_n$ and substituting into the recursion relation $P_n = P_{n-1} + P_{n-2}$ from Proposition 5.31, we get (5.36). □

Example 5.42. As an illustration of Proposition 5.41 we get the number of periodic points in the table below. ◇

n	3	4	5	6	7	8	9	10	11	12	13	14	15	16
$P'(\text{Circ}_n)$	5	8	12	19	30	48	77	124	200	323	522	844	1365	2208

Orbit Equivalence

In, e.g., [90] the concept of *stable isomorphism* is introduced. Two finite dynamical systems are stably isomorphic if they are dynamically equivalent when restricted to their respective periodic points, which is the case if there exists a digraph isomorphism between their periodic orbits. *Orbit equivalence* may therefore be a more descriptive term for this notion.

The notion of orbit equivalence is a little coarse. It is occasionally desirable to distinguish between what we would call *functional orbit equivalence* and *dynamical orbit equivalence*: There is a functional orbit equivalence between two finite dynamical systems if their periodic orbits coincide. There is a dynamical equivalence between two systems if they are dynamically equivalent when restricted to their periodic orbits. The following proposition illustrates the distinction.

Proposition 5.43. *Let Y be a combinatorial graph, let $w \in W_Y'$, let $M = \mathbb{F}_2^n \setminus \{(1, 1, \ldots, 1)\}$, and let $N = \mathbb{F}_2^n \setminus \{(0, 0, \ldots, 0)\}$. We let $\phi = [\textbf{Nor}_Y, w]: M \longrightarrow M$, $\psi = [(1 + \textbf{Nor} + \textbf{Nand})_Y, w]: M \longrightarrow M$ and $\eta = [(\textbf{Nor} + \textbf{Nand})_Y, w]: N \longrightarrow N$. Then we have*
  (i) The dynamical systems ϕ and ψ are functionally orbit equivalent.
  (ii) The dynamical systems ϕ and η are dynamically orbit equivalent.

Proof. Restricted to the periodic points of ϕ the functions nor and $1 + \text{nor} + \text{nand}$ coincide and (i) follows. It is clear that (ii) follows from (i). □

Example 5.44. Figure 4.10 on page 89 shows the phase spaces of the SDS $[\textbf{Nor}_{\text{Circ}_4}, (0, 1, 2, 3)]$ and $[(1 + \textbf{Nor} + \textbf{Nand})_{\text{Circ}_4}, (0, 1, 2, 3)]$. It is easy to see that the orbits are dynamically equivalent. ◇

Problems

5.20. We have seen that threshold SDS have no periodic points of period $p > 1$. This is generally not true for a parallel update order. Give an example of a threshold system updated in parallel that has a periodic orbit of length 2. [**1**]

5.21. A Nor-SDS is an example of inverted threshold SDS. As we know, permutation Nor-SDS never have fixed points. Is this true in general for inverted threshold permutation SDS? Give a proof or a counterexample. [**1**]

5.22. Let $Y = \mathsf{Wheel}_4$ and let $w = W'_Y$. How many periodic points does the SDS $[(\mathbf{Nor} + \mathbf{Nand})_Y, w]$ have? [**1**]

5.23. Let $Y = K_{4,3}$ be the complete, bipartite graph with vertex classes $V_1 = \{1, 2, 3, 4\}$ and $V_2 = \{5, 6, 7\}$ where each vertex $v \in V_1$ has vertex function induced by or: $\mathbb{F}_2^4 \longrightarrow \mathbb{F}_2$ and each vertex $v \in V_2$ has vertex function induced by majority: $\mathbb{F}_2^5 \longrightarrow \mathbb{F}_2$. Show that the induced SDS map $[\mathbf{F}_Y, \pi]$ has no periodic points of period $p > 1$ for any $\pi \in S_Y$. (The graph is shown below.) [**1**]

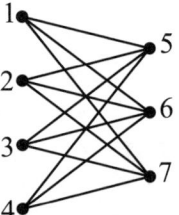

5.24. Figure 5.8 shows a space-time diagram of an SDS map starting at the state $(1, 0, 0, 0, 0)$ at $t = 0$. The graph Y is a connected graph on five vertices.
  (*i*) What state is reached at time $t = 3$, and what type of state is this?

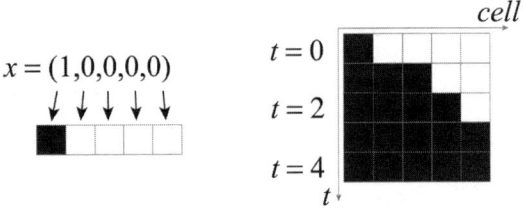

Fig. 5.8. The space-time diagram of Problem 5.24.

  (*ii*) Which of the following SDS-maps can *not* generate this space-time diagram? (There may be more than one correct answer.)
A) $[\mathbf{Nor}_Y, (0, 1, 2, 3, 4)]$ B) $[\mathbf{Majority}_{K_4}, (1, 2, 4, 3)]$
C) $[\mathbf{Nor}_{\mathsf{Circ}_5}, (1, 0, 2, 3, 4)]$ D) $[\mathbf{Majority}_Y, (1, 5, 4, 2, 3)]$
E) $[\mathbf{Or}_Y, (1, 5, 4, 2, 3)]$. [**1+**]

5.25. (Dynamics of [Parity$_{K_n}$, π]) Let $\beta \colon S_n \longrightarrow S_{n+1}$ be the function that maps $\pi = (\pi_1, \ldots, \pi_n)$ (standard form) to the $(n+1)$-*cycle*

$$\beta(\pi) = (\pi_1, \pi_2, \ldots, \pi_n, n+1) .$$

Define $\hat{\mathbb{F}}_2^n \subset \mathbb{F}_2^{n+1}$ by $\hat{\mathbb{F}}_2^n = \{x \in \mathbb{F}_2^{n+1} \mid x_{n+1} = \mathrm{parity}(x_1, x_2, \ldots, x_n)\}$ and the maps $\mathsf{proj} \colon \hat{\mathbb{F}}_2^n \longrightarrow \mathbb{F}_2^n$, $\iota \colon \mathbb{F}_2^n \longrightarrow \hat{\mathbb{F}}_2^n$ and $\sigma \colon \hat{\mathbb{F}}_2^n \longrightarrow \hat{\mathbb{F}}_2^n$ by

$$\mathsf{proj}(x_1, \ldots, x_n, x_{n+1}) = (x_1, \ldots, x_n),$$
$$\iota(x_1, \ldots, x_n) = (x_1, \ldots, x_n, \mathrm{parity}_n(x_1, \ldots, x_n)),$$
$$\sigma_{n+1}(x_1, x_2, \ldots, x_{n+1}) = (x_{n+1}, x_1, \ldots, x_n) .$$

(a) Prove that the set $\hat{\mathbb{F}}_2^n$ in invariant under the permutation action of any $\pi \in S_{n+1}$, and that ι and proj are inverse maps, that is,

$$\mathsf{proj} \circ \iota = \mathrm{id}_{\mathbb{F}_2^n} \quad \text{and} \quad \iota \circ \mathsf{proj} = \mathrm{id}_{\hat{\mathbb{F}}_2^n} . \tag{5.37}$$

(b) Prove that the SDS-map $[\mathbf{Parity}_{K_n}, \pi] \colon \hat{\mathbb{F}}_2^n \longrightarrow \hat{\mathbb{F}}_2^n$ is dynamically equivalent to the permutation action of $\beta(\pi)$ on $\hat{\mathbb{F}}_2^n$. **[3]**

Answers to Problems

5.2. (a) $L_n = 2L_{n-1} - L_{n-2} + L_{n-4}$. Initial values are $L_3 = 2$, $L_4 = 6$, $L_5 = 12$, and $L_6 = 20$. (c) The fixed points can be characterized as all states $x \in \mathbb{F}_2^n$ with no isolated 0's or 1's.

5.3. One needs $f_v = f_{\gamma(v)}$ for all $v \in \mathrm{v}[Y]$ and all $\gamma \in \mathsf{Aut}(Y)$ to have an action. The statement follows easily from Proposition 4.30.

5.4. (a) Easy. (b) All the arguments we used for threshold SDS apply directly to permutation-SDS with monotone vertex functions.

5.5. You need to compute ΔE for the case when x_v is mapped from -1 to 1 and for the case when x_v is mapped from 1 to -1.

5.9. Interesting results should probably be considered for publication.

5.12. Any state containing a vertex state that is 1 cannot be fixed. The only remaining candidate for a fixed point is $x = (0, 0, \ldots, 0)$, but this state is clearly not fixed.

5.13. Consider a permutation SDS $\phi = [\mathbf{F}_Y, \pi]$ induced by nor functions. If, for example, ϕ has a periodic orbit of size 2, then the SDS $[\mathbf{F}_Y, w]$ where w is the concatenation of π with itself clearly has two fixed points.

5.14. The proof of Proposition 5.33 shows that $P_n = L_{n-1} + L_{n-3}$, and from $L_n = F_{n+1}$ it follows that

$$
\begin{aligned}
P_{n+1} - P_n - P_{n-1} &= L_n + L_{n-2} - L_{n-1} - L_{n-3} - L_{n-2} - L_{n-4} \\
&= F_{n+1} - F_n - F_{n-2} - F_{n-3} \\
&= F_n + F_{n-1} - F_n - F_{n-2} - F_{n-3} \\
&= F_{n-1} - (F_{n-2} + F_{n-3}) = F_{n-1} - F_{n-1} = 0 \ .
\end{aligned}
$$

5.16. For a state $x = (x_0, x_1, x_2, x_3, x_4)$ to be mapped to 0 when the update order is $\pi = (1, 2, 0, 3, 4)$, we must have $x_3 = x_4 = 1$. This leaves us with two choices for x_0. If $x_0 = 0$, we must have $x_1 = x_2 = 1$, and if $x_0 = 1$, then x_1 and x_2 are always mapped to 0. Thus,

$$
\begin{aligned}
\phi^{-1}(0,0,0,0,0) = \{&(0,1,1,1,1), (1,0,0,1,1), (1,1,0,1,1), \\
&(1,0,1,1,1), (1,1,1,1,1)\} \ .
\end{aligned}
$$

Similarly, for a point y to be mapped to $(0, 0, 0, 1, 0)$, we see that $y_3 = 0$ and $y_4 = 1$. The last condition follows from the fact that at the time y_4 is to be updated we have that the state of vertex 0 is 0. As before, if $y_0 = 0$, then $y_1 = y_2 = 1$, and if $y_0 = 1$, then y_1 and y_2 are always mapped to 0. Thus, we have

$$
\begin{aligned}
\phi^{-1}(0,0,0,1,0) = \{&(0,1,1,0,1), (1,0,0,0,1), (1,1,0,0,1), \\
&(1,0,1,0,1), (1,1,1,0,1)\} \ .
\end{aligned}
$$

We have $z = \phi(0, 0, 0, 1, 0) = (0, 1, 1, 0, 1)$, and we see that a predecessor y of z must have $y_0 = y_1 = y_2 = 0$. For y_3 to be mapped to 0, we must have $y_3 = 1$ and for y_4 to be mapped to 1, we must have $y_4 = 0$. This gives us $(0, 0, 0, 1, 0)$ as the only preimage of z, and therefore $z = \phi(0, 0, 0, 1, 0)$ has indegree 1. From this it follows that $M(Y, \pi)$ contains at least the points $(0, 0, 0, 0, 0)$ and $(0, 0, 0, 1, 0)$ and thus has cardinality at least 2. The bijection r is the map that assigns to $x \in \phi^{-1}(0, 0, 0, 1, 0)$ the state $r(x)$ obtained from x by setting x_3 to 1 and mapping all other coordinates identically.

5.17. You should consider submitting your answer to a journal.

5.20. Let $Y = \mathrm{Circ}_4$ and let each vertex function be majority$_3$. Using a parallel update scheme we see that $x = (0, 1, 0, 1)$ is mapped to $y = (1, 0, 1, 0)$, which in turn is mapped back to x, and we have our periodic orbit of length 2.

5.21. The minority function is an inverted threshold function. If we take $Y = \mathrm{Circ}_4$, we see that, for example, $(0, 1, 0, 1)$ is a fixed point for $[\mathbf{Minority}_Y, \pi]$ for any permutation update order.

5.24. (i) The state reached at time $t = 3$ is $(1, 1, 1, 1, 1)$, which is a fixed point. (ii) The correct answer is A, B, C, and D. A nor-SDS never have fixed points. That gives A and C. Alternative B is an SDS over a graph with four states so this map cannot have this space-time diagram. For a connected graph a state containing a single vertex state 1 cannot map to a state containing more 1's for a majority SDS. (Why?) The remaining alternative E could have produced the given diagram. (Provide an example graph.)

5.25. (a) If $\pi_{n+1} = n + 1$, the statement clearly holds. Otherwise, assume that $(\pi(x))_{n+1} = x_i$. Then the sum (i.e., parity) of the first n coordinates of $\beta(x)$ is

$$\mathrm{parity}(x_1, \ldots, x_n) + x_1 + x_2 + \cdots + x_{i-1} + x_{i+1} + \cdots + x_n = x_i,$$

and the first part of the lemma follows. The statements in (5.37) are obvious.

(b) The map $\mathrm{parity}_n : \mathbb{F}_2^n \longrightarrow \mathbb{F}_2$ satisfies the functional relation

$$\mathrm{parity}_n(x_1, x_2, \ldots, x_{i-1}, \mathrm{parity}_n(x_1, \ldots, x_n), x_{i+1}, \ldots, x_n)$$
$$= \sum_{j=1, j \neq i}^{n} x_j + \sum_{j=1}^{n} x_j = x_i \quad (5.38)$$

for any $1 \leq i \leq n$. Writing $\overset{i}{\mapsto}$ for the application of Parity_i to a given state $x = (x_1, \ldots, x_n)$, we get through repeated application of (5.38)

$$x = (x_1, x_2, \ldots, x_n) \overset{1}{\mapsto} (\text{parity}_n(x), x_2, x_3, \ldots, x_n)$$
$$\overset{2}{\mapsto} (\text{parity}_n(x), \text{parity}_n(\text{parity}_n(x), x_2, \ldots, x_n), x_3, \ldots, x_n)$$
$$= (\text{parity}_n(x), x_1, x_3, \ldots, x_n)$$
$$\vdots$$
$$\overset{n}{\mapsto} (\text{parity}_n(x), x_1, x_2, \ldots, x_{n-1}) \,.$$

The above computation gives us the commutative diagram

$$
\begin{array}{ccc}
\mathbb{F}_2^n & \xrightarrow{[\mathbf{Parity}_{K_n}, \mathsf{id}]} & \mathbb{F}_2^n \\
\iota \downarrow & & \uparrow \mathsf{proj} \\
\hat{\mathbb{F}}_2^n & \xrightarrow{\sigma_{n+1}} & \hat{\mathbb{F}}_2^n
\end{array}
\tag{5.39}
$$

that is, $[\mathbf{Parity}_{K_n}, \mathsf{id}] = \mathsf{proj} \circ \sigma_{n+1} \circ \iota$. Since ι and proj are inverses, it follows that $[\mathbf{Parity}_{K_n}, \mathsf{id}]$ is dynamically equivalent to the shift map on $\hat{\mathbb{F}}_2^n$. Since $\mathsf{Aut}(K_n) = S_n$, we have $[\mathbf{Parity}_{K_n}, \pi] = \pi \circ [\mathbf{Parity}_{K_n}, \mathsf{id}] \circ \pi^{-1}$ for all $\pi \in S_n$. Consequently, diagram (5.39) can be extended to

$$
\begin{array}{ccc}
\mathbb{F}_2^n & \xrightarrow{[\mathbf{Parity}_{K_n}, \pi]} & \mathbb{F}_2^n \\
\pi^{-1} \downarrow & & \uparrow \pi \\
\mathbb{F}_2^n & \xrightarrow{[\mathbf{Parity}_{K_n}, \mathsf{id}]} & \mathbb{F}_2^n \\
\iota \downarrow & & \uparrow \mathsf{proj} \\
\hat{\mathbb{F}}_2^n & \xrightarrow{\sigma_{n+1}} & \hat{\mathbb{F}}_2^n
\end{array}
\tag{5.40}
$$

Let $\pi \in S_n$ and define $\bar{\pi} \in S_{n+1}$ by $\bar{\pi}_i = \pi_i$ for $1 \le i \le n$ (and thus $\bar{\pi}_{n+1} = n + 1$.) It is straightforward to verify the identities

$$\iota \circ \pi^{-1} = (\bar{\pi})^{-1} \circ \iota \quad \text{and} \quad \pi \circ \mathsf{proj} = \mathsf{proj} \circ \bar{\pi} \,.$$

Consequently, we derive from (5.40) the commutative diagram

$$
\begin{array}{ccc}
\mathbb{F}_2^n & \xrightarrow{[\mathbf{Parity}_{K_n}, \pi]} & \mathbb{F}_2^n \\
\iota \downarrow & & \uparrow \mathsf{proj} \\
\hat{\mathbb{F}}_2^n & \xrightarrow{\beta(\pi)} & \hat{\mathbb{F}}_2^n
\end{array}
\tag{5.41}
$$

where

$$\bar{\pi} \circ \sigma_{n+1} \circ (\bar{\pi})^{-1} = (\pi(1), \pi(2), \ldots, \pi(n), n + 1) = \beta(\pi) \,. \tag{5.42}$$

The identity on the left in (5.42) can be verified by first representing σ_{n+1} as the permutation action of $\bar{\sigma} = (1, 2, \ldots, n+1)$ (using cycle form) and using the properties of group actions: The permutation $\bar{\pi}\bar{\sigma}(\bar{\pi})^{-1}$ maps π_i to π_{i+1}. Again, since ι and proj are inverse maps, we conclude that $[\mathbf{Parity}_{K_n}, \pi]$ is dynamically equivalent to the permutation action of $\beta(\pi)$ on $\hat{\mathbb{F}}_2^n$.

6

Graphs, Groups, and SDS

6.1 SDS with Order-Independent Periodic Points

In this section we show that a certain class of SDS induces a group that encodes the dynamics over periodic points that can be obtained by varying the word update order [93, 113]. Through this construction we can use group theory to prove the existence of certain types of phase-space structures.

In general, neither an SDS nor its Y-local maps are invertible, and therefore we cannot consider the obvious construction: the group generated by the Y-local maps under function composition. Instead we will consider the restriction of an SDS map $[\mathbf{F}_Y, w]$ to its periodic points. If the set of periodic points is independent of the word update order, we can conclude, under mild assumptions on the update word, that the Y-local maps through restrictions induce bijective maps

$$F_{v,Y}|_{\mathsf{P}} : \mathsf{P} \longrightarrow \mathsf{P} \,,$$

where $\mathsf{P} = \mathsf{Per}[\mathbf{F}_Y, w]$ and $F_{v,Y}|_{\mathsf{P}}$ denotes the restriction of $F_{v,Y}$ to P. The group generated by the restriction maps $F_{v,Y}|_{\mathsf{P}}$ encodes the different configurations of periodic points that can obtained by varying the word update order.

The assumption on the update schedule is a technical condition to avoid special situations where some Y-local maps are not being applied. That is, we consider *fair words* over $\mathrm{v}[Y]$ defined by

$$W'_Y = \{w \in W_Y \mid \forall \, v \in \mathrm{v}[Y], \, \exists \, w_i; \, v = w_i\} \,. \tag{6.1}$$

We can now introduce w-independent SDS:

Definition 6.1 (w-independent SDS [93, 113]). An SDS $(Y, \mathbf{F}_Y, w)$ with state space K^n is w-independent if there exists $\mathsf{P} \subset K^n$ such that for all $w \in W'_Y$ we have $\mathsf{Per}[\mathbf{F}_Y, w] = \mathsf{P}$.

Note that in the case of w-independent SDS the set P is the unique maximal subset of K^n such that $[\mathbf{F}_Y, w]|_{\mathsf{P}} : \mathsf{P} \longrightarrow \mathsf{P}$ is bijective. We point out that w-independence does not imply that the periodic orbits are the same for all update orders w. The structure of the SDS phase space critically depends on the update order w.

6.1.1 Preliminaries

We start by analyzing why the periodic points of an SDS generally depend on the update order.

Lemma 6.2. *Let* $[\mathbf{F}_Y, w] = \prod_{i=1}^{k} F_{w_i,Y}$ *be an SDS-map, let* $\mathsf{M} \subset K^n$, *and set*

$$\mathsf{M}_j = \begin{cases} \mathsf{M} & \text{for } j = 1, \\ \prod_{i=1}^{j-1} F_{w_i,Y}(\mathsf{M}) & \text{otherwise.} \end{cases} \tag{6.2}$$

Then we have

$$(\prod_{i=1}^{k} F_{w_i,Y})|_{\mathsf{M}} \text{ is bijective} \iff \forall\, 1 \le j \le k;\; F_{w_j,Y}|_{\mathsf{M}_j} \text{ is bijective}, \tag{6.3}$$

where

$$F_{w_j,Y} \colon \prod_{i=1}^{j-1} F_{w_i,Y}(\mathsf{M}) \longrightarrow \prod_{i=1}^{j} F_{w_i,Y}(\mathsf{M}) . \tag{6.4}$$

The proof of Lemma 6.2 is straightforward and indicates that the question of bijectivity of an SDS restricted to some set $\mathsf{M} \subset K^n$ is generally not reducible to the question of bijectivity of its local functions

$$F_{w_j,Y} \colon \prod_{i=1}^{j-1} F_{w_i,Y}(\mathsf{M}) \longrightarrow \prod_{i=1}^{j} F_{w_i,Y}(\mathsf{M})$$

alone. According to Lemma 6.2, the map $F_{w_j,Y}$ is bijective *restricted* to the set M_j, which reflects the role of the word update order w of the SDS. A consequence of Lemma 6.2 is that the set of periodic points of an SDS $(Y, (F_{w_i,Y})_i, w)$ generally depends on the particular choice of update order w.

Proposition 6.3. *There exist a graph* Y, *a field* K, *and a family* $\mathbf{F}_Y$ *of* Y-*local functions such that the set of periodic points of* $[\mathbf{F}_Y, w]$ *depends on* w.

Proof. Let $K = \mathbb{F}_2$, let $Y = \mathsf{Circ}_4$, and let $F_{i,Y}(x_1, \dots, x_4)$ for $i = 1, \dots, 4$ be Y-local maps induced by the symmetric, Boolean function

$$b \colon \mathbb{F}_2^3 \longrightarrow \mathbb{F}_2, \quad b(x, y, z) = \begin{cases} 1 & \text{for } \mathrm{sum}_{\mathbb{N}}(x, y, z) = 1, \\ 0 & \text{otherwise.} \end{cases}$$

Consider the two words $w = (v_4, v_3, v_2, v_1)$ and $w' = (v_4, v_2, v_3, v_1)$. For the state $(1, 0, 0, 0)$ we obtain

$$(1,0,0,0) \xrightarrow{[\mathbf{F}_Y,w']} (0,1,0,1) .$$

$$(1,0,0,0)$$
$$\swarrow_{[\mathbf{F}_Y,w]} \qquad \searrow$$
$$(0,0,1,1) \longrightarrow (1,1,1,0)$$

Since $(0, 1, 0, 1)$ is a fixed point for $[\mathbf{F}_Y, w]$ and $[\mathbf{F}_Y, w']$, we conclude that $(1, 0, 0, 0)$ is a periodic point for $[\mathbf{F}_Y, w]$ but not for $[\mathbf{F}_Y, w']$. □

6.1.2 The Group $G(Y, \mathbf{F}_Y)$

In Proposition 6.4 we show that a w-independent SDS $(Y, \mathbf{F}_Y, w)$ naturally induces the finite group $G(Y, \mathbf{F}_Y)$. In Theorem 6.5 we show that this group contains information about the structure of the periodic orbits of all phase spaces generated by varying the word update order. In the following we will, by abuse of notation, sometimes write $[\mathbf{F}_Y, w]$ instead of $[\mathbf{F}_Y, w]|_\mathsf{P}$. It is implicitly understood that the map $[\mathbf{F}_Y, w]$ induces the map $[\mathbf{F}_Y, w]|_\mathsf{P}$ by restriction.

Proposition 6.4. *Let Y be a graph, K a finite field, $w \in W_Y'$, and $(Y, \mathbf{F}_Y, w)$ a w-independent SDS. Then for any $v \in \mathrm{v}[Y]$ the local maps $F_{v,Y} : K^n \longrightarrow K^n$ induce the bijections*

$$F_{v,Y}|_\mathsf{P} : \mathsf{P} \longrightarrow \mathsf{P} \ ,$$

and the SDS $(Y, \mathbf{F}_Y, w)$ induces the finite group

$$G(Y, \mathbf{F}_Y) = \langle \{F_{v,Y}|_\mathsf{P} \mid v \in \mathrm{v}[Y]\} \rangle, \tag{6.5}$$

which acts naturally as a permutation group on P.

Proof. By assumption we have $\mathsf{Per}[\mathbf{F}_Y, w] = \mathsf{P}$ for all $w \in W_Y'$. Let $w = (w_1, \ldots, w_k) \in W_Y'$ and $v \in \mathrm{v}[Y]$, and set $w_v = (w_1, \ldots, w_k, v)$. Since $w, w_v \in W_Y'$, we conclude that both the SDS-maps $[\mathbf{F}_Y, w] : \mathsf{P} \longrightarrow \mathsf{P}$ and $[\mathbf{F}_Y, w_v] : \mathsf{P} \longrightarrow \mathsf{P}$ are bijections. Furthermore, we have

$$[\mathbf{F}_Y, w_v] = F_{v,Y} \circ [\mathbf{F}_Y, w] : \mathsf{P} \longrightarrow \mathsf{P}, \tag{6.6}$$

from which follows that $F_{v,Y}|_\mathsf{P} : \mathsf{P} \longrightarrow \mathsf{P}$ is a well-defined bijection. Therefore, the group $G(Y, \mathbf{F}_Y)$ obtained by composition of the maps $F_{v,Y}|_\mathsf{P}$ is well-defined and Proposition 6.4 follows. $\qquad\square$

According to Proposition 6.4, we have the mapping

$$\mathbf{F}_Y = (F_{v_i,Y})_{1 \le i \le n} \mapsto G(Y, \mathbf{F}_Y) = \langle \{F_{v_i,Y}|_\mathsf{P} \mid v_i \in \mathrm{v}[Y]\} \rangle \ , \tag{6.7}$$

which allows us to utilize a group-theoretic framework for analyzing SDS phase spaces. Recall that $\mathsf{Fix}[\mathbf{F}_Y, w]$ denotes the set of fixed points of the SDS $(Y, \mathbf{F}_Y, w)$. An example of how Proposition 6.4 opens the door for group-theoretic arguments is provided by

Theorem 6.5. *Let Y be a graph and let $(Y, \mathbf{F}_Y, w)$ be a w-independent SDS with periodic points P and associated group $G(Y, \mathbf{F}_Y)$. Then we have*
(a) $G(Y, \mathbf{F}_Y) = 1$ if and only if all periodic points of $(Y, \mathbf{F}_Y, w)$ are fixed points.
(b) Suppose $G(Y, \mathbf{F}_Y)$ acts transitively on P, and let p be a prime number such that $|\mathsf{P}| \equiv 0 \mod p$. Then there exists a word $w_0 \in W$ such that

$$(i) \ |\mathsf{Fix}[\mathbf{F}_Y, w_0]| \equiv 0 \mod p, \tag{6.8}$$
$$(ii) \ \text{all periodic orbits of } [\mathbf{F}_Y, w_0] \text{ have length } p. \tag{6.9}$$

In particular, if $[\mathbf{F}_Y, w_0]$ has no fixed points, it has at least one periodic orbit of length p, and if $[\mathbf{F}_Y, w_0]$ has no periodic orbits of length greater than 1, then it has at least p fixed points.

Proof. Ad (a). Obviously, if $G(Y, \mathbf{F}_Y) = 1$, then all local maps restricted to P are the identity, and any SDS $(Y, \mathbf{F}_Y, w)$ only has fixed points as periodic points. Suppose next that $G(Y, \mathbf{F}_Y) \neq 1$. By definition, we conclude from $G(Y, \mathbf{F}_Y) \neq 1$ that there exist $g \in G(Y, \mathbf{F}_Y)$ and $\xi \in \mathsf{P}$ such that $g(\xi) \neq \xi$. We can write $g = \prod_{i=1}^{h} F_{w_i, Y}$ and observe that ξ is not a fixed point of the SDS-map $[\mathbf{F}_Y, (w_1, \dots, w_h))$. Hence, we have shown that $G(Y, \mathbf{F}_Y) \neq 1$ implies that there exists an SDS with periodic points that are not all fixed points and (a) follows.

Ad (b). Since $G(Y, \mathbf{F}_Y)$ acts transitively on P, there exists some $q \in \mathsf{P}$ such that

$$|G(Y, \mathbf{F}_Y)| = |\mathsf{P}||G_q|\,, \tag{6.10}$$

where $G_q = \{g \in G(Y, \mathbf{F}_Y) \mid gq = q\}$, i.e., the subgroup consisting of all elements of $G(Y, \mathbf{F}_Y)$ that fix the periodic point q. Let $k \in \mathbb{N}$ be the highest power for which we have $|\mathsf{P}| \equiv 0 \mod p^k$. Equation (6.10) implies

$$|G(Y, \mathbf{F}_Y)| \equiv 0 \mod p^k,$$

and we can conclude from Sylow's theorems that there exists a subgroup $H < G(Y, \mathbf{F}_Y)$ such that $|H| = p^k$. As a p-group H is solvable, whence there exists a cyclic subgroup $H_p < H < G(Y, \mathbf{F}_Y)$.

Let $g = \prod_{j=1}^{k} F_{w_j}$ be a generator of H_p, that is, $H_p = \langle g \rangle$ and $w_0 = (w_1, \dots, w_k)$. We consider the group action of H_p on P and obtain

$$|\mathsf{P}| = \sum_{\xi \in \Xi} |\langle g \rangle(\xi)|\,, \tag{6.11}$$

where Ξ is a set of representatives of the $\langle g \rangle$-action. We have

$$|\langle g \rangle(\xi)| = [\langle g \rangle : \langle g \rangle_\xi]\,,$$

where $\langle g \rangle_\xi$ is the fixed group of ξ. Since $\langle g \rangle$ is a cyclic group of order p, we have the alternative

$$|\langle g \rangle(\xi)| = \begin{cases} 1 & \text{if and only if } \xi \text{ is fixed by } \langle g \rangle, \\ p & \text{if and only if } \xi \text{ is contained in an } \langle g \rangle\text{-orbit of length } p. \end{cases} \tag{6.12}$$

We conclude from $|\mathsf{P}| \equiv 0 \mod p$ and Eq. (6.11) that

$$|\{\xi \mid \xi \text{ is fixed by } \langle g \rangle\}| \equiv 0 \mod p\,.$$

Furthermore, each nontrivial orbit of $(Y, \mathbf{F}_Y, w_0)$ corresponds exactly to an orbit $\langle g \rangle(\xi)$ for some $\xi \in \mathsf{P}$. The proof of Theorem 6.5 now follows from Eq. (6.12). $\qquad \square$

Example 6.6. Here we will compute the group $G(K_3, \mathbf{Nor})$. There are four periodic points labeled 0 through 3 as given in Table 6.1. Each map Nor_i can

Periodic point	Label	Nor_1	Nor_2	Nor_3
$(0,0,0)$	0	$(1,0,0)$	$(0,1,0)$	$(0,0,1)$
$(1,0,0)$	1	$(0,0,0)$	$(1,0,0)$	$(1,0,0)$
$(0,1,0)$	2	$(0,1,0)$	$(0,0,0)$	$(0,1,0)$
$(0,0,1)$	3	$(0,0,1)$	$(0,0,1)$	$(0,0,0)$

Table 6.1. Periodic points of Nor-SDS over K_3.

be represented as a permutation n_i of the periodic points. Using the labeling from Table 6.1, we get $n_1 = (0,1)$, $n_2 = (0,2)$, and $n_3 = (0,3)$. There are four periodic points, so $G(K_3, \mathbf{Nor})$ (when viewed as a group of permutations) must be a subgroup of S_4, that is, $G(K_3, \mathbf{Nor}) < S_4$. On the other hand, we see that $n_3 n_2 n_1 = (0,1,2,3)$, and since it is known that $S_4 = \langle \{(0,1), (0,1,2,3)\} \rangle$ it follows that $S_4 < G(K_3, \mathbf{Nor})$. We conclude that $G(K_3, \mathbf{Nor})$ is isomorphic to S_4.

What does $G(K_3, \mathbf{Nor}) \cong S_4$ imply? It means we can organize the periodic points in any cycle configuration we like by a suitable choice of the update order word. For instance, we could choose to have a Nor-SDS where the first and last periodic points are fixed points and the remaining two periodic points are contained in a two-cycle. The fact that G is isomorphic to S_4 guarantees that it is possible. It does not tell us how to find the update order, though, but it is easily verified that the update order $w = (1,2,1)$ does the trick.   ◇

In Theorem 6.5 we have seen that a transitive $G(Y, \mathbf{F}_Y)$ action and $|\mathsf{P}| \equiv 0$ mod p allow us to design SDS with specific phase-space properties. We next show that $G(Y, \mathbf{F}_Y)$ acts transitively if the Y-local maps are Nor functions. Recall that an action of a group G on a set X is transitive if for any pair $x, y \in X$ there exists $g \in G$ such that $y = gx$.

Lemma 6.7. *Let Y be a combinatorial graph and let $w \in W_Y'$. The SDS $(Y, \mathbf{Nor}_Y, w)$ is w-independent and $G(Y, \mathbf{Nor}_Y)$ acts transitively on $\mathsf{P} = \mathrm{Per}[\mathbf{Nor}_Y, w]$.*

Proof. Let $p = (p_{v_1}, \ldots, p_{v_n})$ and $p' = (p'_{v_1}, \ldots, p'_{v_n})$ be two periodic points with corresponding independent sets $\beta(p)$ and $\beta(p')$ where

$$\beta(x_{v_1}, \ldots x_{v_n}) = \{v_k \mid x_{v_k} = 1\}$$

(Theorem 5.28). We observe that

$$g = \left[\prod_{v \in \beta(p')} \mathrm{Nor}_v \right] \circ \left[\prod_{v \in \beta(p)} \mathrm{Nor}_v \right]$$

is a well-defined element of G without referencing a particular order within the sets $\beta(p)$ and $\beta(p')$ since for any two $v, v' \in \beta(p)$ and $v, v' \in \beta(p')$ we have

$$\text{Nor}_v \circ \text{Nor}_{v'} = \text{Nor}_{v'} \circ \text{Nor}_v .$$

We proceed by proving $g(p) = p'$. We observe that

$$\left[\prod_{v \in \beta(p)} \text{Nor}_v \right](p) = (0, \ldots, 0), \tag{6.13}$$

$$\left[\prod_{v \in \beta(p')} \text{Nor}_v \right](0, \ldots, 0) = p' , \tag{6.14}$$

from which it follows that

$$g(p) = \left[\prod_{v \in \beta(p')} \text{Nor}_v \right] \circ \left[\prod_{v \in \beta(p)} \text{Nor}_v \right](p) = p',$$

and the proof of Lemma 6.7 is complete.   □

From the proof of Lemma 6.7 we conclude that one possible word w that induces the element

$$g = \left[\prod_{v \in \beta(p')} \text{Nor}_v \right] \circ \left[\prod_{v \in \beta(p)} \text{Nor}_v \right]$$

is given by $w = (w_{v_{j_1}}, \ldots, w_{v_{j_q}}, w_{v_{i_1}}, \ldots, w_{v_{i_r}})$, where $w_{v_{j_h}} \in \beta(p)$ and $w_{v_{i_h}} \in \beta(p')$. Obviously, w is in general not a permutation. Lemma 6.7 and Theorem 6.5 imply

Corollary 6.8. *For any prime p such that $|\text{Per}[\mathbf{Nor}_Y, w]| \equiv 0 \mod p$ there exists an SDS of the form $(Y, \mathbf{Nor}_Y, w)$ with the property $|\text{Fix}[\mathbf{Nor}_Y, w]| \equiv 0 \mod p$, and all periodic orbits have length p.*

Example 6.9. Let Y_0 be the graph on four vertices shown in Figure 6.1. We use nor as vertex functions and derive the following table.

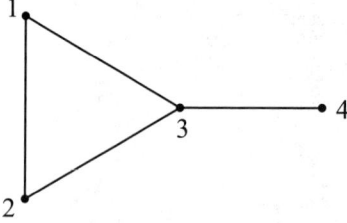

Fig. 6.1. The graph used in Example 6.9.

Label	State	Nor$_1$	Nor$_2$	Nor$_3$	Nor$_4$
1	0000	1000	0100	0010	0001
2	1000	0000	1000	1000	1001
3	1001	0001	1001	1001	1000
4	0100	0100	0000	0100	0101
5	0101	0101	0001	0101	0100
6	0010	0010	0010	0000	0010
7	0001	1001	0101	0001	0000

From this we derive the permutation representations $n_1 = (1,2)(3,7)$, $n_2 = (1,4)(5,7)$, $n_3 = (1,6)$, and $n_4 = (1,7)(2,3)(4,5)$. Each n_i is an element of S_7, so the group $G(Y_0, \mathbf{Nor}_{Y_0})$ must be a subgroup of S_7. However, we note that $n = n_1 n_2 n_3 n_4 = (1,6,5,2,7,4,3)$. Since n and n_3 generate S_7, we must have $S_7 < G(Y_0, \mathbf{Nor}_{Y_0})$, that is, $G(Y_0, \mathbf{Nor}_{Y_0})$ (viewed as a group of permuations) equals S_7.   ◇

6.1.3 The Class of w-Independent SDS

The characterization of w-independent SDS requires one to check the set W'_Y, which makes the verification virtually impossible. In the following we provide an equivalent condition for w-independence that only requires the consideration of the set $S_Y \subset W'_Y$. In other words, the subset of fair words that are permutations completely characterizes w-independence.

Lemma 6.10. *Let $(Y, \mathbf{F}_Y, w)$ be an SDS with state space K^n. If there exists $\mathsf{P} \subset K^n$ such that*

$$\forall w \in S_Y; \; \mathrm{Per}[\mathbf{F}_Y, w] = \mathsf{P} , \tag{6.15}$$

then $(Y, \mathbf{F}_Y, w)$ is w-independent.

Proof. By assumption we have

$$\forall w \in S_Y; \; [\mathbf{F}_Y, w]|_{\mathsf{P}} : \mathsf{P} \longrightarrow \mathsf{P} \quad \text{is bijective}, \tag{6.16}$$

and $\mathsf{P} = \mathrm{Per}[\mathbf{F}_Y, w]$ is maximal and unique with respect to this property. Since the SDS-map $[\mathbf{F}_Y, w] = F_{w_n, Y} \circ \cdots \circ F_{w_1, Y}$ is a finite composition of Y-local maps, we can conclude that $F_{w_1, Y}|_{\mathsf{P}} : \mathsf{P} \longrightarrow F_{w_1, Y}(\mathsf{P})$ is bijective. Since this holds for any $w \in S_Y$, we derive

$$\forall v \in \mathrm{v}[Y]; \; F_{v, Y}|_{\mathsf{P}} : \mathsf{P} \longrightarrow F_{v, Y}(\mathsf{P}) \quad \text{is bijective}. \tag{6.17}$$

Let $v \in \mathrm{v}[Y]$. We choose $w^* \in S_Y$ such that $w_1 = v$, that is,

$$[\mathbf{F}_Y, w^*] = (\prod_{i=2}^{n} F_{w_i, Y}) \circ F_{v, Y} .$$

The next step is to show that $F_{v, Y}(\mathsf{P}) = \mathsf{P}$ holds. Setting $\Phi = \prod_{i=2}^{n} F_{w_i, Y}$, we have by associativity

$$F_{v,Y} \circ (\prod_{i=2}^{n} F_{w_i,Y} \circ F_{v,Y}) = (F_{v,Y} \circ \prod_{i=2}^{n} F_{w_i,Y}) \circ F_{v,Y} . \qquad (6.18)$$

Equation (6.18) can be expressed by the commutative diagram

$$
\begin{array}{ccc}
\mathsf{P} & \xrightarrow{\ \Phi \circ F_{v,Y}\ } & \mathsf{P} \\
{\scriptstyle F_{v,Y}} \downarrow & & \downarrow {\scriptstyle F_{v,Y}} \\
F_{v,Y}(\mathsf{P}) & \xrightarrow{\ F_{v,Y} \circ \Phi\ } & F_{v,Y}(\mathsf{P})
\end{array} \quad ,
$$

from which [in view of Eq. (6.17)] we obtain that

$$F_{v,Y} \circ \Phi \colon F_{v,Y}(\mathsf{P}) \longrightarrow F_{v,Y}(\mathsf{P}) \quad \text{is bijective.}$$

We next observe that $w = (w_2, \ldots, w_n, v) \in S_Y$ and

$$F_{v,Y} \circ \Phi = [\mathbf{F}_Y, w] \ .$$

Using Eq. (6.16) we can conclude that $F_v(\mathsf{P}) \subset \mathsf{P}$ since P is the unique maximal set for which $[\mathbf{F}_Y, w] \colon \mathsf{P} \longrightarrow \mathsf{P}$ is bijective. Since $F_{v,Y}|_{\mathsf{P}}$ is bijective [Eq. (6.17)], the inclusion $F_v(\mathsf{P}) \subset \mathsf{P}$ implies $F_{v,Y}(\mathsf{P}) = \mathsf{P}$. Therefore, we have

$$\forall\, v \in \mathrm{v}[Y]; \ F_{v,Y} \colon \mathsf{P} \longrightarrow \mathsf{P} \quad \text{is bijective.} \qquad (6.19)$$

As a result, $(Y, \mathbf{F}_Y, w)$ is w-independent, and the lemma follows. $\qquad \square$

Let $(Y, \mathbf{F}_Y, w)$ be an SDS. We next show, under mild assumptions on the local maps in terms of the action of $\mathsf{Aut}(Y)$, that for any $\gamma \in \mathsf{Aut}(Y)$ we have $\gamma(\mathsf{Per}[\mathbf{F}_Y, w]) = \mathsf{Per}[\mathbf{F}_Y, \gamma(w)]$. As a consequence, if $(Y, \mathbf{F}_Y, w)$ is w-independent, then $\mathsf{Aut}(Y)$ acts naturally on P by restriction of the natural action $\gamma(x_{v_i}) = (x_{\gamma^{-1}(v_i)})$.

Proposition 6.11. *Let Y be a graph, K a finite field, and $(Y, \mathbf{F}_Y, w)$ an SDS with the property*

$$\forall\, \gamma \in \mathsf{Aut}(Y); \ \forall\, v \in \mathrm{v}[Y]; \quad f_v = f_{\gamma(v)} \ , \qquad (6.20)$$

that is, the vertex functions on any $\langle \gamma \rangle$-orbit are identical. Then we have

$$\forall\, \gamma \in \mathsf{Aut}(Y); \ \mathsf{p} \in \mathsf{Per}[\mathbf{F}_Y, w] \quad \Longrightarrow \quad \gamma(\mathsf{p}) \in \mathsf{Per}[\mathbf{F}_Y, \gamma(w)] \ . \qquad (6.21)$$

Furthermore, if $(Y, \mathbf{F}_Y, w)$ is w-independent, then $\mathsf{Aut}(Y)$ acts on P in a natural way via $\gamma(x_{v_i})_i = (x_{\gamma^{-1}(v_i)})_i$.

Proof. Using the same argument as in the proof of Proposition 4.30 and Eq. (6.20), we have

$$\forall\, \gamma \in \mathsf{Aut}(Y), \ v_i \in \mathrm{v}[Y]; \quad \gamma \circ F_{v_i,Y} \circ \gamma^{-1} = F_{\gamma(v_i),Y} \ , \qquad (6.22)$$

from which it follows that

$$\gamma \circ [\mathbf{F}_Y, w] \circ \gamma^{-1} = [\mathbf{F}_Y, \gamma(w)] \quad \text{where} \quad \gamma(w) = (\gamma(w_1), \ldots, \gamma(w_k)) \ .$$

Hence, we have the commutative diagram

$$
\begin{array}{ccc}
\gamma(\mathsf{P}) & \xrightarrow{\gamma \circ [\mathbf{F}_Y, w] \circ \gamma^{-1}} & \gamma(\mathsf{P}) \\
{\scriptstyle \gamma^{-1}} \downarrow & & \uparrow {\scriptstyle \gamma} \\
\mathsf{P} & \xrightarrow{[\mathbf{F}_Y, w]} & \mathsf{P}
\end{array}
\qquad (6.23)
$$

from which we conclude that if P is a periodic point of $[\mathbf{F}_Y, w]$, then $\gamma(\mathsf{P})$ is a periodic point of $[\mathbf{F}_Y, \gamma(w)]$, proving (6.21).

To prove the second assertion it suffices to show that $\mathsf{Aut}(Y)(\mathsf{P}) = \mathsf{P}$ holds. Clearly, we have $w \in W'_Y$ if and only if $\gamma(w) \in W'_Y$. The commutative diagram (6.23) proves that $[\mathbf{F}_Y, \gamma(w)] : \gamma(\mathsf{P}) \longrightarrow \gamma(\mathsf{P})$ is bijective. Since (1) $\mathsf{P} = \mathsf{Per}[\mathbf{F}_Y, w]$ is the unique, maximal subset of K^n for which $[\mathbf{F}_Y, w] : \mathsf{P} \longrightarrow \mathsf{P}$ with $w \in W'_Y$ is bijective and (2) $w \in W'_Y$ is equivalent to $\gamma(w) \in W'_Y$, we derive

$$\forall \, \gamma \in \mathsf{Aut}(Y); \ \gamma(\mathsf{P}) \subset \mathsf{P} \ .$$

Since γ is an automorphism, we obtain $\gamma(\mathsf{P}) = \mathsf{P}$ and the proof of the proposition is complete.   □

We proceed by showing that the class of *monotone* maps induces w-independent SDS. Let $x = (x_{v_j})_j \in \mathbb{F}_2^n$ and set

$$
x^r = (x^r_{v_j})_j =
\begin{cases}
1 & \text{for } j = r, \\
x_{v_j} & \text{otherwise.}
\end{cases}
$$

A Y-local map $F_{v,Y} : \mathbb{F}_2^n \longrightarrow \mathbb{F}_2^n$ is *monotone* if and only if

$$\forall \, v_j \in \mathrm{v}[Y]; \ r = 1, \ldots, n \quad F_{v,Y}(x)_{v_j} = 1 \quad \Longrightarrow \quad F_{v,Y}(x^r)_{v_j} = 1 \ . \quad (6.24)$$

The SDS $(Y, \mathbf{F}_Y, w)$ is monotone if all $F_{v,Y}$, $v \in \mathrm{v}[Y]$ are monotone local maps.

Proposition 6.12. *Let Y be a graph, $K = \mathbb{F}_2$, and $(Y, \mathbf{F}_Y, w)$ a monotone SDS. Then $(Y, \mathbf{F}_Y, w)$ is w-independent and we have*

$$G(Y, \mathbf{F}_Y) = 1 \ .$$

Proof. It suffices to prove that periodic points of $(Y, \mathbf{F}_Y, w)$ with $w = (w_i)_{1 \leq i \leq k}$ are necessarily fixed points. For this purpose we note that the inverse of $[\mathbf{F}_Y, w]$ restricted to the periodic points is given by $[\mathbf{F}_Y, w^*]$, where

$w^* = (w_{k+1-i})_i$. Suppose $\xi = (\xi_{v_1}, \ldots, \xi_{v_n})$ is a periodic point of $[\mathbf{F}_Y, w]$. We then have

$$\prod_{i=1}^{k} F_{w_{k+1-i}, Y} \circ \prod_{i=1}^{k} F_{w_i, Y}(\xi) = \xi \,.$$

Hence, for an arbitary index $1 \le j \le k$

$$F_{w_j, Y}^2 \left(\prod_{i=1}^{j-1} F_{w_i, Y}(\xi) \right) = \prod_{i=1}^{j-1} F_{w_i, Y}(\xi) \,.$$

By induction over the index j, it follows from (6.24) that $F_{w_j, Y}(\xi) = \xi$ for $1 \le j \le k$. Therefore, all periodic points are necessarily fixed points. The fixed points are independent of update order w, and $G(Y, \mathbf{F}_Y) = 1$. □

6.2 The Class of w-Independent SDS over Circ_n

In this section we study all w-independent SDS over Circ_n that are induced by symmetric Boolean functions. We then compute all groups $G(\mathrm{Circ}_n, \mathbf{F}_{\mathrm{Circ}_n})$ for $n = 4$. In the following we will use the notion of H-class. A state $x \in \mathbb{F}_2^m$ belongs to H-class k if x has exactly k coordinates that are 1, and we write this as $x \in H_{k,m}$.

Theorem 6.13. *For $Y = \mathrm{Circ}_n$, $n \ge 3$, there are exactly 11 symmetric Boolean functions that induce w-independent SDS. They are*

- nor_3 *and* nand_3,
- $(\mathrm{nor}_3 + \mathrm{nand}_3)$ *and* $(1 + \mathrm{nor}_3 + \mathrm{nand}_3)$,
- *the non-constant, monotone maps, i.e.,* and_3, or_3 *and* $\mathrm{majority}_3$,
- *the maps inducing invertible SDS, i.e.,* parity_3 *and* $1 + \mathrm{parity}_3$,
- *the constant maps* $\hat{0}, \hat{1}$.

Proof. Theorem 5.28 and Proposition 5.38 imply that nor_3, nand_3, $(\mathrm{nor}_3 + \mathrm{nand}_3)$, and $(1 + \mathrm{nor}_3 + \mathrm{nand}_3)$ induce w-independent SDS. From Proposition 5.12 we conclude that or_3, and_3, and $\mathrm{majority}_3$ induce w-independent SDS. The case of the constant Boolean functions $\hat{0}_3$ and $\hat{1}_3$ is obvious, but we note that this case also follows from the fact that these SDS are monotone. From Proposition 4.16 we know that parity_3 and $(1 + \mathrm{parity}_3)$ are the only symmetric, Boolean functions that induce invertible SDS, so in particular we have that these maps induce w-independent SDS.

It remains to prove that the five symmetric functions from $\mathrm{Map}(\mathbb{F}_2^3, \mathbb{F}_2)$ that do not appear in the list induce w-dependent SDS. Our strategy is to find points that are periodic for one update order and transient for other update orders.

Consider the Boolean function

$$b_1 : \mathbb{F}_2^3 \longrightarrow \mathbb{F}_2, \quad b_1(x, y, z) = \begin{cases} 1 & \text{for } (x, y, z) \in H_{1,3}, \\ 0 & \text{otherwise.} \end{cases}$$

Let $Y = \mathsf{Circ}_4$, and consider the two words $w = (3, 2, 1, 0)$, $w' = (2, 0, 3, 1)$, and the state $(0, 0, 0, 1)$. We compute

and observe that $(1, 0, 1, 0)$ is a fixed point for $[\mathbf{F}_Y, w]$ and $[\mathbf{F}_Y, w']$, respectively. Therefore, $(0, 0, 0, 1)$ is a periodic point for $[\mathbf{F}_Y, w]$ but not for $[\mathbf{F}_Y, w']$. We conclude that b_1 induces w-dependent SDS.

For the function

$$b_2 : \mathbb{F}_2^3 \longrightarrow \mathbb{F}_2, \quad b_2(x, y, z) = \begin{cases} 0 & \text{for } (x, y, z) \in H_{2,3}, \\ 1 & \text{otherwise} , \end{cases}$$

we have

$$\mathsf{inv} \circ b_1 \circ \mathsf{inv} = b_2 , \tag{6.25}$$

which implies that b_2 induces w-dependent SDS.

Next let

$$b_3 : \mathbb{F}_2^3 \longrightarrow \mathbb{F}_2, \quad b_3(x, y, z) = \begin{cases} 1 & \text{for } (x, y, z) \in H_{0,3} \cup H_{1,3}, \\ 0 & \text{otherwise.} \end{cases}$$

For $Y = \mathsf{Circ}_4$, $w = (0, 1, 2, 3)$ and $w' = (0, 2, 1, 3)$, we have

and observe that $(1, 0, 1, 0)$ is a fixed point for $(\mathsf{Circ}_4, \mathbf{F}_Y, w')$. Since $(1, 0, 0, 0)$ is a periodic point for $(\mathsf{Circ}_4, \mathbf{F}_Y, w)$ and a transient point for $(\mathsf{Circ}_4, \mathbf{F}_Y, w')$, it follows that b_3 induces w-dependent SDS.

Next we consider

$$b_4 : \mathbb{F}_2^3 \longrightarrow \mathbb{F}_2, \quad b_4(x, y, z) = \begin{cases} 1 & \text{for } (x, y, z) \in H_{2,3}, \\ 0 & \text{otherwise,} \end{cases}$$

and take $Y = \mathsf{Circ}_4$, $w = (0, 1, 2, 3)$, and $w' = (0, 1, 3, 2)$. For the SDS $(\mathsf{Circ}_4, \mathbf{F}_Y, w)$ all periodic points are fixed points. Explicitly, we have

$$\mathrm{Per}[\mathbf{F}_Y, w] = \{(0,0,0,0), (0,0,1,1), (1,0,0,1), (1,1,0,0), (0,1,1,0)\} \,.$$

In contrast, the SDS $(\mathrm{Circ}_4, \mathbf{F}_Y, w')$ has two additional orbits of length 2. For the state $(1,1,1,1)$ we obtain

$$[\mathbf{F}_Y, w'](1,1,1,1) = (0,1,0,1) \quad \text{and} \quad [\mathbf{F}_Y, w'](0,1,0,1) = (1,1,1,1) \,,$$

which prove that $(1,1,1,1)$ is a transient point for $w = (0,1,2,3)$ and a periodic point for $w' = (0,1,3,2)$; hence, b_4 induces w-dependent SDS. Since the function

$$\mathsf{b}_5 \colon \mathbb{F}_2^3 \longrightarrow \mathbb{F}_2, \quad \mathsf{b}_5(x,y,z) = \begin{cases} 0 & \text{for } (x,y,z) \in H_{1,3}, \\ 1 & \text{otherwise} \end{cases}$$

satisfies

$$\mathrm{inv} \circ \mathsf{b}_4 \circ \mathrm{inv} = \mathsf{b}_5,$$

it also induces w-dependent SDS for $n = 4$, and the proof of the theorem is complete. $\qquad\square$

6.2.1 The Groups $G(\mathrm{Circ}_4, \mathbf{F}_{\mathrm{Circ}_4})$

Here we compute the group $G(Y, \mathbf{F}_Y)$ for all w-independent SDS over Circ_4.

Proposition 6.14. *Let $Y = \mathrm{Circ}_4$, and let G_b denote the group generated by the Y-local maps induced by b over Y restricted to the periodic points. We then have*

- $G_{\mathrm{nor}} \cong A_7$ *and* $G_{\mathrm{nand}} \cong A_7$,
- $G_{\mathrm{nor+nand}} \cong A_7$ *and* $G_{1+\mathrm{nor+nand}} \cong A_7$,
- $G_{\mathrm{or}} = 1$, $G_{\mathrm{and}} = 1$ *and* $G_{\mathrm{majority}} = 1$,
- $G_{\mathrm{parity}} \cong G_{1+\mathrm{parity}} \cong \mathrm{GAP}(96, 227)$,
- $G_{\hat{0}} = 1$ *and* $G_{\hat{1}} = 1$.

Here A_7 is the alternating group on seven elements, and $\mathrm{GAP}(96, 227)$ is the (unique) group with GAP index $(96, 227)$; see [114].

Proof. The SDS induced by nor functions has seven periodic points, which we label as $0 \leftrightarrow (0,0,0,0)$, $1 \leftrightarrow (1,0,0,0)$, $2 \leftrightarrow (0,1,0,0)$, $3 \leftrightarrow (0,0,1,0)$, $4 \leftrightarrow (1,0,1,0)$, $5 \leftrightarrow (0,0,0,1)$, and $6 \leftrightarrow (0,1,0,1)$. Rewriting the corresponding maps Nor_i for $0 \leq i \leq 3$ as permutations n_i of S_7 (using cycle form) gives $n_0 = (0,1)(3,4)$, $n_1 = (0,2)(5,6)$, $n_2 = (0,3)(1,4)$, and $n_3 = (0,5)(2,6)$. Next, note that the group A_7 has a presentation

$$\langle x, y \mid x^3 = y^5 = (xy)^7 = (xy^{-1}xy)^2 = (xy^{-2}xy^2) = 1 \rangle$$

and that $a = (0,1,2)$ and $b = (2,3,4,5,6)$ are two elements of S_7 that will generate A_7. Now, $a' = n_2(n_0 n_3 n_1)^2 = (0,4,1,6,3)$ and $b' = (n_3 n_2)^2 (n_2 n_1)^2 = $

$(2, 5, 3)$, and after relabeling of the periodic points using the permutation $(0, 3, 2)(1, 5)$ we transform a' into a and b' into b. With G_{nor} viewed as a permutation group we therefore have $A_7 \leq G_{\text{nor}}$. Since every generator n_i is even, we also have $G_{\text{nor}} \leq A_7$, proving the statement for G_{nor}. Since nor and nand induce dynamically equivalent SDS, it follows that

$$G_{\text{nand}} \cong G_{\text{nor}} \cong A_7 .$$

The proof for $G_{\text{nor+nand}}$ and $G_{1+\text{nor+nand}}$ is completely analogous, so we only give the labeling of the periodic points and the corresponding generator relations as $0 \leftrightarrow (0, 0, 0, 0)$, $1 \leftrightarrow (1, 0, 1, 0)$, $2 \leftrightarrow (1, 1, 1, 0)$, $3 \leftrightarrow (0, 1, 0, 1)$, $4 \leftrightarrow (1, 1, 0, 1)$, $5 \leftrightarrow (1, 0, 1, 1)$, $6 \leftrightarrow (0, 1, 1, 1)$, and $7 \leftrightarrow (1, 1, 1, 1)$. The generators are $n_0 = (3, 4)(6, 7)$, $n_1 = (1, 2)(5, 7)$, $n_2 = (3, 6)(4, 7)$, and $n_3 = (1, 5)(2, 7)$. If we simply relabel the periodic points using the permutation $(0, 7)(1, 6)(2, 5)(3, 4)$, the generators are mapped into the generators of G_{nor}, which, along with equivalence, proves

$$G_{\text{nor+nand}} \cong G_{1+\text{nor+nand}} \cong A_7 .$$

Since monotone SDS only have fixed points as periodic points, the corresponding groups are trivial (Proposition 6.12).

The final cases are G_{parity} and $G_{1+\text{parity}}$. In both cases all points are periodic, so we simply use the decimal encoding of each point as its label using (4.17), which in the case of G_{parity} (viewed as a permutation group) gives us the generators $n_0 = (2, 3)(6, 7)(8, 9)(12, 13)$, $n_1 = (1, 3)(4, 6)(9, 11)(12, 14)$, $n_2 = (2, 6)(3, 7)(8, 12)(9, 13)$, and $n_3 = (1, 9)(3, 11)(4, 12)(6, 14)$. (Note that there are four fixed points.)

A straightforward but tedious computation shows that G_{parity} has order $96 = 2^5 \cdot 3$. From [115, 116] it is known that there are 231 groups of order 96. Explicit computations show that G_{parity} is non-Abelian since, for example, $n_0 n_2 \neq n_2 n_0$, that it has 16 Sylow 3-subgroups, and that its center and its Frattini subgroups are trivial. These properties uniquely identify G_{parity} as the group with GAP [114] index $(96, 227)$.

The four generators for $G_{1+\text{parity}}$ (viewed as a permutation group) are

$$n_0 = (0, 1)(4, 5)(10, 11)(14, 15),$$
$$n_1 = (0, 2)(5, 7)(8, 10)(13, 15),$$
$$n_2 = (0, 4)(1, 5)(10, 14)(11, 15),$$
$$n_3 = (0, 8)(2, 10)(5, 13)(7, 15) .$$

Using the relabeling

$$\begin{pmatrix} 0 & 1 & 2 & 4 & 5 & 7 & 8 & 10 & 11 & 13 & 14 & 15 \\ 3 & 2 & 1 & 7 & 6 & 4 & 11 & 9 & 8 & 14 & 13 & 12 \end{pmatrix},$$

the generators for $G_{1+\text{parity}}$ are transformed into the generators for G_{parity}; hence, $G_{1+\text{parity}} \cong G_{\text{parity}}$, and the proof is complete. □

6.1. Write a program to identify the maps $f \colon \mathbb{F}_2^3 \longrightarrow \mathbb{F}_2$ that do not induce w-independent SDS over Circ_n. Prove that the remaining maps induce w-independent SDS.   [3+C]

6.3 A Presentation of S_{35}

We conclude this chapter by showing that the symmetric group S_{35} is isomorphic to the group of $(Q_2^3, (\mathbf{Nor} + \mathbf{Nand})_{Q_2^3}, w)$.

Proposition 6.15. *Let $Y = Q_2^3$ and let $\pi \in S_8$; then any SDS of the form $(Q_2^3, (\mathbf{Nor} + \mathbf{Nand})_{Q_2^3}, \pi)$ has precisely 35 periodic points of period $2 \le p \le 17$ and precisely 1 fixed point. Furthermore, we have*

$$ G(Q_2^3, (\mathbf{Nor} + \mathbf{Nand})_{Q_2^3}) \cong S_{35} . $$

Proof. From Proposition 5.38 we know that the periodic points of an arbitrary SDS induced by $(\mathrm{nor}_k + \mathrm{nand})_k$ are w-independent. A straightforward analysis of the phase space of the SDS

$$ [(\mathbf{Nor} + \mathbf{Nand})_{Q_2^3}, (0, 1, 2, 3, 4, 5, 6, 7)] $$

shows that

- there exists exactly one fixed point,
- exactly 35 points of period $p \ge 2$.

The second part of the first statement ($p \le 17$) follows by inspection of the SDS phase spaces induced by all $\Delta(Q_2^3) = 54$ representatives of the $\mathsf{Aut}(Y)$-action on $S_8/\sim_Y$.

We consider the periodic points as binary numbers and order them in the natural order. Then we obtain for the restrictions of the local maps F_{i,Q_2^3}, $0 \le i \le 7$, to the periodic points of period $p \ge 2$:

$$
\begin{aligned}
g_0 &= (0,1)(5,6)(12,13)(15,16)(17,18)(20,21)(22,23)(24,25)(26,27) , \\
g_1 &= (0,2)(3,4)(7,8)(9,10)(17,19)(26,28)(29,30)(31,32)(33,34) , \\
g_2 &= (0,3)(2,4)(7,9)(8,10)(12,14)(26,29)(28,30)(31,33)(32,34) , \\
g_3 &= (0,5)(1,6)(7,11)(12,15)(13,16)(17,20)(18,21)(22,24)(23,25) , \\
g_4 &= (0,7)(2,8)(3,9)(4,10)(5,11)(26,31)(28,32)(29,33)(30,34) , \\
g_5 &= (0,12)(1,13)(3,14)(5,15)(6,16)(17,22)(18,23)(20,24)(21,25) , \\
g_6 &= (0,17)(1,18)(2,19)(5,20)(6,21)(12,22)(13,23)(15,24)(16,25) , \\
g_7 &= (0,26)(1,27)(2,28)(3,29)(4,30)(7,31)(8,32)(9,33)(10,34) .
\end{aligned}
\tag{6.26}
$$

Since there are 35 periodic points of period $p \ge 2$, it is clear that $G < S_{35}$. We next observe that a 35-cycle can be generated as follows:

$$\alpha_1 = g_6\,g_4\,g_7\,g_2\,g_1\,g_5\,g_7\,g_3\,g_2\,g_6\,g_4\,g_0\,g_1\,g_6\,g_1\,g_7\,g_2\,g_1\,g_5\,g_7\,g_3\,g_2\,g_6\,g_4\,g_0\,g_5$$
$$= (1, 13, 0, 4, 24, 8, 16, 9, 27, 31, 21, 29, 33, 11, 6, 17, 10, 20,$$
$$28, 22, 34, 30, 14, 12, 2, 15, 5, 19, 25, 3, 18, 26, 32, 23, 7) \qquad (6.27)$$

Next, a 2-cycle can be generated from

$$\alpha_2 = g_4\,g_7\,g_1\,g_2\,g_6\,g_7\,g_3\,g_1\,g_5\,g_4\,g_0$$
$$= (1, 3, 0, 25, 34, 32, 19, 11, 22, 6, 4, 2, 5, 23, 33, 31, 17,$$
$$16, 10, 8, 20, 13, 9, 12, 21, 14, 7, 24, 27, 29, 26, 18, 15)(28, 30) \qquad (6.28)$$

as $\alpha' = \alpha_2^{33} = (28, 30)$. Since $\gcd(4, 35) = 1$, we see that $\beta' = \alpha_1^4$ is a 35-cycle where 28 and 30 are consecutive elements. Since it is known that $\alpha = (1, 2)$ and $\beta = (1, 2, \ldots, 35)$ generate S_{35}, and since we can transform α' and β' into α and β by relabeling, we conclude that G is isomorphic to S_{35}. □

By the above proposition, an SDS of the form $(Q_2^3, (\mathbf{Nor} + \mathbf{Nand})_{Q_2^3}, \pi)$ with $\pi \in S_8$ has a maximal orbit length 17. With arbitrary words as update orders, additional periodic orbits can be obtained:

Corollary 6.16. *For any* $1 \le p \le 35$ *there exists a word* w *such that* $(Q_2^3, (\mathbf{Nor} + \mathbf{Nand})_{Q_2^3}, w)$ *has a periodic orbit of length* p. *In particular, for any bijection* β *over a set of 35 elements there exists some* w *such that*

$$\beta = [(\mathbf{Nor} + \mathbf{Nand})_{Q_2^3}, w]|_P .$$

Problems

6.2. Determine the periodic points for Nor-SDS over Line_2, and give permutation representations n_0 and n_1 of the restriction of Nor_0 and Nor_1 to the set of periodic points. What is the group $G_2 = G(\mathsf{Line}_2, (\mathrm{Nor})) = \langle\{n_0, n_1\}\rangle$? Interpret your result in terms of periodic orbit structure and update orders.
[1]

6.3. What is the group $G_3 = G(\mathsf{Line}_3, (\mathrm{Nor}_v)_v)$?
[2]

6.4. Show that $G_4 = G(\mathsf{Circ}_4, (\mathrm{Nor})) \cong A_7$, the alternating group on seven letters. *Hint.* The group A_7 has a presentation

$$\langle x, y \,|\, x^3 = y^5 = (xy)^7 = (xy^{-1}xy)^2 = (xy^{-2}xy^2) = 1 \rangle.$$

Check that $(0, 1, 2)$ and $(2, 3, 4, 5, 6)$ are two such generators, and use this to solve the problem.
[2]

6.5. Show that $G_5 = G(\mathsf{Circ}_5, (\mathrm{Nor})) = S_{11}$. *Hint.* Use the fact that S_n is generated by $(0, 1)$ and $(0, 1, 2, \ldots, n-1)$.
[2]

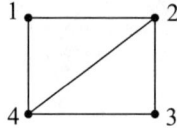

Fig. 6.2. The square with a diagonal edge.

6.6. Let Y be the graph shown in Figure 6.2, let the vertex functions be induced by $(\text{nor}_k)_{k=3}^4$, and let w be a fair word over the vertex set. Let m be the number of periodic points of the SDS $(Y, \mathbf{Nor}_Y, w)$. (i) What is m? (ii) Give the periodic points.

Label the periodic points 1 through m by viewing them as binary numbers such that they are given in increasing order. For each $v \in \mathrm{v}[Y]$ let Nor'_v be the restriction of Nor_v to the periodic points and let n_v be the permutation encoding of Nor'_v based on your labeling of the periodic points. (iii) What are the n_v's? (iv) Explain why the group $G = \langle\{n_v\}\rangle$ is well-defined.

(v) What is the group G?

(vi) Interpret your answer in (v) in the context of update orders and periodic orbit structure.   **[2]**

6.7.   (Subgroups of $G(K_n, \mathbf{Parity}_{K_n})$ [117]) From Problem 5.25, and as shown in its solution, the invertible permutation SDS-map $[\mathbf{Parity}_{K_n}, \pi]$ is dynamically equivalent to the permutation action of the $(n+1)$-cycle

$$\beta(\pi) = (\pi(1), \pi(2), \ldots, \pi(n), n+1)$$

on $\hat{\mathbb{F}}_2^n$. Refer to Problem 5.25 and its solution for notation and definitions.

(a) Show that

$$H_n = \langle\{[\mathbf{Parity}_{K_n}, \pi] \mid \pi \in S_{K_n}\}\rangle \tag{6.29}$$

is a subgroup of $G(K_n, (\text{Parity}_i)_i)$. Find a representation of the identity element in H_n in terms of the generators of H_n.

(b) For $n \geq 3$ show that

(i) the n-cycles in S_n generate A_n when n is odd,
(ii) the n-cycles in S_n generate S_n when n is even.

(c) Define the map $\phi\colon H_n \longrightarrow S_{n+1}$ for n odd and $\phi\colon H_n \longrightarrow A_{n+1}$ for n even by $\phi([\mathbf{Parity}_{K_n}, \pi]) = \beta(\pi)$ with $\pi \in S_n$, and $\phi([\mathbf{Parity}_{K_n}, \pi^k] \circ \cdots \circ [\mathbf{Parity}_{K_n}, \pi^1]) = \phi(\pi^k) \cdots \phi(\pi^1)$ with $\pi^i \in S_n$, $1 \leq i \leq k$. Show that ϕ is a well-defined group homomorphism.

(d) Argue that any $(n+1)$-cycle can be represented as $\beta(\pi)$ for some $\pi \in S_n$. Use the results from (b) and (c) to conclude that the map ϕ is an isomorphism, and hence that

$$H_n \cong \begin{cases} S_{n+1}, & n \text{ odd}, \\ A_{n+1}, & n \text{ even}. \end{cases} \qquad (6.30)$$

[3]

6.8. The construction of H_n in Problem 6.7 can be done over any graph Y. Determine the order of the group H generated by permutation SDS induced by parity functions in the case of $Y = \text{Circ}_4$. Verify that $|H|$ divides $|G(\text{Circ}_4, (\textbf{Parity})_{\text{Circ}_4})|$. Identify the group H.   [2C+]

6.9. Let ϕ be a w-independent SDS over a graph Y with Y-local functions $(F_{v,Y})_v$ and periodic points $P \subset K^n$. Let $F'_{v,Y}$ denote the restriction of $F_{v,Y}$ to P. Argue that

$$H(Y, (F_{v,Y})_v) = \langle \{ [(F_{v,Y})_v, \pi] : P \longrightarrow P \mid \pi \in S_Y \} \rangle$$

is a well-defined subgroup of $G = G(Y, (F_{v,Y})_v)$. What is H if $Y = \text{Circ}_4$ for SDS induced by nor functions?   [2C+]

Answers to Problems

6.2. The periodic points are $(0,0)$, $(1,0)$, and $(0,1)$. With the chosen labeling

periodic point	label	Nor_0	Nor_1
$(0,0)$	0	$(1,0)$	$(0,1)$
$(0,1)$	1	$(0,1)$	$(0,0)$
$(1,0)$	2	$(0,0)$	$(1,0)$

in the table we have $n_0 = (0,2)$ and $n_1 = (0,1)$ as permutation representations of Nor_0 and Nor_1, respectively. Clearly, we have $G_2 \leq S_3$. Since $n_0 n_1 = (0,1,2)$ and $n_1 = (0,1)$ and since we know $(0,1)$ and $(0,1,2)$ generate S_3, we also have $G_2 \geq S_3$, so $G_2 = S_3$.

6.3. Using the labeling from the table below, we have $n_0 = (0,1)$, $n_1 = (0,2)$,

periodic point	label	Nor_0	Nor_1	Nor_2
$(0,0,0)$	0	$(1,0,0)$	$(0,1,0)$	$(0,0,1)$
$(1,0,0)$	1	$(0,0,0)$	$(1,0,0)$	$(1,0,0)$
$(0,1,0)$	2	$(0,1,0)$	$(0,0,0)$	$(0,1,0)$
$(0,0,1)$	3	$(0,0,1)$	$(0,0,1)$	$(0,0,0)$

and $n_3 = (0,3)$. From $n_2 n_1 n_0 = (0,1,2,3)$ and $S_4 = \langle \{(0,1),(0,1,2,3)\} \rangle \leq G_3 \leq S_4$, we see that $G_3 = S_4$.

6.4. Here we use the labeling $0 \leftrightarrow (0,0,0,0)$, $1 \leftrightarrow (1,0,0,0)$, $2 \leftrightarrow (0,1,0,0)$, $3 \leftrightarrow (0,0,1,0)$, $4 \leftrightarrow (1,0,1,0)$, $5 \leftrightarrow (0,0,0,1)$, and $6 \leftrightarrow (0,1,0,1)$ of the periodic points. You can verify that $n_0 = (0,1)(3,4)$, $n_1 = (0,2)(5,6)$, $n_2 = (0,3)(1,4)$, and $n_3 = (0,5)(2,6)$. Note that, for example, $a = n_2(n_0 n_3 n_1)^2 = (0,4,1,6,3)$ and $b = (n_3 n_2)^2 (n_2 n_1)^2 = (2,5,3)$. If we relabel the periodic points by $\begin{pmatrix} 0\ 1\ 2\ 3\ 4\ 5\ 6 \\ 3\ 5\ 0\ 2\ 4\ 1\ 6 \end{pmatrix}$, we get $a = (0,1,2)$ and $b = (2,3,4,5,6)$. By the hint we know that G_4 contains A_7. However, since every generator of G_4 is an even permutation and since G_4 is contained in S_7, we must have $G_4 = A_7$.

6.5. There are 11 periodic points, which we initially label as in the table: You can now verify that $n_0 n_1 n_4 = (0,9,1)(2,8)(3,10,4)(5,7,6)$. By using the transpositions $(n_1 n_0 n_2)^3 = (1,3)$, $(n_2 n_1 n_3)^3 = (2,5)$ and $(n_3 n_2 n_4)^3 = (3,8)$, we construct the 11-cycle

$$a = (n_3 n_2 n_4)^3 (n_2 n_1 n_3)^3 (n_1 n_0 n_2)^3 (n_0 n_1 n_4) = (0,9,8,5,7,6,2,3,10,4,1).$$

We also have $b = (n_4 n_0 n_1)^3 = (0,9)$. Using the problem hint, it is now easy to see that $G_5 = S_{11}$.

Label	Point	Label	Point
0	$(0,0,0,0,0)$	1	$(1,0,0,0,0)$
2	$(0,1,0,0,0)$	3	$(0,0,1,0,0)$
4	$(1,0,1,0,0)$	5	$(0,0,0,1,0)$
6	$(1,0,0,1,0)$	7	$(0,1,0,1,0)$
8	$(0,0,0,0,1)$	9	$(0,1,0,0,1)$
10	$(0,0,1,0,1)$		

Challenge: In terms of the number of generators, find a minimal 11-cycle.

6.7. (a) Since parity-SDS are invertible, and since everything is finite, every element of H_n can be written as $[\mathbf{Parity}_{K_n}, w]$ for some finite, fair word w. Thus, H_n is a subset of $G(K_n, \mathbf{Parity}_{K_n})$, and H_n is a group by construction. Let $\pi = (1, 2, \ldots, n)$, i.e., the identity permutation. The inverse of $[\mathbf{Parity}_{K_n}, \pi]$ is $[\mathbf{Parity}_{K_n}, \pi^*]$; thus, the identity element in H_n has a representation in terms of generators as $[\mathbf{Parity}_{K_n}, \pi^*] \circ [\mathbf{Parity}_{K_n}, \pi]$.

(b) Let C_n denote the subgroup of S_n generated by the n-cycles. We always have (using cycle-form)

$$(a, b, c) = (a, c, b, \alpha_{n-3}, \alpha_{n-4}, \ldots, \alpha_1)(a, c, \alpha_1, \alpha_2, \ldots, \alpha_{n-3}, b) ,$$

and since A_n is generated by the three-cycles it follows that $A_n \leq C_n$. When $n \equiv 1 \mod 2$, every element of C_n is an even permutation and consequently $C_n = A_n$, giving the first statement. When $n \equiv 0 \mod 2$, we also have

$$(1, 2) = (1, 2, 3, \ldots, n)^2 (1, n, n - 2, \ldots, 4, 2, n - 1, n - 3, \ldots, 3) .$$

The fact that $(1, 2)$ and $(1, 2, 3, \ldots, n)$ generate S_n shows that $S_n = C_n$. Alternatively, $S_n : A_n = 2$ and since C_n contains an odd permutation when $n \equiv 0 \mod 2$, we deduce from $A_n \leq C_n \leq S_n$ that $C_n = S_n$ in this case.

(c) For $1 \leq i \leq l$ set $h_i = [\mathbf{Parity}_{K_n}, \pi^i]$ with $\pi^i \in S_n$, and let $h = h_l \circ \cdots \circ h_1$. Using the second commutative diagram (5.41) from the solution of Problem 5.25, and using the fact that the maps $\iota \colon \mathbb{F}_2^n \longrightarrow \hat{\mathbb{F}}_2^n$ and $\mathsf{proj} \colon \hat{\mathbb{F}}_2^n \longrightarrow \mathbb{F}_2^n$ (defined in Problem 5.25) are inverses of one another, we obtain

$$\phi(h) = \phi(h_l)\phi(h_{l-1}) \cdots \phi(h_2)\phi(h_1) \tag{6.31}$$
$$= [\iota \circ h_l \circ \mathsf{proj}]\,[\iota \circ h_{l-1} \circ \mathsf{proj}] \cdots$$
$$\cdots [\iota \circ h_2 \circ \mathsf{proj}]\,[\iota \circ h_1 \circ \mathsf{proj}]$$
$$= \iota \circ h \circ \mathsf{proj} .$$

Thus, if an element $g \in H_n$ has two different representations, say $g = \prod g_i = \prod g_i'$, it is clear from (6.31) that $\phi(\prod g_i) = \phi(\prod g_i')$. The map ϕ is thus well-defined. It is a homomorphism by construction. Note that $\beta(\pi)\beta(\pi^*) = \mathsf{id}$, as it should.

(d) An $(n+1)$-cycle element of S_{n+1} can always be shifted cyclically so that the element $n+1$ occurs in the last position — it still represents the same permutation. In light of this, it is clear that any $(n+1)$-cycle of S_{n+1} has a cycle representation $\beta(\pi)$ for some $\pi \in S_n$. From (6.31) it follows that $\phi(g) = \phi(g')$ if and only if $g = g'$ for any $g, g' \in H_n$. The map ϕ is thus an injection. From (b) it follows that $\phi(H_n) = A_{n+1}$ for odd n and $\phi(H_n) = S_{n+1}$ for even n. Thus, ϕ is surjective. We conclude that ϕ is an isomorphism.

6.8. $|H| = 48$, and $48 \mid 96 = |G|$.

6.9. The subgroup H is isomorphic to A_7 and thus equals G.

7

Combinatorics of Sequential Dynamical Systems over Words

In Chapter 4 we introduced SDS over permutations, that is, SDS for which each Y-local map is applied exactly once. A combinatorial result of SDS over permutations developed in Chapter 4 based on Eq. (3.15) allowed us to identify identical SDS via the acyclic orientations of the base graph Y through

$$\mathcal{O}_Y \colon S_k/\sim_Y \longrightarrow \mathsf{Acyc}(Y)\,, \tag{7.1}$$

where we identify S_Y with S_k, the symmetric group over k letters, and where $\sigma_1 \sim_Y \sigma_2$ if and only if they can be transformed into each other by successive transpositions of consecutive letters that are pairwise non-adjacent Y-vertices. Let us recall how this equivalence relation $\sim_Y$ ties to SDS: Two local maps F_v and $F_{v'}$ commute if v and v' are not adjacent since in this case $F_{v'}(x_{v_1}, \ldots, x_{v_n})$ does not depend on x_v and $F_v(x_{v_1}, \ldots, x_{v_n})$ does not depend on $x_{v'}$ in the coordinate functions corresponding to v' and v, respectively. As a result two SDS-maps are identical if their underlying permutation update orders belong to the same $\sim_Y$-equivalence class.

In this chapter we generalize SDS over permutations to SDS over general words as in [118, 119]. This allows us to analyze and model much broader classes of systems. For instance, SDS over words can be used to study discrete event simulations where agents are typically updated multiple times [121]. We will simplify notation as follows: If v_i is a vertex in Y and $\{v_i, v_j\}$ is an edge of Y, we write $v_i \in Y$ and $\{v_i, v_j\} \in Y$, respectively.

This chapter is organized into two sections. In the first section we derive an analogue of Eq. (7.1) by introducing a new combinatorial object: the undirected, loop-free graph $G(w, Y)$, which has vertex set $\{1, \ldots, k\}$ and edge set $\{\{r, s\} \mid \{w_s, w_r\} \in Y\}$. It is evident that $G(w, Y)$ is much too "fine" since it uses the indices of the word instead of its letters. The key idea consists of identifying a suitable equivalence relation over acyclic orientations of $G(w, Y)$ in order to obtain the invariance of the resulting class under transpositions of non-adjacent letters in w. We will show that this equivalence relation $\sim_w$ over acyclic orientations is induced by a new group action: the subgroup of

$G(w, Y)$-automorphisms that fix the word w denoted $\mathsf{Fix}(w)$. Obviously, the fixed group of a permutation-word is trivial, and accordingly $\mathsf{Fix}(w)$ did not appear in the framework of permutations-SDS. The orbits of $\mathsf{Fix}(w)$ allow us to generalize Eq. (7.1) to SDS over words as follows:

$$\mathcal{O}'_Y : W_k / \sim_Y \longrightarrow \dot{\bigcup}_{\varphi \in \Phi} [\mathsf{Acyc}(G(\varphi, Y)) / \sim_\varphi] , \qquad (7.2)$$

where Φ is a set of representatives of the natural S_k-action on W_k (words of length k) given by

$$\sigma \cdot w = (w_{\sigma^{-1}(1)}, \dots, w_{\sigma^{-1}(k)}) .$$

In analogy with permutation-SDS, the above correspondence is not only of combinatorial interest but also relevant for SDS-maps since for $w \sim_Y w'$ the SDS-maps of $(Y, \mathbf{F}_Y, w)$ and $(Y, \mathbf{F}_Y, w')$ are identical.

In the second section we introduce a generalized equivalence relation over words. Let $\mathsf{A}(w)$ be the automorphism group of $G(w, Y)$, and let $\mathsf{N}(w)$ be the normalizer of $\mathsf{Fix}(w)$ in $\mathsf{A}(w)$, that is,

$$\mathsf{N}(w) = \{\alpha \in \mathsf{A}(w) \mid \alpha \mathsf{Fix}(w) \alpha^{-1} = \mathsf{Fix}(w)\} .$$

Then the short exact sequence

$$1 \longrightarrow \mathsf{Fix}(w) \longrightarrow \mathsf{N}(w) \longrightarrow \mathsf{Aut}(Y)$$

(Theorem 7.6) allows one to define a new equivalence relation over words denoted by $\sim_{\mathsf{N}(w)}$. This relation is directly induced by the group $\mathsf{N}(w)$ and arises in the context of the question of whether it is possible to replace $\mathsf{Fix}(w)$ by a larger group G of $G(w, Y)$-automorphisms. As in the case of $\mathsf{Fix}(w)$ the group G should give rise to a new equivalence relation "$\sim_G$" that has the property that $w \sim_G w'$ implies that the SDS-maps associated to $(Y, \mathbf{F}_Y, w)$ and $(Y, \mathbf{F}_Y, w')$ are equivalent. The main result of this section is that $G = \mathsf{N}(w)$ induces such an equivalence relation $\sim_{\mathsf{N}(w)}$. Explicitly $\sim_{\mathsf{N}(w)}$ has the properties that

(P1) $\mathcal{O}_Y^{\mathsf{N}(\varphi)} : S_k(\varphi) / \sim_{\mathsf{N}(\varphi)} \longrightarrow [\mathsf{Acyc}(\varphi) / \sim_{\mathsf{N}(\varphi)}] ,$

$\mathcal{O}_Y^{\mathsf{N}(\varphi)}([\sigma \cdot \varphi]_{\mathsf{N}(\varphi)}) = [\mathcal{O}_Y(\sigma)]_{\mathsf{N}(\varphi)}$

is a bijection and

(P2) $w \sim_{\mathsf{N}(\varphi)} w' \Longrightarrow [\mathbf{F}_Y, w] \sim [\mathbf{F}_Y, w'] .$

The equivalence relation $\sim_{\mathsf{N}(w)}$ can differ significantly from $\sim_Y$. In this chapter we will show in Lemma 7.21 that $w \sim_{\mathsf{N}(\varphi)} w'$ implies that there exist $g, g' \in \mathsf{N}(\varphi)$ such that $\vartheta(g) \circ w \sim_Y \vartheta(g') \circ w'$, where $\vartheta : \mathsf{N}(w) \longrightarrow \mathsf{Aut}(Y)$ is given by $\vartheta(\alpha)(w_i) = w_{\alpha^{-1}(i)}$ (Theorem 7.6). This result connects the actions of the groups $\mathsf{A}(w)$ and $\mathsf{Aut}(w)$.

7.1 Combinatorics of SDS over Words

7.1.1 Dependency Graphs

Let us begin by introducing two basic group actions. First, we have S_k, the symmetric group over k letters, $\{1, \ldots, k\}$, which acts on the set of all words of length k, denoted W_k, via $\sigma \cdot w = (w_{\sigma^{-1}(1)}, \ldots, w_{\sigma^{-1}(k)})$. The orbits of this action induce the partition

$$W_k = \dot{\bigcup}_{\varphi \in \Phi} S_k(\varphi) \,, \tag{7.3}$$

where Φ is a corresponding set of representatives. Second, the automorphism group of Y acts on W_k via $\gamma \circ w = (\gamma(w_1), \ldots, \gamma(w_k))$ and $\circ$ has by definition the property

$$\gamma(w_s) = (\gamma \circ w)_s \,.$$

Lemma 7.1. *Let $\sigma \in S_k$ and $\gamma \in \mathsf{Aut}(Y)$. Then we have*

$$\gamma \circ (\sigma \cdot w) = \sigma \cdot (\gamma \circ w) \,. \tag{7.4}$$

Proof. To prove the lemma, we compute

$$\begin{aligned}
\gamma \circ (\sigma \cdot w) &= (\gamma(w_{\sigma^{-1}(1)}), \ldots, \gamma(w_{\sigma^{-1}(k)})) \\
&= ((\gamma \circ w)_{\sigma^{-1}(1)}, \ldots, (\gamma \circ w)_{\sigma^{-1}(k)}) \\
&= \sigma \cdot (\gamma \circ w) \,.
\end{aligned}$$

$\square$

We next define the dependency graph of a word w and a combinatorial graph Y.

Definition 7.2. A word $w \in W_k$ and a combinatorial graph Y naturally induce the combinatorial graph $G(w, Y)$ with vertex set $\{1, \ldots, k\}$ and edge set $\{\{r, s\} \mid \{w_s, w_r\} \in Y\}$. We call $G(w, Y)$ the *dependency graph* of w and Y.

Example 7.3. Let $w = (v_1, v_2, v_1, v_2, v_3)$ and $Y = v_1 \text{——} v_2 \text{——} v_3$. Then we have

$$G(w, Y) = \begin{matrix} 1 \text{——} 2 \\ \times \diagdown \\ 3 \text{——} 4 \text{——} 5 \end{matrix}$$

In the following we will use the notation $\mathsf{A}(w)$ for the group of graph automorphisms of $G(w, Y)$, and we denote the set of acyclic orientations of $G(w, Y)$ by $\mathsf{Acyc}(w)$. For $w \in W_k$ we set $\mathsf{Fix}(w) = \{\rho \in S_k \mid \rho \cdot w = w\}$. $\diamond$

Proposition 7.4. *Let Y be a combinatorial graph, $w \in W_k$, $\gamma \in \mathsf{Aut}(Y)$, and $\sigma \in S_k$. Then*

$$\sigma \colon G(w, Y) \longrightarrow G(\sigma \cdot w, Y), \quad r \mapsto \sigma(r) \tag{7.5}$$

is a graph isomorphism. In particular, $\mathsf{Fix}(w)$ is a subgroup of $\mathsf{A}(w)$ and

$$\mathsf{A}(\sigma \cdot w) = \sigma\,\mathsf{A}(w)\,\sigma^{-1} \quad \text{and} \quad \mathsf{Fix}(\sigma \cdot w) = \sigma\,\mathsf{Fix}(w)\,\sigma^{-1}. \tag{7.6}$$

For any $\gamma \in \mathsf{Aut}(Y)$ we have $G(w, Y) \cong G(\gamma \circ w, Y)$ and $\mathsf{Fix}(\gamma \circ w) = \mathsf{Fix}(w)$.

Proof. We set $w' = (w_{\sigma(1)}, \ldots, w_{\sigma(k)}) = \sigma^{-1} \cdot w$ and show that $\sigma \colon G(w', Y) \longrightarrow G(w, Y)$ is an isomorphism of graphs. Let $\{r, s\} \in G(w', Y)$. By definition of $w' = (w_{\sigma(1)}, \ldots, w_{\sigma(k)})$, we have $w_{\sigma(h)} = w'_h$ for $h = 1, \ldots, k$ and obtain

$$\{w'_s, w'_r\} \in Y \quad \Longleftrightarrow \quad \{w_{\sigma(s)}, w_{\sigma(r)}\} \in Y\,;$$

hence, $\{\sigma(s), \sigma(r)\} \in G(w, Y)$ and (7.5) follows. Next we prove that $\mathsf{Fix}(w)$ is a subgroup of $\mathsf{A}(w)$. Let $\rho \in \mathsf{Fix}(w)$, that is, $\rho \cdot w = w$ and we immediately observe $\rho \colon G(w, Y) \longrightarrow G(\rho \cdot w, Y) = G(w, Y)$, and $\rho \in \mathsf{A}(w)$. In order to prove (7.6) we consider the diagrams

with $\alpha \in \mathsf{A}(w)$ and $\beta \in \mathsf{A}(\sigma \cdot w)$, respectively. It is clear that each α induces a unique $G(\sigma \cdot w, Y)$-automorphism via $\sigma\alpha\sigma^{-1}$, and similarly each β its respective $G(w, Y)$-automorphism via $\sigma^{-1}\beta\sigma$, and (7.6) follows. According to Lemma 7.1, we have $\rho \cdot (\gamma \circ w) = \gamma \circ (\rho \cdot w)$, and $\mathsf{Fix}(w) = \mathsf{Fix}(\gamma \circ w)$. Finally, we observe that $\{w_s, w_r\} \in Y$ is equivalent to $\{\gamma(w_s), \gamma(w_r)\} \in Y$, from which $G(w, Y) \cong G(\gamma \circ w, Y)$ follows. $\square$

7.1. Let w and w' be the words $w = (v_1, v_2, v_3, v_1)$ and $w' = (v_1, v_1, v_3, v_2)$, and let $Y = v_1 \rule{0.8em}{0.4pt} v_3 \rule{0.8em}{0.4pt} v_2$. Draw $G(w', Y)$ and $G(w', Y)$ and show that $G(w', Y) \cong G(w, Y)$. [1]

7.2. Let $Y = K_3$ be the complete graph with vertex set $\{v_1, v_2, v_3\}$, $w = (v_1, v_1, v_2, v_3)$, $w' = (v_3, v_2, v_2, v_1)$, and $w'' = (v_2, v_2, v_3, v_1)$. Show that $G(w, Y) \cong G(w', Y) \cong G(w'', Y)$ (Proposition 7.4). [1]

One immediate consequence of Proposition 7.4 is that for permutations the graph $G(w, Y)$ can naturally be identified with Y.

Corollary 7.5. *Let $w \in S_Y$. Then we have*

$$G(w, Y) \cong Y\,. \tag{7.7}$$

Proof. We have $\{r, s\} \in G(w, Y)$ if and only if $\{w_r, w_s\} \in Y$. In view of Proposition 7.4, we may without loss of generality assume that $w = \mathsf{id} = (v_1, v_2, v_3, \ldots, v_n)$ and derive $G(w, Y) \cong G((v_1, \ldots, v_n), Y) = Y$. $\square$

7.1.2 Automorphisms

In this section we study the normalizer of $\mathsf{Fix}(w)$ in $\mathsf{A}(w)$. We prove a short exact sequence that relates it to the groups $\mathsf{Fix}(w)$ and $\mathsf{Aut}(Y)$ (Theorem 7.6). Before we state the main result, let us have a closer look at $G(w, Y)$-automorphisms.

Observation 1. We first present a $G(w, Y)$-automorphism α that is not contained in $\mathsf{Fix}(w)$. Let $Y = v_1 \text{------} v_2$ and set $w = (v_1, v_1, v_2, v_2)$. Then $\alpha = (1, 4)(2, 3)$ is an automorphism of $G(w, Y)$. Furthermore, we have $\alpha \cdot w = (v_2, v_2, v_1, v_1)$ and $\alpha \notin \mathsf{Fix}(w)$. That is,

$$G(w, Y) = \begin{array}{cc} 1 & 2 \\ \big| \times \big| \\ 3 & 4 \end{array} \longrightarrow \begin{array}{cc} 4 & 3 \\ \big| \times \big| \\ 2 & 1 \end{array} = G(\alpha \cdot w, Y) = G(w, Y) \,.$$

Observation 2. Second, we show that $\mathsf{Fix}(w)$ is in general not a normal subgroup of $\mathsf{A}(w)$. For this purpose, let $Y = v_1 \quad v_2 \text{------} v_3 \quad v_4$ and $w = (v_1, v_1, v_2, v_3, v_4, v_4)$. Then we have $\mathsf{Fix}(w) = \langle (1, 2), (5, 6) \rangle$. We set $\alpha = (1, 5)(3, 4)$, that is, $\alpha \cdot w = (v_4, v_1, v_3, v_2, v_1, v_4)$ and observe

$$\alpha \in \mathsf{A}(w); \quad \alpha \mathsf{Fix}(w)\alpha^{-1} = \mathsf{Fix}(\alpha \cdot w) = \langle (6, 1), (5, 2) \rangle \neq \mathsf{Fix}(w) \,.$$

Since we will use $G(w, Y)$-automorphisms to obtain equivalence classes of acyclic $G(w, Y)$-orientations, we first study $\mathsf{Fix}(w)$ in $\mathsf{A}(w)$ and set

$$\mathsf{N}(w) = \{\alpha \in \mathsf{A}(w)) \mid \alpha \mathsf{Fix}(w)\alpha^{-1} = \mathsf{Fix}(w)\} \,. \tag{7.8}$$

Recall that $\langle \gamma \rangle$ is the cyclic group generated by γ and $\langle \gamma \rangle(v_j) = \{\gamma^h v_j \mid h \in \mathbb{N}\}$ denotes the orbit of $\langle \gamma \rangle$ that contains v_j.

Theorem 7.6. *Let $G(w, Y)$ be the dependency graph of the fair word w and Y. Then there exists a group homomorphism*

$$\vartheta \colon \mathsf{N}(w) \longrightarrow \mathsf{Aut}(Y), \quad \vartheta(\alpha)(w_i) = w_{\alpha^{-1}(i)}, \tag{7.9}$$

and we have the short exact sequence

$$1 \longrightarrow \mathsf{Fix}(w) \longrightarrow \mathsf{N}(w) \longrightarrow \mathsf{Aut}(Y) \,, \tag{7.10}$$

or equivalently, $\mathsf{Ker}(\vartheta) = \mathsf{Fix}(w)$. Furthermore, we have

$$\mathsf{Im}(\vartheta) = \{\gamma \in \mathsf{Aut}(Y) \mid \forall\, r \in \mathbb{N}_k; \; \forall\, w_s \in \langle \gamma \rangle(w_r); |\mathsf{Fix}(w)(r)| = |\mathsf{Fix}(w)(s)|\} \,. \tag{7.11}$$

Proof. Let $\alpha \in \mathsf{N}(w)$ and set

$$\vartheta(\alpha)(x_i) = x_{\alpha^{-1}(i)}, \quad x_i \in Y \,.$$

In particular, we have $\vartheta(\alpha)(w_i) = w_{\alpha^{-1}(i)}$. By definition of $G(w, Y)$, we have

$$\{r, s\} \in G(w, Y) \iff \{w_r, w_s\} \in Y,$$

and accordingly obtain

$$\{\alpha(i), \alpha(j)\} \in G(w, Y) \iff \{w_{\alpha^{-1}(i)}, w_{\alpha^{-1}(j)}\} \in Y . \tag{7.12}$$

We conclude from Eq. (7.12) that for any $\alpha \in N(w)$, $\vartheta(\alpha)$ induces mappings such that

$$
\begin{array}{ccc}
i & \xrightarrow{\ \alpha\ } & \alpha(i) \\
\downarrow & & \downarrow \\
w_i & \xrightarrow{\ \vartheta(\alpha)\ } & w_{\alpha^{-1}(i)} \ ,
\end{array}
\qquad
\begin{array}{ccc}
\{i, j\} & \xrightarrow{\ \alpha\ } & \{\alpha(i), \alpha(j)\} \\
\downarrow & & \downarrow \\
\{w_i, w_j\} & \xrightarrow{\ \vartheta(\alpha)\ } & \{w_{\alpha^{-1}(i)}, w_{\alpha^{-1}(j)}\}
\end{array}
$$

are commutative diagrams.

Claim. For any $\alpha \in N(w)$ the mapping

$$\vartheta(\alpha) : Y \longrightarrow Y, \quad \vartheta(\alpha)(w_i) = w_{\alpha^{-1}(i)}$$

is well-defined and an automorphism of Y.

We first show that $\vartheta(\alpha)$ is well-defined. By assumption every Y-vertex w_i is contained in w, and we conclude that $\vartheta(\alpha)$ is defined over Y. By construction, $\vartheta(\alpha)$ maps Y-edges into Y-edges. For arbitrary $\rho \in \mathrm{Fix}(w)$ we have the following situation:

$$
\begin{array}{ccc}
\rho(i) & \xrightarrow{\ \alpha\ } & \alpha(\rho(i)) \\
\downarrow & & \downarrow \\
w_{\rho(i)} & \xrightarrow{\ \vartheta(\alpha)\ } & w_{\alpha^{-1}(\rho(i))}
\end{array}
$$

Since $\alpha \in N(w) = \{\alpha \in A(w) \mid \alpha \mathrm{Fix}(w) \alpha^{-1} = \mathrm{Fix}(w)\}$, we have

$$\forall \, \rho \in \mathrm{Fix}(w), \ \exists \, \rho' \in \mathrm{Fix}(w); \quad \rho' \alpha^{-1} = \alpha^{-1} \rho ,$$

from which we derive $w_{\alpha^{-1}(\rho(i))} = w_{\rho' \alpha^{-1}(i)}$ and $w_{\alpha^{-1}(\rho(j))} = w_{\rho' \alpha^{-1}(j)}$. Furthermore, for $\rho', \rho \in \mathrm{Fix}(w)$ and $r \in \mathbb{N}_k$ we have $w_{\rho(r)} = w_r$ and $w_{\rho'(r)} = w_r$, respectively, that is,

$$w_{\rho(i)} = w_i, \ w_{\rho(j)} = w_j, \ w_{\rho'(\alpha^{-1}(i))} = w_{\alpha^{-1}(i)}, \text{ and } w_{\rho'(\alpha^{-1}(j))} = w_{\alpha^{-1}(j)} .$$

Accordingly, we have shown

$$\forall \, \rho \in \mathrm{Fix}(w), \ \vartheta(\alpha)(w_{\rho(i)}) = \vartheta(\alpha)(w_i), \ \vartheta(\alpha)(\{w_{\rho(i)}, w_{\rho(j)}\}) = \vartheta(\alpha)(\{w_i, w_j\}) ,$$

which proves that $\vartheta(\alpha)$ is well-defined over Y.

Next we show injectivity. Note that $\vartheta(\alpha)(w_r) = \vartheta(\alpha)(w_s)$ is equivalent to $w_{\alpha^{-1}(r)} = w_{\alpha^{-1}(s)}$, that is, there exists some $\rho' \in \mathsf{Fix}(w)$ such that $\rho'\alpha^{-1}(r) = \alpha^{-1}(s)$. Since α is in the normalizer of $\mathsf{Fix}(w)$, $\rho'\alpha^{-1}(r) = \alpha^{-1}(s)$ guarantees that $\alpha^{-1}(\rho(r)) = \alpha^{-1}(s)$, and since α^{-1} is bijective we conclude $\rho(r) = s$. Hence, $\vartheta(\alpha)$ is injective and the claim follows.

Claim. The map $\vartheta \colon \mathsf{N}(w) \longrightarrow \mathsf{Aut}(Y)$ is a group homomorphism.

To prove this we observe $\vartheta(\alpha_2\alpha_1)(w_i) = w_{(\alpha_2\alpha_1)^{-1}(i)} = w_{\alpha_1^{-1}\alpha_2^{-1}(i)}$. We next set $y_i = \vartheta(\alpha_1)(w_i)$ for $i = 1, \ldots, k$ and compute

$$
\begin{aligned}
\vartheta(\alpha_2) \circ \vartheta(\alpha_1)(w_i) &= \vartheta(\alpha_2)(\vartheta(\alpha_1)(w_i)) \\
&= y_{\alpha_2^{-1}(i)} \\
&= \vartheta(\alpha_1)(w_{\alpha_2^{-1}(i)}) \\
&= w_{\alpha_1^{-1}\alpha_2^{-1}(i)} \\
&= \vartheta(\alpha_2\alpha_1)(w_i) \, ,
\end{aligned}
$$

proving the claim.

Next we prove that $\mathsf{Fix}(w) = \mathsf{Ker}(\vartheta)$. For $\rho \in \mathsf{Fix}(w)$ we obtain $\vartheta(\rho)(w_i) = w_{\rho^{-1}(i)} = w_i$, and $\mathsf{Fix}(w) \subset \mathsf{Ker}(\vartheta)$. Now let $\beta \in \mathsf{Ker}(\vartheta)$, i.e., $\vartheta(\beta)(w_i) = w_{\beta^{-1}(i)} = w_i$ for $i \in \mathbb{N}_k$, which is equivalent to $\beta \cdot w = w$, that is, $\beta \in \mathsf{Fix}(w)$.

Claim. $\mathsf{Im}(\vartheta) = \{\gamma \in \mathsf{Aut}(Y) \mid \forall \, r \in \mathbb{N}_k; \ \forall \, w_s \in \langle\gamma\rangle(w_r); |\mathsf{Fix}(w)(r)| = |\mathsf{Fix}(w)(s)|\}$.

To prove the claim we consider $\gamma \in \mathsf{Aut}(Y)$. By assumption, every Y-vertex v_i is contained in w at least once, and we may choose, modulo $\mathsf{Fix}(w)$, some index $a \in \mathbb{N}_k$ such that

$$
w_a = \gamma(w_i) \, .
$$

In order to define $\alpha_\gamma \in \mathsf{N}(w)$, we consider the diagrams below and define α_γ in two steps.

$$
\begin{array}{ccc}
i \xmapsto{\ \alpha_\gamma\ } \alpha_\gamma(i) & \qquad & \rho(i) \xmapsto{\ \alpha_\gamma\ } \alpha_\gamma(\rho(i)) \\
\Big\downarrow \qquad\qquad \Big\downarrow & & \Big\downarrow \qquad\qquad\qquad \Big\downarrow \\
w_i \xmapsto{\ \gamma\ } \gamma(w_i) = w_a & & w_{\rho(i)} \xmapsto{\ \gamma\ } \gamma(w_{\rho(i)}) = \gamma(w_i) = w_a
\end{array}
$$

Step 1. By assumption we can select a subset of indices $V = \{k_1, \ldots, k_n\} \subset \mathbb{N}_k$ such that $\{w_{k_1}, \ldots, w_{k_n}\} = Y$ and define

$$
\forall \, s \in \mathbb{N}_n, \qquad\qquad \alpha_\gamma^{-1}(k_s) = a(k_s) \, .
$$

Step 2. In view of the diagram on the right we compute $\gamma(w_{\rho(k_s)}) = w_{a(k_s)}$, that is, $w_{\alpha_\gamma^{-1}(\rho(k_s))} = w_{\alpha_\gamma^{-1}(k_s)}$. Therefore, we define

$$
\forall \, s \in \mathbb{N}_n, \ \forall \, \rho \in \mathsf{Fix}(w), \qquad \alpha_\gamma^{-1}(\rho(k_s)) = \rho(a(k_s)) \, . \tag{7.13}
$$

Claim. Suppose any two Y-vertices that belong to the same $\langle \gamma \rangle$-orbit have the same multiplicity in w. Then $\alpha_\gamma \in \mathsf{N}(w)$ and $\vartheta(\alpha_\gamma) = \gamma$.

In view of (7.13) we observe that α_γ is bijective if and only if any two Y-vertices that belong to the same $\langle \gamma \rangle$-orbit have the same multiplicity in w, i.e., $|\mathsf{Fix}(w)(k_s)| = |\mathsf{Fix}(w)(a(k_s))|$. We consider the diagram

$$\{\rho(k_r), \rho(k_s)\} \xmapsto{\quad \alpha_\gamma \quad} \{\alpha_\gamma(\rho(k_r)), \alpha_\gamma(\rho(s))\}$$

$$\downarrow \qquad\qquad\qquad\qquad\qquad \downarrow$$

$$\{w_{k_r}, w_{k_s}\} \xmapsto{\;\gamma\;} \{w_{a(k_r)}, w_{a(k_s)}\} = \{w_{\alpha_\gamma^{-1}(\rho(k_r))}, w_{\alpha_\gamma^{-1}(\rho(k_s))}\} \,,$$

from which we conclude that α_γ maps $G(w, Y)$-edges into $G(w, Y)$-edges. We observe

$$\forall\, s \in \mathbb{N}_n, \ \forall\, \rho, \rho_1 \in \mathsf{Fix}(w); \qquad \alpha_\gamma \rho_1 \alpha_\gamma^{-1}(\rho(k_s)) = \rho_1(\rho(k_s)),$$

and finally compute

$$\vartheta(\alpha_\gamma)(w_{\rho(k_s)}) = w_{\alpha_\gamma^{-1}(\rho(k_s))} = w_{a(k_s)} = \gamma(w_{k_s}) = \gamma(w_{\rho(k_s)}) \,,$$

proving the claim, and the proof of the theorem is complete. $\qquad\square$

Corollary 7.7. *Suppose w is a fair word over Y and that for any $w_s \in \mathsf{Aut}(Y)(w_r)$ the elements w_s and w_r have the same multiplicity in w. Then we have the long exact sequence*

$$1 \longrightarrow \mathsf{Fix}(w) \longrightarrow \mathsf{N}(w) \longrightarrow \mathsf{Aut}(Y) \longrightarrow 1 \,. \tag{7.14}$$

Equivalently, $\mathsf{Ker}(\vartheta) = \mathsf{Fix}(w)$ and ϑ is surjective.

Proof. Equation (7.14) follows immediately from Theorem 7.6 since then, by assumption, any two Y-vertices that belong to the same $\mathsf{Aut}(Y)$-orbit have the same multiplicity and we have $\mathsf{Im}(\vartheta) = \mathsf{Aut}(Y)$. $\qquad\square$

7.1.3 Words

We begin this section by endowing the set of words of length k, denoted W_k, with a graph structure. As in the case of permutation-word Section 4.2, the following notion of adjacency is a consequence of the fact that two local maps indexed by non-adjacent Y-vertices commute, that is, $F_{v_i} \circ F_{v_j} = F_{v_j} \circ F_{v_i}$ if either $v_j = v_i$ or $\{v_i, v_j\} \notin Y$.

Let U_k be the graph over words of length k defined as follows: Two different words $w, w' \in W_k$ are adjacent in U_k if and only if there exists some index $1 \le i < k$ such that

$$\forall\, j \ne i, i+1; \ w_j = w'_j, \ w_i = w'_{i+1}, \ w_{i+1} = w'_i \ \wedge \ \{w_i, w_{i+1}\} \notin Y \,. \tag{7.15}$$

That is, two words w and w' are adjacent in U_k if and only if they can be transformed into each other by flipping exactly one pair of consecutive letters $\{w_i, w_{i+1}\}$ such that $\{w_i, w_{i+1}\}$ is not an edge in Y.

7.3. Identify the components of the graph W_3 over $Y = v_1 \;\text{------}\; v_2 \;\text{------}\; v_3$.

[1+]

As a result two words within a given component of U_k induce not only equivalent but identical SDS-maps:

Proposition 7.8. *Let* $(Y, \mathbf{F}_Y, w)$ *and* $(Y, \mathbf{F}_Y, w')$ *be two* SDS. *Then we have*

$$w \sim_Y w' \quad \Longrightarrow \quad [\mathbf{F}_Y, w] = [\mathbf{F}_Y, w'] . \tag{7.16}$$

7.4. Prove Proposition 7.8. **[2]**

Two words $w, w' \in S_k(\varphi)$ are called $\sim_Y$ equivalent if they belong to the same U_k-component. We denote the $\sim_Y$-equivalence class by $[w] = \{w' \mid w' \sim_Y w\}$. We proceed by showing that for any two $\sim_Y$-nonequivalent words $w, w' \in S_k(\varphi)$ there exists some family of Y-local maps $\mathbf{F}_Y$ such that $[\mathbf{F}_Y, w]$ and $[\mathbf{F}_Y, w']$ are different as mappings.

Proposition 7.9. *Let* Y *be a graph with non-empty edge set and let* $w, w' \in S_k(\varphi)$. *Then we have*

$$w \nsim_Y w \quad \Longrightarrow \quad \exists\, (F_{v_i}); \quad [\mathbf{F}_Y, w] \neq [\mathbf{F}_Y, w'] . \tag{7.17}$$

7.5. Prove Proposition 7.9. **[3]**

7.1.4 Acyclic Orientations

Next we present some results on acyclic orientations which we need for the proof of Theorem 7.17 ahead. Any subgroup of $G(w, Y)$-automorphisms, $H < A(w)$, acts on the acyclic orientations of $G(w, Y)$ via

$$h \bullet \mathcal{O}(\{r, s\}) = h(\mathcal{O}(\{h^{-1}(r), h^{-1}(s)\})) . \tag{7.18}$$

In this section we will utilize this action for the particular case of $H = \mathsf{Fix}(w)$ in order to obtain equivalence classes of acyclic orientations of $G(w, Y)$.

Definition 7.10. *Let* $\mathcal{O}$ *and* $\mathcal{O}'$ *be two* $G(w, Y)$-orientations. *We call* $\mathcal{O}$ *and* $\mathcal{O}'$ *equivalent and write* $\mathcal{O} \sim_w \mathcal{O}'$ *if and only if we have*

$$\exists\, \rho \in \mathsf{Fix}(w); \; \rho(\mathcal{O}(\{r, s\})) = \mathcal{O}'(\{\rho(r), \rho(s)\}), \tag{7.19}$$

or equivalently,

$$\exists\, \rho \in \mathsf{Fix}(w); \quad \mathcal{O}' = \rho \bullet \mathcal{O} ,$$

and we have the commutative diagram

$$
\begin{array}{ccc}
r & \xrightarrow{\;\rho\;} & \rho(r) \\
\Big\downarrow{\scriptstyle\mathcal{O}} & & \Big\downarrow{\scriptstyle\mathcal{O}'} \\
s & \xrightarrow[\;\rho\;]{} & \rho(s)
\end{array}\; .
$$

We denote the equivalence class of $\mathcal{O}$ with respect to $\sim_w$ by $[\mathcal{O}]_w$.

7.6. Let $Y = v_1 \relbar\joinrel\relbar v_2 \relbar\joinrel\relbar v_3$ and $w = (v_1, v_2, v_1, v_2, v_3)$. Determine $G(w, Y)$ and the equivalence class $[\mathcal{O}]_w$.   [2]

In Lemma 7.13 we will show how to map words $w \in S_k(\varphi)$, for a fixed representative $\varphi \in \Phi$, into $\sim_\varphi$-equivalence classes of acyclic orientations of $G(\varphi, Y)$. For the construction of this mapping the following class (Section 3.1.5) of acyclic orientations will be central.

Definition 7.11. Let $\sigma \in S_k$ and $w \in W$; then we set

$$\mathcal{O}_Y(\sigma)(\{r, s\}) = \begin{cases} (r, s) & \text{iff} \quad \sigma(r) < \sigma(s), \\ (s, r) & \text{iff} \quad \sigma(r) > \sigma(s). \end{cases}$$

We continue by proving basic properties of $\mathcal{O}_Y$-orientations.

Lemma 7.12. Let $\sigma', \sigma \in S_k$ and $\lambda \in A(w)$ be such that $\sigma'\lambda = \sigma$. Then we have

$$\lambda(\mathcal{O}_Y(\sigma)(\{r, s\})) = \mathcal{O}_Y(\sigma')(\{\lambda(r), \lambda(s)\}) . \tag{7.20}$$

In particular, for $\rho \in \mathsf{Fix}(w)$ and $\mathcal{O}_Y(\sigma'\rho), \mathcal{O}_Y(\sigma') \in \mathsf{Acyc}(G(w, Y))$ we have

$$\mathcal{O}_Y(\sigma'\rho) \sim_w \mathcal{O}_Y(\sigma') ,$$

and furthermore

$$[\mathcal{O}_Y(\sigma')]_w = \{\mathcal{O}_Y(\sigma'\rho) \mid \rho \in \mathsf{Fix}(w)\} . \tag{7.21}$$

Proof. For $\{r, s\} \in G(w, Y)$ we compute

$$\mathcal{O}_Y(\sigma)(\{r, s\}) = (r, s) \quad\Longleftrightarrow\quad \sigma(r) < \sigma(s),$$
$$\mathcal{O}_Y(\sigma')(\{\lambda(r), \lambda(s)\}) = (\lambda(r), \lambda(s)) \quad\Longleftrightarrow\quad \sigma(r) = \sigma'\lambda(r) < \sigma'\lambda(s) = \sigma(s),$$

from which we conclude

$$\lambda(\mathcal{O}_Y(\sigma)(\{r, s\})) = \mathcal{O}_Y(\sigma')(\{\lambda(r), \lambda(s)\}) .$$

By Definition 7.10, $\mathcal{O}_Y(\sigma'\rho) \sim_w \mathcal{O}_Y(\sigma')$ follows from Eq. (7.20) with $\lambda = \rho$ and $\sigma = \sigma'\rho$.

In view of $\mathcal{O}_Y(\sigma'\rho) \sim_w \mathcal{O}_Y(\sigma')$ it suffices in order to prove Eq. (7.21):

$$[\mathcal{O}_Y(\sigma')]_w \subset \{\mathcal{O}_Y(\sigma'\rho) \mid \rho \in \mathsf{Fix}(w)\} .$$

Let $\mathcal{O}_Y \in [\mathcal{O}_Y(\sigma')]_w$. Using Eq. (7.20), we obtain

$$\exists \rho \in \mathsf{Fix}(w); \quad \rho(\mathcal{O}_Y(\{r, s\})) = \mathcal{O}_Y(\sigma')(\{\rho(r), \rho(s)\})$$
$$= \rho(\mathcal{O}_Y(\sigma'\rho)(\{r, s\})) .$$

Since $\rho \in \mathsf{Fix}(w)$ is a $G(w, Y)$-automorphism, we conclude that $\mathcal{O}_Y = \mathcal{O}_Y(\sigma'\rho)$ holds, and the proof of the lemma is complete. $\square$

7.1.5 The Mapping $\mathcal{O}_Y$

The orbits of the S_k-action $\sigma \cdot w = (w_{\sigma^{-1}(1)}, \ldots, w_{\sigma^{-1}(k)})$ induce the partition $W_k = \dot{\bigcup}_{\varphi \in \Phi} S_k(\varphi)$ where Φ is a set of representatives. The set W_k is the disjoint union of its S_k orbits, and any w is contained in exactly one orbit $S_k(\varphi)$ where $w = \sigma \cdot \varphi$, for $\sigma \in S_k$.

Lemma 7.13. *For any $\varphi \in \Phi$ we have the the surjective mapping*

$$\mathcal{O}_Y : S_k(\varphi) \longrightarrow [\mathsf{Acyc}(\varphi)/\sim_\varphi], \quad \sigma \cdot \varphi \mapsto \mathcal{O}_Y(\sigma \cdot \varphi) = [\mathcal{O}_Y(\sigma)]_\varphi . \quad (7.22)$$

Proof. We first show that $\mathcal{O}_Y : S_k(\varphi) \longrightarrow [\mathsf{Acyc}(\varphi)/\sim_\varphi]$ is well-defined. Suppose we have $\sigma \cdot \varphi = \sigma' \cdot \varphi$. We set $\rho = \sigma'^{-1}\sigma$ and have $\rho \cdot \varphi = \varphi$. For $\rho \in \mathsf{Fix}(\varphi)$ we obtain by Lemma 7.12 that $\mathcal{O}_Y(\sigma) \sim_\varphi \mathcal{O}_Y(\sigma')$ and

$$[\mathcal{O}_Y(\sigma)]_\varphi = [\mathcal{O}_Y(\sigma')]_\varphi .$$

We next prove that $\mathcal{O}_Y : S_k(\varphi) \longrightarrow [\mathsf{Acyc}(\varphi)/\sim_\varphi]$ is surjective.
Claim. For any $\mathcal{O} \in \mathsf{Acyc}(\varphi)$ there exists some $\sigma \in S_k$ with the property $\mathcal{O} = \mathcal{O}_Y(\sigma)$.

Since $\mathcal{O}$ is acyclic, there exists some $\sigma \in S_k$ such that

$$\forall\, \{r, s\} \in G(\varphi, Y), \quad \mathcal{O}(\{r, s\}) = (r, s); \qquad \sigma(r) < \sigma(s) , \quad (7.23)$$

which proves $\mathcal{O} = \mathcal{O}_Y(\sigma)$. $\qquad\square$

In the following we investigate under which conditions for $\sigma, \sigma' \in S_k$ $\mathcal{O}_Y(\sigma \cdot \varphi) = \mathcal{O}_Y(\sigma' \cdot \varphi)$, holds. In Section 7.1.7 this will allow us to prove the bijection between equivalence classes of words and $\sim_\varphi$-equivalence classes of acyclic orientations of $G(\varphi, Y)$.

Lemma 7.14. *Suppose $\sigma \cdot \varphi, \sigma' \cdot \varphi \in W_k$. Then we have*

$$\sigma \cdot \varphi \sim_Y \sigma' \cdot \varphi \quad \Longrightarrow \quad \mathcal{O}_Y(\sigma \cdot \varphi) = \mathcal{O}_Y(\sigma' \cdot \varphi) . \quad (7.24)$$

Proof. We set $w = \sigma \cdot \varphi$ and $w' = \sigma' \cdot \varphi$. By induction on the U_k-distance between w and w', we may without loss of generality assume that w and w' are adjacent in U_k, that is, we have the following situation:

$$\tau \cdot w = w', \quad \tau = (i, i+1), \quad \{w_i, w_{i+1}\} \notin Y .$$

Claim. Without loss of generality we may assume $\tau\sigma = \sigma'$.

We have $\sigma'^{-1}\tau\sigma \cdot \varphi = \varphi$, and $\rho = \sigma'^{-1}\tau\sigma \in \mathsf{Fix}(\varphi)$ together with Lemma 7.12 implies for σ' and $\sigma'\rho$: $\mathcal{O}_Y(\sigma') \sim_\varphi \mathcal{O}_Y(\sigma'\rho)$. Thus, we obtain $[\mathcal{O}_Y(\sigma')]_\varphi = [\mathcal{O}_Y(\sigma'\rho)]_\varphi$ and the claim follows.
Claim. Suppose $\tau\sigma = \sigma'$ holds, then we obtain $\mathcal{O}_Y(\sigma) = \mathcal{O}_Y(\sigma')$.

By definition, we have for $\mathcal{O}_Y(\sigma), \mathcal{O}_Y(\sigma') \in \mathsf{Acyc}(\varphi)$

$$\mathcal{O}_Y(\sigma)(\{r, s\}) = (r, s) \quad \Longleftrightarrow \quad \sigma(r) < \sigma(s),$$
$$\mathcal{O}_Y(\tau\sigma)(\{r, s\}) = (r, s) \quad \Longleftrightarrow \quad \tau\sigma(r) < \tau\sigma(s) .$$

Claim. $\{\sigma^{-1}(i), \sigma^{-1}(i+1)\} \notin G(\varphi, Y)$.

We have the following commutative diagram of graph isomorphisms:

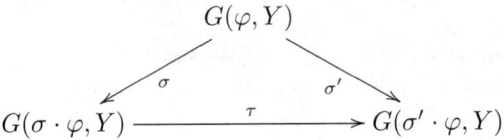

By definition of $G(\varphi, Y)$, we have

$$\{\sigma^{-1}(i), \sigma^{-1}(i+1)\} \in G(\varphi, Y) \quad \Longleftrightarrow \quad \{\varphi_{\sigma^{-1}(i)}, \varphi_{\sigma^{-1}(i+1)}\} \in Y .$$

Since $\sigma \cdot \varphi = w$, we obtain $w_i = \varphi_{\sigma^{-1}(i)}$ and hence $\{\varphi_{\sigma^{-1}(i)}, \varphi_{\sigma^{-1}(i+1)}\} = \{w_i, w_{i+1}\} \notin Y$, and the claim follows.

Obviously, $\sigma^{-1}(i), \sigma^{-1}(i + 1)$ are the only two indices for which $i = \sigma(\sigma^{-1}(i)) < \sigma(\sigma^{-1}(i+1)) = i+1$ and $i+1 = \tau\sigma(\sigma^{-1}(i)) > \tau\sigma(\sigma^{-1}(i+1)) = i$ holds, and

$$\forall \{r, s\} \in G(\varphi, Y) : \quad \{ \sigma(r) < \sigma(s) \iff \tau\sigma(r) < \tau\sigma(s) \} . \tag{7.25}$$

Equation (7.25) is equivalent to

$$\forall \{r, s\} \in G(\varphi, Y); \quad \mathcal{O}_Y(\sigma)(\{r, s\}) = \mathcal{O}_Y(\sigma')(\{r, s\}) ;$$

thus,

$$\mathcal{O}_Y(\sigma \cdot \varphi) = [\mathcal{O}_Y(\sigma)]_\varphi = [\mathcal{O}_Y(\sigma')]_\varphi = \mathcal{O}_Y(\sigma' \cdot \varphi) ,$$

and the lemma follows. $\square$

We give an illustration of Lemma 7.14:

Example 7.15. Let $\varphi = (v_1, v_2, v_3, v_2)$, $\tau \cdot \varphi = (v_1, v_3, v_2, v_2)$ where $\tau = (2, 3)$ and $Y = v_1 \rule{0.5cm}{0.4pt} v_2 \qquad v_3$. Then we have $\mathcal{O}_Y(\tau)(\{1, 2\}) = (1, 2)$ since $\tau(1) = 1 < 3 = \tau(2)$ and $\mathcal{O}_Y(\tau)(\{1, 4\}) = (1, 4)$ since $\tau(1) = 1 < 4 = \tau(4)$.

$$\mathcal{O}_Y(\mathsf{id}) = \begin{array}{ccc} 1 & \longrightarrow & 2 \\ \downarrow & & \\ 4 & & 3 \end{array} \quad \text{and} \quad \mathcal{O}_Y(\tau) = \begin{array}{ccc} 1 & \longrightarrow & 2 \\ \downarrow & & \\ 4 & & 3 \end{array} .$$

$\diamond$

7.1.6 A Normal Form Result

In this section we prove a lemma that will be instrumental in the proof of our main correspondence, which is Theorem 7.17. Its proof is based on a construction related to the Cartier–Foata normal form in partially commutative monoids [68].

Lemma 7.16. *Let* $\sigma, \sigma' \in S_k$, $w \in W_k$, *and* $\mathcal{O}_Y(\sigma), \mathcal{O}_Y(\sigma') \in \mathsf{Acyc}(w)$. *Then we have*

$$\mathcal{O}_Y(\sigma) = \mathcal{O}_Y(\sigma') \quad \Longrightarrow \quad \sigma \cdot w \sim_Y \sigma' \cdot w . \tag{7.26}$$

Proof. By definition, w_j has index $\sigma(j)$ in $\sigma \cdot w$ and index $\sigma'(j)$ in $\sigma' \cdot w$, respectively. We observe that $\mathcal{O}_Y(\sigma) = \mathcal{O}_Y(\sigma')$ is equivalent to

$$\forall \{i,j\} \in G(w,Y), \quad (\, \sigma(i) < \sigma(j) \,) \quad \Longleftrightarrow \quad (\, \sigma'(i) < \sigma'(j) \,) . \tag{7.27}$$

Now let $\sigma(j_1) = 1$. By definition, w_{j_1} has index 1 in $\sigma \cdot w$. According to Eq. (7.27), there is no $\{i, j_1\} \in G(w, Y)$ with $\sigma'(i) < \sigma'(j_1)$, and as a result there exists no w_h in position $s < \sigma'(j_1)$ in $\sigma' \cdot w$ such that $\{w_h, w_{j_1}\} \in Y$. Hence, we can move w_{j_1} in $\sigma' \cdot w$ to first position by successive transpositions of consecutive, non-adjacent letters. Setting $\sigma_1 \cdot w = (w_{j_1}, w_{\sigma'^{-1}(1)}, \dots, w_{\sigma'^{-1}(k)})$, we obtain

$$\exists\, \sigma_1 \in S_k; \ [\sigma_1(j_1) = \sigma(j_1) = 1] \wedge [\sigma' \cdot w \sim_Y \sigma_1 \cdot w] . \tag{7.28}$$

We observe further that $\mathcal{O}_Y(\sigma) = \mathcal{O}_Y(\sigma_1)$ holds, that is,

$$\forall \{i,j\} \in G(w,Y), \quad (\, \sigma(i) < \sigma(j) \,) \quad \Longleftrightarrow \quad (\, \sigma_1(i) < \sigma_1(j) \,) . \tag{7.29}$$

We proceed by induction. By the induction hypothesis, we have for $\sigma_m \cdot w = (w_{j_1}, w_{j_2}, \dots, w_{j_m}, \dots)$,

$$\exists\, \sigma_m \in S_k; \ \forall\, r \in \mathbb{N}_m; \ [\sigma_m(j_r) = \sigma(j_r) = r] \wedge [\sigma' \cdot w \sim_Y \sigma_m \cdot w]$$

and $\mathcal{O}_Y(\sigma) = \mathcal{O}_Y(\sigma_m)$ or, equivalently,

$$\forall \{i,j\} \in G(w,Y), \quad \sigma(i) < \sigma(j) \,) \quad \Longleftrightarrow \quad (\, \sigma_m(i) < \sigma_m(j) \,) . \tag{7.30}$$

Let $\sigma(j_{m+1}) = m + 1$. If there exists some index $\sigma_m(i)$ with the property $\sigma_m(i) < \sigma_m(j_{m+1})$ and $\{i, j_{m+1}\} \in G(w, Y)$, we obtain from Eq. (7.30): $\sigma(i) < \sigma(j_{m+1}) = m + 1$, i.e., $\in \{j_1, \dots, j_m\}$. In view of $\sigma_m(j_r) = \sigma(j_r) = r$ for $1 \le r \le m$, we derive $1 \le \sigma_m(i) \le m$. Hence, we can move $w_{j_{m+1}}$ in $\sigma_m \cdot w$ to position $m + 1$ by successive transpositions of consecutive, non-adjacent letters. Accordingly, we have for $\sigma_{m+1} \cdot w = (w_{j_1}, \dots, w_{j_{m+1}}, \dots)$

$$\exists\, \sigma_{m+1} \in S_k; \ \forall\, r \in \mathbb{N}_{m+1}; [\sigma_{m+1}(j_r) = \sigma(j_r) = r] \wedge [\sigma' \cdot w \sim_Y \sigma_{m+1} \cdot w]$$

and

$$\mathcal{O}_Y(\sigma) = \mathcal{O}_Y(\sigma_{m+1}) ,$$

and the lemma follows. □

7.1.7 The Bijection

Now we are prepared to combine our results in order to prove

Theorem 7.17. *Let $k \in \mathbb{N}$ and Φ be a set of representatives of the S_k-action on W_k. Then for each $w \in W_k$ there exist some $\sigma_w \in S_k$ and $\varphi_w \in \Phi$ such that $w = \sigma_w \cdot \varphi_w$ and we have the bijection*

$$\mathcal{O}'_Y : W_k / \sim_Y \longrightarrow \dot{\bigcup}_{\varphi \in \Phi} [\mathsf{Acyc}(\varphi) / \sim_\varphi], \quad \text{where} \quad \mathcal{O}'_Y([w]_\varphi) = \mathcal{O}_Y(\sigma_w \cdot \varphi_w) .$$
$$(7.31)$$

Proof. According to Lemma 7.13, we have the well-defined and surjective mapping

$$\mathcal{O}_Y : S_k(\varphi) \longrightarrow [\mathsf{Acyc}(\varphi) / \sim_\varphi], \quad \sigma \cdot \varphi \mapsto \mathcal{O}_Y(\sigma \cdot \varphi) = [\mathcal{O}_Y(\sigma)]_\varphi,$$

and according to Lemma 7.14, we have for $\sigma \cdot \varphi, \sigma' \cdot \varphi \in W_k$

$$\sigma \cdot \varphi \sim_Y \sigma' \cdot \varphi, \quad \Longrightarrow \quad \mathcal{O}_Y(\sigma \cdot \varphi) = \mathcal{O}_Y(\sigma' \cdot \varphi) .$$

Since $W_k = \dot{\bigcup}_{\varphi \in \Phi} S_k(\varphi)$, we have the mapping over $\sim_Y$-equivalence classes

$$\mathcal{O}'_Y : W_k / \sim_Y \longrightarrow \dot{\bigcup}_{\varphi \in \Phi} [\mathsf{Acyc}(\varphi) / \sim_\varphi], \quad \mathcal{O}'_Y([w]_\varphi) = \mathcal{O}_Y(\sigma_w \cdot \varphi_w) .$$

According to Lemmas 7.14 and 7.13, for any fixed representative φ the mapping

$$\mathcal{O}'_Y|_{S_k(\varphi)} : S_k(\varphi) / \sim_Y \longrightarrow [\mathsf{Acyc}(\varphi) / \sim_\varphi], \quad [\sigma \cdot \varphi] \mapsto \mathcal{O}_Y(\sigma \cdot \varphi)$$

is surjective. In view of $\mathsf{U}_k = \dot{\bigcup}_{\varphi \in \Phi} S_k(\varphi)$, we conclude that $\mathcal{O}'_Y$ is surjective. It remains to prove that $\mathcal{O}'_Y$ is injective.
Claim. Let $w = \sigma \cdot \varphi$ and $w' = \sigma' \cdot \varphi$. Then we have

$$\sigma \cdot \varphi \not\sim_Y \sigma' \cdot \varphi \quad \Longrightarrow \quad \mathcal{O}_Y(\sigma \cdot \varphi) \neq \mathcal{O}_Y(\sigma' \cdot \varphi) . \quad (7.32)$$

Let $\mathcal{O}_Y(\sigma), \mathcal{O}_Y(\sigma') \in \mathsf{Acyc}(\varphi)$ be representatives for $\mathcal{O}_Y(\sigma \cdot \varphi)$ and $\mathcal{O}_Y(\sigma' \cdot \varphi)$; respectively. By Proposition 7.4 we have the following commutative diagram:

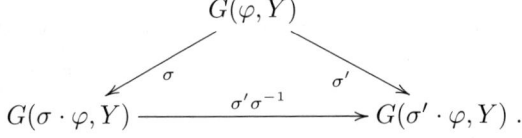

Suppose now $[\mathcal{O}_Y(\sigma)]_\varphi = [\mathcal{O}_Y(\sigma')]_\varphi$, that is, $\mathcal{O}_Y(\sigma) \sim_\varphi \mathcal{O}_Y(\sigma')$. Then there exists some $\rho \in \mathsf{Fix}(\varphi)$ such that $\rho(\mathcal{O}_Y(\sigma)(\{r, s\})) = \mathcal{O}_Y(\sigma')(\{\rho(r), \rho(s)\})$. According to Lemma 7.12, we obtain

$$\mathcal{O}_Y(\sigma')(\{\rho(r), \rho(s)\}) = \rho\left(\mathcal{O}_Y(\sigma'\rho)(\{r, s\})\right),$$

and since ρ is an $G(\varphi, Y)$-automorphism, $\mathcal{O}_Y(\sigma) = \mathcal{O}_Y(\sigma'\rho)$. In view of $\rho \cdot \varphi = \varphi$, Lemma 7.16 implies

$$\sigma \cdot \varphi \sim_Y (\sigma'\rho) \cdot \varphi = \sigma' \cdot \varphi,$$

which is a contradiction. Thus, we have proved $[\mathcal{O}_Y(\sigma)]_\varphi \neq [\mathcal{O}_Y(\sigma')]_\varphi$, which is exactly Eq. (7.32), and the proof of the theorem is complete. $\qquad\square$

We proceed by revisiting the bijection

$$\mathcal{O}_Y : S_k/\sim_Y \longrightarrow \mathsf{Acyc}(Y)$$

of Eq. (3.15) from Chapter 3. In the context of Theorem 7.17 the result becomes a corollary:

Corollary 7.18. *Let w be a permutation and identify $S_k(\mathsf{id})$ with S_k. We have the bijection*

$$\mathcal{O}_Y : S_k/\sim_Y \longrightarrow \mathsf{Acyc}(Y). \tag{7.33}$$

7.7. Prove Corollary 7.18. [2]

7.2 Combinatorics of SDS over Words II

7.2.1 Generalized Equivalences

We call two $G(w, Y)$-orientations $\mathcal{O}$ and $\mathcal{O}'$ G-equivalent and write $\mathcal{O} \sim_G \mathcal{O}'$ if and only if there exists some $g \in G$ such that $\mathcal{O} = g \bullet \mathcal{O}'$ holds. The G-equivalence class of $\mathcal{O}$ with respect to $\sim_G$ is denoted by $[\mathcal{O}]_G$. As in Section 7.1 we have

$$\forall \sigma \in S_k, \qquad \mathcal{O}_Y(\sigma)(\{r, s\}) = \begin{cases} (r, s) & \text{iff} \quad \sigma(r) < \sigma(s), \\ (s, r) & \text{iff} \quad \sigma(r) > \sigma(s), \end{cases} \tag{7.34}$$

and for $\sigma', \sigma \in S_k$, $\lambda \in \mathsf{A}(w)$ such that $\sigma'\lambda = \sigma$ Lemma (7.12) guarantees

$$\lambda(\mathcal{O}_Y(\sigma)(\{r, s\})) = \mathcal{O}_Y(\sigma')(\{\lambda(r), \lambda(s)\}). \tag{7.35}$$

Lemma 7.19. *Suppose $\sigma', \sigma, \lambda \in S_k$ and $\lambda \in \mathsf{A}(w)$ such that $\sigma'\lambda = \sigma$. Then for $g \in \mathsf{N}(w)$ and $\mathcal{O}_Y(\sigma'g), \mathcal{O}_Y(\sigma') \in \mathsf{Acyc}(w)$ and*

$$\mathcal{O}_Y(\sigma'g) \sim_{\mathsf{N}(w)} \mathcal{O}_Y(\sigma')$$

holds. Furthermore, we have

$$[\mathcal{O}_Y(\sigma')]_{\mathsf{N}(w)} = \{\mathcal{O}_Y(\sigma'g) \mid g \in \mathsf{N}(w)\}. \tag{7.36}$$

Proof. By definition $\mathcal{O}_Y(\sigma'g) \sim_{N(w)} \mathcal{O}_Y(\sigma')$ follows directly from (7.35) upon setting $\lambda = g$ and $\sigma = \sigma'g$. In view of $\mathcal{O}_Y(\sigma'g) \sim_{N(w)} \mathcal{O}_Y(\sigma')$, it suffices in order to prove Eq. (7.36):

$$[\mathcal{O}_Y(\sigma')]_{N(w)} \subset \{\mathcal{O}_Y(\sigma'g) \mid g \in N(w)\} .$$

Let $\mathcal{O} \in [\mathcal{O}_Y(\sigma')]_{N(w)}$. Using Eq. (7.35) we obtain

$$\exists \, g \in N(w); \quad g(\mathcal{O}(\{r, s\})) = \mathcal{O}_Y(\sigma')(\{g(r), g(s)\}) = g(\mathcal{O}_Y(\sigma'g)(\{r, s\})) .$$

Since $g \in N(w)$ is a $G(w, Y)$-automorphism, we conclude that $\mathcal{O} = \mathcal{O}_Y(\sigma'g)$ and the proof of the lemma is complete. $\qquad\square$

Let U_k be the graph over W_k [Eq. (7.15)]. We set Φ' to be the set of words of length k in which each Y-vertex occurs at least once (Φ' is needed to satisfy the conditions of Theorem 7.6) since only words contained in Φ' yield Y-automorphisms via Theorem 7.6. It is clear that Φ' equipped with this notion of adjacency forms a subgraph of U_k since flips of consecutive coordinates preserve Φ'.

We now introduce the equivalence relation $\sim_{N(\varphi)}$ by

$$\sigma \cdot \varphi \sim_{N(\varphi)} \sigma \cdot \varphi \quad \Longleftrightarrow \quad (\exists \, g, g \in N(\varphi); \sigma g \cdot \varphi \sim_Y \sigma'g' \cdot \varphi) , \qquad (7.37)$$

and refer to $[w] = \{w' \mid w' \sim_Y w\}$ and $[w]_{N(\varphi)} = \{w' \mid w' \sim_{N(\varphi)} w\}$ as the equivalence classes of w with respect to $\sim_Y$ and $\sim_{N(\varphi)}$, respectively.

Remark 7.20. In this notation the equivalence relation $\sim_Y$ equals $\sim_{Fix(w)}$. Indeed, we observe

$$\sigma \cdot \varphi \sim_{Fix(\varphi)} \sigma' \cdot \varphi \quad \Longleftrightarrow \quad \exists \, \rho, \rho' \in Fix(\varphi); \sigma\rho \cdot \varphi \sim_Y \sigma\rho' \cdot \varphi ,$$

where $\sigma\rho \cdot \varphi = \sigma \cdot \varphi$ and $\sigma\rho' \cdot \varphi = \sigma \cdot \varphi$. In particular we have $S_k(\varphi)/ \sim_{Fix(\varphi)} = S_k(\varphi)/ \sim_Y$. Replacing $N(w)$ by $Fix(w)$ in Eq. (7.37), we obtain $[w] = [w]_{Fix(w)}$.

The following result shows how the equivalence relation $\sim_{N(\varphi)}$ relates to Y-automorphisms. As mentioned earlier, a result of the action of Y-automorphisms is that $\sim_{N(w)}$ and $\sim_{Fix(w)}$ can differ significantly:

Let K_n be the complete graph, over n vertices, and permutation-words. Clearly, $Fix(w) = 1$ and $N(w) \cong S_n$ and there is exactly one $\sim_{N(w)}$-equivalence class of words in contrast to $\sim_Y$, where [using Eq. (7.1)] each equivalence class contains exactly one element. In case of $K_2 \cong v_1 \text{——} v_2$, for instance, we have exactly the two permutation-words (v_1, v_2) and (v_2, v_1). Since $\{v_1, v_2\}$ is a K_2-edge, we have $(v_1, v_2) \not\sim_Y (v_2, v_1)$ but $[(v_1, v_2)]_{N((v_1, v_2))} = \{(v_1, v_2), (v_2, v_1)\}$ since the map $g\colon K_2 \longrightarrow K_2$, where $g(v_1) = v_2$ and $g(v_2) = v_1$ is a K_2-automorphism and $g \circ (v_1, v_2) = (gv_1, gv_2) = (v_2, v_1)$ holds.

Lemma 7.21. *Let $\varphi \in \Phi'$ and $w, w' \in S_k(\varphi)$. Then we have*

$$w \sim_{\mathsf{N}(\varphi)} w' \quad \Longleftrightarrow \quad \exists\, g, g' \in \mathsf{N}(\varphi);\ \vartheta(g) \circ w \sim_Y \vartheta(g') \circ w'. \tag{7.38}$$

Furthermore, $\sim_{\mathsf{N}(\varphi)}$ is independent of the choice of representative in the orbit $S_k(\varphi)$:

$$\forall\, w, w' \in S_k(\varphi),\ \lambda \in S_k;\quad w \sim_{\mathsf{N}(\varphi)} w' \quad \Longleftrightarrow \quad w \sim_{\mathsf{N}(\lambda \cdot \varphi)} w'. \tag{7.39}$$

Proof. By definition, $w = \sigma \cdot \varphi \sim_{\mathsf{N}(\varphi)} \sigma' \cdot \varphi = w'$ is equivalent to $\sigma g \cdot \varphi \sim_Y \sigma' g' \cdot \varphi$ for some $g, g' \in \mathsf{N}(\varphi)$. Using Theorem 7.6 we obtain

$$\sigma g \cdot \varphi = \sigma \cdot (\vartheta(g) \circ \varphi) \quad \text{and} \quad \sigma' g' \cdot \varphi = \sigma' \cdot (\vartheta(g') \circ \varphi)$$

and derive, using the compatibility of the two group actions,

$$\sigma \cdot \vartheta(g) \circ \varphi = \vartheta(g) \circ \sigma \cdot \varphi \quad \text{and} \quad \sigma' \cdot \vartheta(g') \circ \varphi = \vartheta(g') \circ \sigma' \cdot \varphi.$$

Hence, we have

$$w \sim_{\mathsf{N}(\varphi)} w' \quad \Longleftrightarrow \quad \vartheta(g) \circ w \sim_Y \vartheta(g') \circ w'.$$

We next show that

$$\forall\, w, w' \in S_k(\varphi),\ \lambda \in S_k;\quad w \sim_{\mathsf{N}(\varphi)} w' \quad \Longleftrightarrow \quad w \sim_{\mathsf{N}(\lambda \cdot \varphi)} w'. \tag{7.40}$$

Indeed, with $w = \sigma \cdot \varphi$ and $w' = \sigma' \cdot \varphi$ we have by definition of $\sim_{\mathsf{N}(\varphi)}$

$$\sigma \cdot \varphi \sim_{\mathsf{N}(\varphi)} \sigma' \cdot \varphi \quad \Longleftrightarrow \quad \exists\, g, g' \in \mathsf{N}(\varphi);\ \sigma g \cdot \varphi \sim_Y \sigma' g' \cdot \varphi.$$

Since $\mathsf{N}(\lambda \cdot \varphi) = \lambda \mathsf{N}(\varphi) \lambda^{-1}$, we observe that $\sigma \cdot \varphi \sim_{\lambda \mathsf{N}(\varphi) \lambda^{-1}} \sigma' \varphi$ is equivalent to

$$\exists\, g, g' \in \mathsf{N}(\varphi);\ (\sigma \lambda^{-1})(\lambda g \lambda^{-1}) \cdot (\lambda \cdot \varphi) \sim_Y (\sigma' \lambda^{-1})(\lambda g' \lambda^{-1}) \cdot (\lambda \cdot \varphi). \tag{7.41}$$

Equation (7.41) is immediately identified as $\sigma g \cdot \varphi \sim_Y \sigma' g' \cdot \varphi$, and Eq. (7.39) follows, completing the proof of the lemma. □

7.2.2 The Bijection (P1)

In this section we prove a bijection between $\mathsf{N}(\varphi)$-equivalence classes of words and $\mathsf{N}(\varphi)$-orbits of acyclic orientations.

Theorem 7.22. *Let $k \in \mathbb{N}$, $\varphi \in \Phi'$, the set of all words that contain each Y-vertex at least once and $\mathsf{N}(\varphi)$ the normalizer of $\mathsf{Fix}(\varphi)$ in $\mathsf{A}(\varphi)$. Then we have the bijection*

$$\mathcal{O}_Y^{\mathsf{N}(\varphi)} : S_k(\varphi) / \sim_{\mathsf{N}(\varphi)} \longrightarrow \left[\mathsf{Acyc}(\varphi) / \sim_{\mathsf{N}(\varphi)} \right],$$

$$\text{where} \quad \mathcal{O}_Y^{\mathsf{N}(\varphi)}([\sigma \cdot \varphi]_{\mathsf{N}(\varphi)}) = [\mathcal{O}_Y(\sigma)]_{\mathsf{N}(\varphi)}.$$

Proof. We begin by showing that there exists the surjective mapping

$$\tilde{\mathcal{O}}_Y^{\mathsf{N}(\varphi)} : S_k(\varphi) \longrightarrow \left[\mathsf{Acyc}(\varphi)/\sim_{\mathsf{N}(\varphi)}\right], \quad \tilde{\mathcal{O}}_Y^{\mathsf{N}(\varphi)}(\sigma \cdot \varphi) = [\mathcal{O}_Y(\sigma)]_{\mathsf{N}(\varphi)}. \quad (7.42)$$

We first prove that $\tilde{\mathcal{O}}_Y^{\mathsf{N}(\varphi)}$ is well-defined. Suppose we have $\sigma \cdot \varphi = \sigma' \cdot \varphi$. We set $\rho = \sigma'^{-1}\sigma$ and have $\rho \cdot \varphi = \varphi$. Hence, we have $\rho \in \mathsf{Fix}(\varphi) \subset \mathsf{N}(\varphi)$ and obtain from Lemma 7.19:

$$\mathcal{O}_Y(\sigma) \sim_{\mathsf{N}(\varphi)} \mathcal{O}_Y(\sigma'), \quad \text{i.e.,} \quad [\mathcal{O}_Y(\sigma)]_{\mathsf{N}(\varphi)} = [\mathcal{O}_Y(\sigma')]_{\mathsf{N}(\varphi)}.$$

Lemma 7.21 shows that $\sim_{\mathsf{N}(\varphi)}$ is independent of the choice of representative of $\varphi \in S_k(\varphi)$; hence, $\tilde{\mathcal{O}}_Y^{\mathsf{N}(\varphi)}$ is well-defined. Next we show that $\tilde{\mathcal{O}}_Y^{\mathsf{N}(\varphi)}$ is surjective. For this purpose we observe that for any $\mathcal{O} \in \mathsf{Acyc}(\varphi)$ there exists some $\sigma \in S_k$ with the property $\mathcal{O} = \mathcal{O}_Y(\sigma)$. Clearly, since $\mathcal{O}$ is acyclic there exists some $\sigma \in S_k$ such that

$$\forall \{r, s\} \in G(\varphi, Y), \quad \mathcal{O}(\{r, s\}) = (r, s); \quad \sigma(r) < \sigma(s), \quad (7.43)$$

which proves $\mathcal{O} = \mathcal{O}_Y(\sigma)$, and $[\mathcal{O}]_{\mathsf{N}(\varphi)} = [\mathcal{O}_Y(\sigma)]_{\mathsf{N}(\varphi)}$ follows. We proceed by establishing independence of the choice of representatives within $[\sigma \cdot \varphi]_{\mathsf{N}(\varphi)}$:

$$\sigma \cdot \varphi \sim_{\mathsf{N}(\varphi)} \sigma' \cdot \varphi \implies \tilde{\mathcal{O}}_Y^{\mathsf{N}(\varphi)}(\sigma \cdot \varphi) = \tilde{\mathcal{O}}_Y^{\mathsf{N}(\varphi)}(\sigma' \cdot \varphi). \quad (7.44)$$

By definition of the equivalence relation $\sim_{\mathsf{N}(\varphi)}$ [Eq. (7.37)], we have

$$\sigma \cdot \varphi \sim_{\mathsf{N}(\varphi)} \sigma' \cdot \varphi \iff \exists\, g, g' \in \mathsf{N}(\varphi); \quad \sigma g \cdot \varphi \sim_Y \sigma' g' \cdot \varphi.$$

According to Lemma 7.14 (using induction on the U_k-distance of words), we have

$$\sigma \cdot \varphi \sim_Y \sigma' \cdot \varphi \implies [\mathcal{O}_Y(\sigma)]_{\mathsf{Fix}(\varphi)} = [\mathcal{O}_Y(\sigma')]_{\mathsf{Fix}(\varphi)}, \quad (7.45)$$

and using (7.45) we observe that $\sigma \cdot \varphi \sim_{\mathsf{N}(\varphi)} \sigma'\varphi$ implies

$$\exists\, g, g' \in \mathsf{N}(\varphi); \quad \tilde{\mathcal{O}}_Y^{\mathsf{N}(\varphi)}(\sigma g \cdot \varphi) = [\mathcal{O}_Y(\sigma g)]_{\mathsf{N}(\varphi)}$$
$$= [\mathcal{O}_Y(\sigma' g')]_{\mathsf{N}(\varphi)}$$
$$= \tilde{\mathcal{O}}_Y^{\mathsf{N}(\varphi)}(\sigma' g' \cdot \varphi).$$

Lemma 7.19 guarantees $\mathcal{O}_Y(\sigma g) \sim_{\mathsf{N}(\varphi)} \mathcal{O}_Y(\sigma)$ and $\mathcal{O}_Y(\sigma' g') \sim_{\mathsf{N}(\varphi)} \mathcal{O}_Y(\sigma')$, and we obtain

$$\tilde{\mathcal{O}}_Y^{\mathsf{N}(\varphi)}(\sigma \cdot \varphi) = [\mathcal{O}_Y(\sigma)]_{\mathsf{N}(\varphi)}$$
$$= [\mathcal{O}_Y(\sigma g)]_{\mathsf{N}(\varphi)}$$
$$= [\mathcal{O}_Y(\sigma' g')]_{\mathsf{N}(\varphi)}$$
$$= [\mathcal{O}_Y(\sigma')]_{\mathsf{N}(\varphi)}$$
$$= \tilde{\mathcal{O}}_Y^{\mathsf{N}(\varphi)}(\sigma' \cdot \varphi),$$

and Eq. (7.44) is proved. Therefore, we have for any $\varphi \in \Phi'$ the surjective mapping

$$\mathcal{O}_Y^{\mathsf{N}(\varphi)} : S_k(\varphi)/\sim_{\mathsf{N}(\varphi)} \longrightarrow \left[\mathsf{Acyc}(\varphi)/\sim_{\mathsf{N}(\varphi)}\right], \quad \mathcal{O}_Y^{\mathsf{N}(\varphi)}([\sigma \cdot \varphi]_{\mathsf{N}(\varphi)}) = [\mathcal{O}_Y(\sigma)]_{\mathsf{N}(\varphi)}.$$

It remains to prove injectivity

Let $\mathcal{O}_Y(\sigma g), \mathcal{O}_Y(\sigma' g') \in \mathsf{Acyc}(\varphi)$. Then, according to Lemma 7.16, we have

$$\mathcal{O}_Y(\sigma g) = \mathcal{O}_Y(\sigma' g') \quad \Longrightarrow \quad \sigma g \cdot \varphi \sim \sigma' g' \cdot \varphi,$$

which is equivalent to

$$\mathcal{O}_Y(\sigma g) = \mathcal{O}_Y(\sigma' g') \quad \Longrightarrow \quad \sigma \cdot \varphi \sim_{\mathsf{N}(\varphi)} \sigma' \cdot \varphi. \tag{7.46}$$

Suppose we have $w = \sigma \cdot \varphi$ and $w' = \sigma' \cdot \varphi$. Then the following implication holds:

$$\sigma \cdot \varphi \not\sim_{\mathsf{N}(\varphi)} \sigma' \cdot \varphi \quad \Longrightarrow \quad \mathcal{O}_Y^{\mathsf{N}(\varphi)}([\sigma \cdot \varphi]_{\mathsf{N}(\varphi)}) \neq \mathcal{O}_Y^{\mathsf{N}(\varphi)}([\sigma' \cdot \varphi]_{\mathsf{N}(\varphi)}). \tag{7.47}$$

Let $\mathcal{O}_Y(\sigma), \mathcal{O}_Y(\sigma') \in \mathsf{Acyc}(\varphi)$ be representatives for $\mathcal{O}_Y^{\mathsf{N}(\varphi)}([\sigma \cdot \varphi]_{\mathsf{N}(\varphi)})$ and $\mathcal{O}_Y^{\mathsf{N}(\varphi)}([\sigma' \cdot \varphi]_{\mathsf{N}(\varphi)})$, respectively. We will prove Eq. (7.32) by contradiction using (7.46). Suppose we have

$$\mathcal{O}_Y^{\mathsf{N}(\varphi)}([\sigma \cdot \varphi]_{\mathsf{N}(\varphi)}) = \mathcal{O}_Y^{\mathsf{N}(\varphi)}([\sigma' \cdot \varphi]_{\mathsf{N}(\varphi)}), \quad \text{i.e.,} \quad \mathcal{O}_Y(\sigma) \sim_{\mathsf{N}(\varphi)} \mathcal{O}_Y(\sigma').$$

Then there exists some $g \in \mathsf{N}(\varphi)$ such that

$$g\left(\mathcal{O}_Y(\sigma)(\{r, s\})\right) = \mathcal{O}_Y(\sigma')(\{g(r), g(s)\}).$$

According to Lemma 7.19, we have

$$\mathcal{O}_Y(\sigma')(\{g(r), g(s)\}) = g\left(\mathcal{O}_Y(\sigma' g)(\{r, s\})\right),$$

and since g is a $G(\varphi, Y)$-automorphism, $\mathcal{O}_Y(\sigma) = \mathcal{O}_Y(\sigma' g)$ follows. Equation (7.46) guarantees

$$\mathcal{O}_Y(\sigma) = \mathcal{O}_Y(\sigma' g) \quad \Longrightarrow \quad \sigma \cdot \varphi \sim_{\mathsf{N}(\varphi)} \sigma' \cdot \varphi,$$

which is a contradiction. Thus, we have proved that $[\sigma \cdot \varphi]_{\mathsf{N}(\varphi)} \neq [\sigma' \cdot \varphi]_{\mathsf{N}(\varphi)}$ implies $[\mathcal{O}_Y(\sigma)]_{\mathsf{N}(\varphi)} \neq [\mathcal{O}_Y(\sigma')]_{\mathsf{N}(\varphi)}$, and the proof of the theorem is complete. □

Corollary 7.23. *Let $k \in \mathbb{N}$ and Φ be a set of representatives of the S_k-action on W_k. Then we have the bijection*

$$\mathcal{O}'_Y : W_k/\sim_Y \longrightarrow \dot{\bigcup}_{\varphi \in \Phi} \left[\mathsf{Acyc}(\varphi)/\sim_{\mathsf{Fix}(\varphi)}\right],$$

$$\text{where} \quad \mathcal{O}'_Y([w]_{\mathsf{Fix}(\varphi)}) = \mathcal{O}_Y^{\mathsf{Fix}(\varphi)}([\sigma \cdot \varphi]_{\mathsf{Fix}(\varphi)}).$$

Proof. We first observe that in the case of $\mathsf{Fix}(w)$ the condition that φ contains each Y-vertex at least once becomes obsolete. In complete analogy with Theorem 7.22, we derive for fixed $\varphi \in \Phi$ the bijection

$$\mathcal{O}_Y^{\mathsf{Fix}(\varphi)} : S_k(\varphi)/ \sim_{\mathsf{Fix}(\varphi)} \longrightarrow \left[\mathsf{Acyc}(\varphi)/ \sim_{\mathsf{Fix}(\varphi)} \right] .$$

Since $W_k = \dot{\bigcup}_{\varphi \in \Phi} S_k(\varphi)$, each $w \in W_k$ is contained in exactly one orbit $S_k(\varphi)$, and $\mathcal{O}_Y'$ is well-defined. Since the equivalence relation $\sim_{\mathsf{Fix}(w)}$ equals $\sim_Y$, Corollary 7.23 follows from Theorem 7.22.   □

7.2.3 Equivalence (P2)

In this section we address (P2), that is, we prove that $w \sim_{\mathsf{N}(w)} w'$ implies the equivalence equivalence of the SDS-maps $[\mathbf{F}_Y, w] \sim [\mathbf{F}_Y, w']$. We recall (Definition 4.28, Chapter 4) that two SDS-maps $[\mathbf{F}_Y, w]$ and $[\mathbf{F}_Y, w']$ are equivalent if and only if there exists a bijection β such that

$$[\mathbf{F}_Y, w'] = \beta \circ [\mathbf{F}_Y, w] \circ \beta^{-1} .$$

Hence, (P2) is equivalent to the statement that, up to equivalence of dynamical systems, an SDS-map depends only on the combinatorial equivalence class $\sim_{\mathsf{N}(\varphi)}$ of its underlying word $[w]_{\mathsf{Fix}(w)}$.

Theorem 7.24. *Let* $(Y, \mathbf{F}_Y, w)$ *be an* SDS *with the properties that the vertex functions* $f_v \colon K^{d(v)+1} \longrightarrow K$ *are symmetric, and that for any* $\gamma \in \mathsf{Aut}(Y)$, $v_j \in \langle \gamma \rangle(v_i)$ *we have* $f_{v_j} = f_{v_i}$. *Furthermore, let* $\varphi \in \Phi'$, $\mathsf{N}(\varphi)$ *be the normalizer of* $\mathsf{Fix}(\varphi)$ *in* $\mathsf{A}(w)$, *and* $w, w' \in S_k(\varphi)$. *Then we have*

$$w \sim_{\mathsf{N}(\varphi)} w' \quad \Longrightarrow \quad [\mathbf{F}_Y, w] \sim [\mathbf{F}_Y, w'] . \tag{7.48}$$

Proof. According to Eq. (4.5) of Chapter 4, a Y-local map is a mapping $F_{v_i} \colon K^n \longrightarrow K^n$,

$$F_{v_i}(x) = (x_1, \ldots, x_{v_{i-1}}, f_{v_i}(x[v_i]), x_{v_{i+1}}, \ldots, x_{v_n}),$$

where $x[v] = (x_{n[v](1)}, \ldots, x_{n[v](d(v)+1)})$ and $n[v] \colon \{1, 2, \ldots, d(v)+1\} \longrightarrow \mathrm{v}[Y]$ (Section 4.1). Since $v_j \in \langle \gamma \rangle(v_i)$ holds for any $\gamma \in \mathsf{Aut}(Y)$, we derive

$$\forall \, \gamma \in \mathsf{Aut}(Y), \; \forall v_j \in \langle \gamma \rangle(v_i); \quad F_{v_i} = F_{v_j} , \tag{7.49}$$

where $\langle \gamma \rangle(v_i)$ denotes the orbit of the cyclic group $\langle \gamma \rangle$ containing v_i. Lemma 7.21 guarantees

$$w \sim_{\mathsf{N}(\varphi)} w' \quad \Longleftrightarrow \quad \exists \, g, g \in \mathsf{N}(\varphi); \; \vartheta(g) \circ w \sim_Y \vartheta(g') \circ w' ,$$

where $\vartheta \colon \mathsf{N}(w) \longrightarrow \mathsf{Aut}(Y)$ is given by $\vartheta(\alpha)(w_i) = w_{\alpha^{-1}(i)}$ (Theorem 7.6). For two non-adjacent Y-vertices w_i and w_{i+1} we observe that

$$F_{w_i} \circ F_{w_{i+1}} = F_{w_{i+1}} \circ F_{w_i} \tag{7.50}$$

since the Y-local functions F_{w_i} and $F_{w_{i+1}}$ depend only on the states of their nearest neighbors. By induction on the U_k-distance between $\vartheta(g) \circ w$ and $\vartheta(g') \circ w'$, we conclude from Eq. (7.50) that

$$\vartheta(g) \circ w \sim_Y \vartheta(g') \circ w' \implies [\mathbf{F}_Y, \vartheta(g) \circ w] = [\mathbf{F}_Y, \vartheta(g') \circ w'] . \tag{7.51}$$

We proceed by showing

$$[\mathbf{F}_Y, w] \sim [\mathbf{F}_Y, \vartheta(g) \circ w] \quad \text{and} \quad [\mathbf{F}_Y, w'] \sim [\mathbf{F}_Y, \vartheta(g') \circ w'] .$$

Let x_{v_i} be the state of the vertex v_i of Y. The group $\mathsf{Aut}(Y)$ acts naturally on $(x_{v_1}, \ldots, x_{v_n})$ via

$$\gamma \cdot (x_{v_1}, \ldots, x_{v_n}) = (x_{\gamma^{-1}(v_1)}, \ldots, x_{\gamma^{-1}(v_n)}) . \tag{7.52}$$

Claim.

$$\vartheta(g) \circ [\mathbf{F}_Y, w] \circ \vartheta(g)^{-1} = [\mathbf{F}_Y, \vartheta(g) \circ w], \quad \text{i.e.,} \quad [\mathbf{F}_Y, w] \sim [\mathbf{F}_Y, \vartheta(g) \circ w]. \tag{7.53}$$

We set $\gamma = \vartheta(g)$ and first prove what amounts to a version of the claim for a single Y-local function F_{v_i},

$$\forall \gamma \in \mathsf{Aut}(Y), \ v_i \in Y; \quad \gamma \circ F_{v_i} \circ \gamma^{-1} = F_{\gamma(v_i)} . \tag{7.54}$$

To prove this we imitate the proof of Proposition 4.30:

$$\gamma \circ F_{v_i} \circ \gamma^{-1}((x_{v_j})) = \gamma \cdot (F_{v_i}(\gamma^{-1} \cdot (x_{v_j})))$$

and for arbitrary $\gamma \in \mathsf{Aut}(Y)$, we have $\gamma(B_1(v_i)) = B_1(\gamma(v_i))$. In view of

$$(\gamma^{-1} \cdot (x_{v_j}))_{v_i} = x_{\gamma(v_i)} \quad \text{and} \quad (\gamma \cdot (y_{v_j}))_{\gamma(v_i)} = y_{v_i},$$

we derive

$$\gamma \cdot (F_{v_i}(\gamma^{-1} \cdot (x_{v_j}))) = \gamma \cdot (x_{\gamma(v_1)}, \ldots, \underbrace{f_{v_i}((x_{\gamma(v_k)})_{v_k \in B_1(v_i)})}_{v_i \text{th-position}}, \ldots, x_{\gamma(v_n)})$$

$$= (x_{v_1}, \ldots, \underbrace{f_{v_i}((x_{\gamma(v_k)})_{v_k \in B_1(v_i)})}_{\gamma(v_i) \text{th position}}, \ldots, x_{v_n}),$$

$$F_{\gamma(v_i)}((x_{v_j})) = (x_{v_1}, \ldots, \underbrace{f_{\gamma(v_i)}((x_{\gamma(v_k)})_{\gamma(v_k) \in B_1(\gamma(v_i))})}_{\gamma(v_i) \text{th-position}}, \ldots, x_{v_n}) .$$

Equation (7.54) now follows from the fact that the functions $f_v \colon K^{d(v)+1} \longrightarrow K$ are symmetric, Eq. (7.49), and

$$f_{v_i}(x_{\gamma(v_s)} \mid \gamma(v_s) \in B_1(\gamma(v_i))) = f_{v_i}(x_{\gamma(v_s)} \mid v_s \in B_1(v_i)) .$$

Obviously, Eq. (7.53) follows by composing the corresponding local maps according to the word w as

$$\vartheta(g) \circ \left(\prod_{i=1}^{k} F_{w_i, Y} \right) \circ \vartheta(g)^{-1} = \prod_{i=1}^{k} \left(\vartheta(g) \circ F_{w_i} \circ \vartheta(g)^{-1} \right) = \prod_{i=1}^{k} F_{\vartheta(g)(w_i)} \,,$$

and the claim follows. Accordingly, we obtain

$$[\mathbf{F}_Y, w] \sim [\mathbf{F}_Y, \vartheta(g) \circ w] = [\mathbf{F}_Y, \vartheta(g') \circ w'] \sim [\mathbf{F}_Y, w'] \quad \text{i.e.} \quad [\mathbf{F}_Y, w] \sim [\mathbf{F}_Y, w'] \,,$$

and the proof of the theorem is complete.   □

7.2.4 Phase-Space Relations

Next we will generalize Theorem 4.47 of Section 4.4.3 originally proved in the context of permutation-SDS to word-SDS.

Let Y and Z be connected combinatorial graphs and let $\mathsf{h} \colon Y \longrightarrow Z$ be a graph morphism. For a given word $w' = (w'_1, \ldots, w'_r)$ we set $\mathsf{h}^{-1}(w'_j) = (v_{j_1}, \ldots, v_{j_{s(j)}})$, where $j_i < j_{i+1}$, and observe that h and w' induce the family

$$\left(v_{1_1}, \ldots, v_{1_{s(1)}}, v_{2_1}, \ldots, v_{2_{s(2)}}, \ldots, v_{r_1}, \ldots, v_{r_{s(r)}} \right) \,.$$

We set $w_{t + \sum_{q < j} s(q)} = v_{j_t}$ and obtain the word

$$\mathsf{h}^{-1}(w') = (w_1, \ldots, w_{\sum_{q=1}^{r} s(q)}) \,. \tag{7.55}$$

We now observe that h induces a morphism between dependency graphs

$$\mathsf{h}_1 \colon G(\mathsf{h}^{-1}(w'), Y) \longrightarrow G(w', Z), \quad \text{where } h_1(i) \text{ satisfies} \quad w'_{h_1(i)} = \mathsf{h}(w_i) \,. \tag{7.56}$$

The relation between the dependency graphs $G(\mathsf{h}^{-1}(w'), Y)$ and $G(w', Z)$ in (7.56) motivates the study of phase-space relations between the SDS $(Y, \mathbf{F}_Y, \mathsf{h}^{-1}(w'))$ and $(Z, \mathbf{F}_Z, w)$.

Lemma 7.25. *Let Y and Z be connected combinatorial graphs, and $\mathsf{h} \colon Y \longrightarrow Z$ a surjective graph morphism. Further let $w' = (w'_1, \ldots, w'_{k'}) \in W_{k'}$ be a word over Z, and $(Y, \mathbf{F}_Y, \mathsf{h}^{-1}(w'))$ and $(Z, \mathbf{F}_Z, w)$ two SDS. Furthermore we introduce*

$$\mathsf{H} \colon K^{|Z|} \longrightarrow K^{|Y|}, \quad \mathsf{H}(x)_t = x_{\mathsf{h}(t)} \,.$$

Suppose that we have the commutative diagram

$$
\begin{array}{ccc}
K^{|\mathsf{v}[Z]|} & \xrightarrow{\;\;\mathsf{H}\;\;} & K^{|\mathsf{v}[Y]|} \\
\downarrow{\scriptstyle F_{Z, w'_j}} & & \downarrow{\scriptstyle \prod_{w_{js} \in \mathsf{h}^{-1}(w'_j)} F_{Y, w_{js}}} \\
K^{|\mathsf{v}[Z]|} & \xrightarrow{\;\;\mathsf{H}\;\;} & K^{|\mathsf{v}[Y]|}
\end{array}
\tag{7.57}
$$

i.e., we have

$$H \circ F_{Z,w'_j} = \prod_{w_{j_s} \in h^{-1}(w'_j)} F_{Y,w_{j_s}} \circ H \,.$$

Then

$$H \colon \Gamma(Z, \mathbf{F}_Z, w') \longrightarrow \Gamma(Y, \mathbf{F}_Y, h^{-1}(w'))$$

is a digraph-morphism.

Proof. We first observe that $h^{-1}(w'_j)$ is a Y-independence set since Z is loop-free by assumption. Hence, the product of local maps

$$\prod_{w_{j_s} \in h^{-1}(w'_j)} F_{Y,w_{j_s}}$$

is independent of the ordering of its factors. We next claim that we have the commutative diagram

$$(7.58)$$

By definition of $h^{-1}(w')$ [Eq. (7.55)] and since $\prod_{w_{j_s} \in h^{-1}(w'_j)} F_{Y,w_{j_s}}$ is independent of the ordering of its factors, whence

$$[\mathbf{F}_Y, h^{-1}(w')] = \prod_{j=1}^{k'} \left[\prod_{w_{j_s} \in h^{-1}(w'_j)} F_{Y,w_{j_s}} \right].$$

According to Eq. (7.57), we obtain by induction, composing the local maps $F_{Y,w_{j_s}}$,

$$\prod_{j=1}^{k'} \left[\prod_{w_{j_s} \in c^{-1}(w'_j)} F_{Y,w_{j_s}} \right] \circ H = H \circ \prod_{j=1}^{k'} F_{Z,w'_j} \,,$$

whence Eq. (7.58), and the proof of the lemma is complete. $\qquad\square$

In this context it is of interest to analyze under which conditions the local maps of the SDS $(Y, \mathbf{F}_Y, h^{-1}(w'))$ and $(\mathbf{F}_Z, w')$ satisfy

$$H \circ F_{Z,w'_j} = \prod_{w_{j_s} \in h^{-1}(w'_j)} F_{Y,w_{j_s}} \circ H \,.$$

We next show that locally bijective graph morphisms c induce such a relation between the SDS $(Y, \mathbf{F}_Y, c^{-1}(w'))$ and $(Z, \mathbf{F}_Z, w)$ if the local functions associated to the Z-vertex w'_j and the Y-vertices $w_{j_s} \in c^{-1}(w'_j)$ are identical and induced by symmetric vertex functions f_v.

Theorem 7.26. *Let Y and Z be connected combinatorial graphs, $\mathsf{c} \colon Y \longrightarrow Z$ be a locally bijective graph morphism, and $w' = (w'_1, \ldots, w'_{k'}) \in W_{k'}$ a word over Z. Suppose the local functions of the SDS $(Y, \mathbf{F}_Y, \mathsf{c}^{-1}(w'))$ and $(Z, \mathbf{F}_Z, w)$ are induced by symmetric vertex functions and satisfy*

$$\forall\, w_{j_s} \in \mathsf{c}^{-1}(w'_j); \quad F_{Y, w_{j_s}} = F_{Z, w'_j} . \tag{7.59}$$

Then there exists an injective digraph morphism

$$\mathsf{C} \colon \Gamma(Z, \mathbf{F}_Z, w') \longrightarrow \Gamma(Y, \mathbf{F}_Y, \mathsf{c}^{-1}(w')), \quad \text{where} \quad \mathsf{C}(x)_t = x_{\mathsf{c}(t)} .$$

Proof. We first observe that Lemma 4.45 implies that $\mathsf{c} \colon Y \longrightarrow Z$ is surjective and prove the theorem in two steps. First, we show that we have the commutative diagram

$$
\begin{array}{ccc}
K^{|\mathsf{v}[Z]|} & \xrightarrow{\ \ \mathsf{C}\ \ } & K^{|\mathsf{v}[Y]|} \\
\Big\downarrow{\scriptstyle F_{Z, w'_j}} & & \Big\downarrow{\scriptstyle \prod_{w_{j_s} \in \mathsf{c}^{-1}(w'_j)} F_{Y, w_{j_s}}} \\
K^{|\mathsf{v}[Z]|} & \xrightarrow{\ \ \mathsf{C}\ \ } & K^{|\mathsf{v}[Y]|}
\end{array}
\tag{7.60}
$$

or equivalently

$$\mathsf{C} \circ F_{Z, w'_j} = \prod_{w_{j_s} \in \mathsf{c}^{-1}(w'_j)} F_{Y, w_{j_s}} \circ \mathsf{C} ,$$

and second, we apply Lemma 7.25. We first analyze $\prod_{w_{j_s} \in \mathsf{c}^{-1}(w'_j)} F_{Y, w_{j_s}} \circ \mathsf{C}$. The map $F_{Y, w_{j_s}}(\mathsf{C}(x))$ updates the state of w_{j_s} as a function of $\mathsf{C}(x)_v, v \in B_{1,Y}(w_{j_s})$. Since $\mathsf{C}(x)_v = x_{\mathsf{c}(v)}$, we have

$$(\mathsf{C}(x)_v \mid v \in B_{1,Y}(w_{j_s})) = (x_{\mathsf{c}(v)} \mid v \in B_{1,Y}(w_{j_s})) ,$$

and local bijectivity implies

$$\mathsf{c}(B_{1,Y}(w_{j_s})) = B_{1,Z}(w'_j) .$$

As Z is by assumption a loop-free graph, $\mathsf{c}^{-1}(w'_j)$ is a Y-independence set. Accordingly, we have a well-defined mapping

$$F_{Y, w'_j} = \left[\prod_{w_{j_s} \in \mathsf{c}^{-1}(w'_j)} F_{Y, w_{j_s}} \right] ,$$

since the product is independent of the ordering of its factors. The local map F_{Y, w'_j} updates all Y-vertices $w_{j_s} \in \mathsf{c}^{-1}(w'_j)$ based on $(x_{\mathsf{c}(v)} \mid \mathsf{c}(v) \in B_{1,Z}(w'_j))$ to the state $F_{Y, w_{j_s}}(\mathsf{C}(x))_{w_{j_s}}$.

Next we compute $\mathsf{C} \circ F_{Z,w'_j}(x)$. By definition, $F_{Z,w'_j}(x)$ updates the state of the Z-vertex w'_j as a function of $(x_{v'} \mid v' \in B_{1,Z}(w'_j))$ and we obtain

$$(\mathsf{C} \circ F_{Z,w'_j}(x))_{w_{j_s}} = F_{Z,w'_j}(x)_{w'_j} ,$$

i.e., $\mathsf{C} \circ F_{Z,w'_j}(x)$ updates the states of every Y-vertex $w_{j_s} \in \mathsf{c}^{-1}(w'_j)$ to the state $F_{Z,w'_j}(x)_{w'_j}$ and the diagram in Eq. (7.60) is indeed commutative. From Lemma 7.25 we have the commutative diagram

$$
\begin{array}{ccc}
K^{|\mathsf{v}[Z]|} & \xrightarrow{\;\;\mathsf{c}\;\;} & K^{|\mathsf{v}[Y]|} \\
{\scriptstyle [\mathbf{F}_Z,w']}\Big\downarrow & & \Big\downarrow{\scriptstyle [\mathbf{F}_Y,\mathsf{c}^{-1}(w')]} \\
K^{|\mathsf{v}[Z]|} & \xrightarrow{\;\;\mathsf{c}\;\;} & K^{|\mathsf{v}[Y]|} ,
\end{array}
$$

and the theorem is proved.   □

Problems

7.8. Let $Y = \mathsf{Circ}_4$, $w = (0,1,0,2,3)$, and $w' = (0,0,1,2,3)$. Derive the graphs $G(w,Y)$ and $G(w',Y)$.   [1]

Answers to Problems

7.1. For the words $w = (v_1, v_2, v_3, v_1)$ and $w' = (v_1, v_1, v_3, v_2)$, and the graph $Y = v_1 \underline{\hspace{1cm}} v_3 \underline{\hspace{1cm}} v_2$ with $\tau = (2,4)$, we have $w_\tau = w'$,

$$G(w,Y) = \begin{array}{cc} 1 & 2 \\ & \\ 4 \underline{\hspace{0.5cm}} 3 \end{array} , G(w',Y) = \begin{array}{cc} 1 & 4 \\ & \\ 2 \underline{\hspace{0.5cm}} 3 \end{array} ,$$

and $\tau \colon G(w',Y) \longrightarrow G(w,Y)$ is a graph isomorphism.

7.2. In view of $\mathsf{Aut}(Y) = S_3$, we obtain $w' = \gamma \circ \sigma \cdot w$ and $w'' = \gamma' \circ w$, and

$$G(w,Y) = \begin{array}{cc} 4 \underline{\hspace{0.5cm}} 2 \\ | \diagdown | \\ 1 \underline{\hspace{0.5cm}} 3 \end{array} , \; G(w',Y) = \begin{array}{cc} 1 \underline{\hspace{0.5cm}} 2 \\ | \diagdown | \\ 3 \underline{\hspace{0.5cm}} 4 \end{array} , \; G(w'',Y) = \begin{array}{cc} 4 \underline{\hspace{0.5cm}} 2 \\ | \diagdown | \\ 1 \underline{\hspace{0.5cm}} 3 \end{array} ,$$

where $\sigma' = \langle 1,3,4 \rangle \langle 2 \rangle \colon G(w,Y) \longrightarrow G(w',Y)$ is a graph isomorphism and $G(w,Y) = G(w'',Y)$.

7.4. (Proof of Proposition 7.8) Obviously, $w, w' \in W_k$ and $w \sim_Y w'$ implies $w, w' \in S_k(\varphi)$. For two non-adjacent Y-vertices w_i, w_{i+1} we observe

$$F_{w_i} \circ F_{w_{i+1}} = F_{w_{i+1}} \circ F_{w_i} ,$$

from which we immediately conclude using induction on the U_k-distance between w and w':

$$\forall \, w \sim_Y w' \quad \Longrightarrow \quad [\mathbf{F}_Y, w] = [\mathbf{F}_Y, w'] .$$

7.5. (Proof of Proposition 7.9) We set $w = \sigma \cdot \varphi$ and $w' = \sigma' \cdot \varphi$. Since $w \not\sim_Y w'$, Lemma 7.16 guarantees

$$\forall \, \rho, \rho' \in \mathsf{Fix}(\varphi); \quad \mathcal{O}_Y(\sigma\rho) \neq \mathcal{O}_Y(\sigma'\rho') . \tag{7.61}$$

Let $\sigma(j_1) = 1$ and let t be the minimal position of φ_{j_1} in $w' = \sigma' \cdot \varphi$. In case of $t = 1$, we are done; otherwise we try to move φ_{j_1} to first position by successively transposing consecutive indices of Y-independent letters. In case we were able to move φ_{j_1} to the first position, we continue the procedure with φ_{j_2} and proceed inductively. In view of Eq. (7.61) we have

$$\forall \, \rho, \rho' \in \mathsf{Fix}(\varphi), \, \exists \, \{i,j\} \in G(\varphi, Y), \quad \mathcal{O}_Y(\sigma\rho)(\{i,j\}) \neq \mathcal{O}_Y(\sigma'\rho')(\{i,j\}) ,$$

and our inductively defined procedure must fail. Let us assume it fails after exactly r steps, that is, we have

$$w' \sim_Y (\varphi_{j_1}, \varphi_{j_2}, \ldots, \varphi_{j_r}, \ldots, \varphi_j, \ldots, \varphi_i, \ldots) = w'' \, ,$$

and there exists some φ_j preceding φ_i in w'' such that $\{\varphi_i, \varphi_j\} \in Y$. We now define the family of Y-local maps $\mathbf{F}_Y$ as follows:

$$F_{\varphi_j}(x_{v_1}, \ldots, x_{v_n})_{\varphi_j} = x_{\varphi_j} + 1,$$

$$F_{\varphi_i}(x_{v_1}, \ldots, x_{v_n})_{\varphi_i} = \begin{cases} x_{\varphi_j} & \text{for } x_{\varphi_j} \le m, \\ \max\{x_{\varphi_i}, m\} + 1 & \text{for } x_{\varphi_j} > m, \end{cases}$$

$$F_{\varphi_s}(x_{v_1}, \ldots, x_{v_n})_{\varphi_s} = 0 \quad \text{for } s \ne i, j \, .$$

Suppose the word $(\varphi_{j_1}, \varphi_{j_2}, \ldots, \varphi_{j_r})$ contains φ_j exactly q times. We choose $m = q$ and claim

$$([\mathbf{F}_Y, w](0, 0, \ldots, 0, 0))_{\varphi_i} + 1 \le ([\mathbf{F}_Y, w'](0, 0, \ldots, 0, 0))_i \, .$$

We have the following situation:

$$w = (\varphi_{j_1}, \ldots, \varphi_{j_r}, \varphi_i, \ldots, \varphi_j, \ldots), \tag{7.62}$$

$$w' \sim_Y (\varphi_{j_1}, \varphi_{j_2}, \ldots, \varphi_{j_r}, \ldots, \varphi_j, \ldots, \varphi_i, \ldots) = w'' \, . \tag{7.63}$$

Let us first compute $([\mathbf{F}_Y, w](0, 0, \ldots, 0, 0))_{\varphi_i}$. We observe that φ_i being at index $r + 1$ updates into state $x_{\varphi_i} = q$, regardless of u_1, the number of times $(\varphi_{j_1}, \varphi_{j_2}, \ldots, \varphi_{j_r})$ contains φ_i. Let u be the number of times φ_i appears in the word w. In view of Eqs. (7.62) and (7.63), we observe that w exhibits at most $[u - u_1 - 1]$ φ_i-updates under the condition $x_{\varphi_j} > q$, and we obtain

$$([\mathbf{F}_Y, w](0, 0, \ldots, 0, 0))_{\varphi_i} \le [q + (u - u_1 - 1)] \, .$$

Next we compute $([\mathbf{F}_Y, w'](0, 0, \ldots, 0, 0))_{\varphi_i}$. By assumption φ_j precedes φ_i in w'', and φ_i has some index $s > r+1$ and updates into the state $q+1$, regardless of how many positions $r + 1 \le l \le s$, φ_j occurred in w''. Accordingly, we compute, in view of Eq. (7.62), Eq. (7.63) and Proposition 7.8:

$$([\mathbf{F}_Y, w'](0, 0, \ldots, 0, 0))_{\varphi_i} = [q + (u - u_1)] \, ,$$

and the proof is complete.

7.6. For $Y = v_1 \, \relbar\joinrel\relbar \, v_2 \, \relbar\joinrel\relbar \, v_3 \, ,$ $w = (v_1, v_2, v_1, v_2, v_3)$, and $\rho = (3, 1)(2, 4)$ we have

$$G(w, Y) = \quad \begin{array}{c} 1 \relbar\joinrel\relbar 2 \\ \times \\ 3 \relbar\joinrel\relbar 4 \relbar\joinrel\relbar 5 \end{array} \quad \text{and with} \quad \mathcal{O} = \quad \begin{array}{c} 1 \longrightarrow 2 \\ \times \\ 3 \longleftarrow 4 \longrightarrow 5 \end{array} \, .$$

The equivalence class $[\mathcal{O}]_w$ of $\mathcal{O} \in \mathsf{Acyc}(G(w, Y))$ is given by

$$[\mathcal{O}]_w = \left\{ \begin{array}{c} \text{(diagrams)} \end{array} \right\}.$$

$$[\mathcal{O}]_w = \left\{\begin{array}{cc} 1 \longrightarrow 2 \quad 3 \longrightarrow 2 \\ \times \qquad \times \\ 3 \longleftarrow 4 \longrightarrow 5\,, \quad 1 \longleftarrow 4 \longrightarrow 5\,, \\[2em] 1 \longrightarrow 4 \quad 3 \longrightarrow 4 \\ \times \qquad \times \\ 3 \longleftarrow 2 \longrightarrow 5\,, \quad 1 \longleftarrow 2 \longrightarrow 5 \end{array}\right\}.$$

7.7. (Proof of Corollary 7.18) By Corollary 7.5 we have $G(w, Y) \cong Y$. With $\varphi = \mathrm{id} = (1, 2, \ldots, n)$ we note that $\mathrm{Fix}(\mathrm{id}) = 1$, and

$$[\mathcal{O}_Y(\sigma)]_\varphi = \{\mathcal{O}_Y(\sigma)\}\,,$$

that is, the φ-equivalence class consists exclusively of the acyclic orientation induced by σ itself.

8

Outlook

In the previous chapters we gave an account of the SDS theory developed so far. This final chapter describes a collection of ongoing work and possible directions for future research in theory and applications of SDS. This is, of course, not an exhaustive list, and some of what follows is in an early stage. The material presented here reflects, however, what we currently consider interesting and important and what has already been identified as useful for many application areas.

8.1 Stochastic SDS

When modeling systems one is often confronted with phenomena that are known only at the level of distributions or probabilities. An example is the modeling of biological systems where little data are available or where we only know the empirical statistical distributions. Another example is a physical device that occasionally fails such as a data transmission channel exposed to electromagnetic fields. In this case we typically have an error rate for the channel, and coding theory is used as a framework. In fact, there are also situations where it is possible to show that certain deterministic systems can be simulated by stochastic models such that the corresponding stochastic model is computationally more tractable than the original system. Sampling methods like Monte Carlo simulations [122] are good examples of this. Accordingly, stochasticity can be an advantageous attribute of the model even if it is not an inherent system property.

For SDS there are many ways by which probabilistic elements can be introduced, and in this section we discuss some of these along with associated research questions.

8.1.1 Random Update Order

Stochastic update orders emerge in the context of, for example, *discrete event simulations*. A discrete event simulation is a system basically organized as follows: It consists of a set of agents that are mutually linked through the edges of an interaction graph and where each agent initially has a list of time-stamped tasks to execute at given points in time. When an agent executes a certain task, it may affect the execution of tasks of its neighbors. For this reason an *event* is sent from the agent to its neighbors whenever it has processed a task. Hence, neighboring agents have to respond to this event, and this may cause new tasks to be created and executed, which in turn may cause additional events to be passed around. All the tasks are executed in chronological order. From an implementation point of view this computation can be organized through, for instance, a queue of tasks. The tasks are executed in order, and the new tasks spawned by an event are inserted into the queue as appropriate.

In general, there is no formal guarantee that such a computation will terminate. If events following tasks trigger too many new tasks, the queue will just continue to grow and it will become impossible (at least in practice) to complete all tasks. Moreover, if the time progression is too slow, there is no guarantee that the computation will advance to some prescribed time.

8.1. What reasons do you see for organizing computations using the discrete event simulation setup? Note that the alternative (a time-stepped simulation) would often be to have every agent check its list of tasks at every time step of the computation. If only a small fraction of the agents execute tasks at every time step, it seems like this could lead to a large amount of redundant testing and processing.   [1]

Distributed, Discrete Event Simulations

For efficient computation discrete event simulations are frequently implemented on multiprocessor architectures. In this case each processor (CPU) will be responsible for a subset of the agents and their tasks. Since some CPUs may have fewer agents or a lighter computational load, it can easily happen that the CPU's local times advance at different rates. The resulting CPU misalignment in time can cause synchronization problems. As tasks trigger events, and events end up being passed between CPUs, we can encounter the situation where a given CPU receives events with a time stamp that is "in the past" according to its current time. If the CPU has been idle in the interim, this presents no problem. However, if it has executed tasks that would have been affected by this new event, the situation is more involved.

One way to ensure correct computation order is through *roll-back*; see, e.g., [1, 2]. In this approach each CPU keeps track of its history of tasks, events, and states. When an event from "the past" appears and it spawns new tasks, then time is "rolled back" and the computation starts up at the correct point in time. It is not hard to see that at least in theory this allows one to compute the tasks in the "right" order. However, it is also evident that this scheme can have some significant drawbacks as far as bookkeeping, processor memory usage, and computation speed are concerned.

Local Update Orders

As an alternative to costly global synchronization caused by, e.g., roll-back, [123] has discussed the following approach: Each CPU is given a set of neighbor CPUs, where neighbor could mean being adjacent in the current computing architecture. This set is typically small compared to the total number of processors. Additionally, there is the notion of *time horizon*, which is some time interval Δt.

Starting at time t_0 no processor is allowed to compute beyond the time horizon and there is a *global synchronization barrier* at $t_0 + \Delta t$. Throughout the time interval Δt each processor will only synchronize with its neighbors, for instance, through roll-back. The idea is that this local synchronization combined with a suitable choice of time horizon leads to a global system update that satisfies mutual dependencies. Obviously, this update is bound to execute faster, and it is natural to ask how closely it matches the results one would get by using some roll-back-induced global update. This is precisely where SDS with stochastic update orders offer conceptual insight.

An SDS Model

From the discussion above we conclude that the update derived from local computations will have tasks computed in an order π' that could possibly be different from the a priori update, π. Of course, the actual set of tasks executed could also be different, but we will restrict ourselves to the case where the set of tasks remains the same. We furthermore stipulate that the extent to which the update orders π and π' differ will be a function of the choice of the synchronization of the neighbors and the size of the time horizon.

SDS provide a natural setting to study this problem: Let $[\mathbf{F}_Y, \pi]$ be an SDS-map. If we are given another update order π' such that $d(\pi, \pi') < k$ for some suitable measure of permutation distance, then we ask: How similar or different are the phase spaces $\Gamma[\mathbf{F}_Y, \pi]$ and $\Gamma[\mathbf{F}_Y, \pi']$? To proceed it seems natural to introduce a probability space $\mathcal{P}$ of update orders and an induced probability space of SDS. As described in Section 2.2, this leads to Markov chains or what we may call a probabilistic phase space, the probabilistic phase space being the weighted combination of the phase spaces $\Gamma[\mathbf{F}_Y, \sigma]$ with $\sigma \in \mathcal{P}$.

Example 8.1. We consider the dependency graph Star_3 with vertex function induced by the function $\mathbb{I}_{1,k} \colon \mathbf{F}_2^k \longrightarrow \mathbf{F}_2$, which returns 1 if exactly one of its inputs is 1, and returns 0 otherwise. We obtain two SDS maps ϕ and ψ by using the update orders $(0, 1, 2, 3)$, respectively $(1, 0, 2, 3)$. If we choose the probability $p = 1/2$, then we obtain a probabilistic phase space as shown in Figure 8.1.   ◇

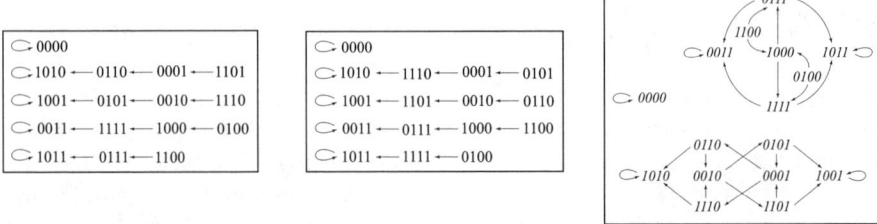

Fig. 8.1. The probabilistic phase space for Example 8.1 (shown on the right) induced by the two deterministic phase spaces ϕ (left) and ψ (middle). For simplicity the weights of the edges have been omitted.

One particular way to define a distance measure on permutation update orders is through acyclic orientations as in Section 3.1.3. The distance between two permutations π and π' is then the number of edges for which the acyclic orientations $\mathcal{O}_Y(\pi)$ and $\mathcal{O}_Y(\pi')$ differ. This distance measure captures how far apart the corresponding components are in the update graph. Assume that we have π as reference permutation. We may construct a probability space $\mathcal{P} = \mathcal{P}(\pi)$ by taking all possible permutation update orders and giving each permutation a probability inversely proportional to the distance to the reference permutation π. Alternatively, we may choose the probability space to consist of all permutations of distance less than k, say, to π and assign them uniform probability.

Random updates should be studied systematically for the specific classes of SDS (Chapter 5). For instance, for SDS induced by threshold functions and linear SDS w-independent SDS are particularly well suited since in this case we have a fixed set of periodic points for all update orders. If we restrict ourselves to the periodic points, it is likely that we can reduce the size of the Markov chain significantly.

From Section 5.3 we know that all periodic points of threshold SDS are fixed points. One question in the context of random updates is then to ask which sets Ω of fixed points can be reached from a given initial state x (see Proposition 4.11 in this context). Note that the choice of update order may affect the transients starting at x. Let $\omega_\pi(x)$ be the fixed point reached under system evolution using the update order π starting at x. The size of the set $\Omega = \cup_{\pi \in \mathcal{P}} \omega_\pi(x)$ is one possible measure for the degree of update order instability. Clearly, this question of stability is relevant to, for example, discrete event simulations. See also Problem 5.11 for an example of threshold systems that exhibit update order instability.

8.1.2 SDS over Random Graphs

In some applications the graph Y may not be completely known or may change at random as time progresses, as, for instance, in stationary radio networks where noise is present. Radios that are within broadcast range may send and receive data. A straightforward way to model such networks is to let radios or antennas correspond to the vertices of a graph and to connect each antenna pair that is within communication range of one another. Noise or other factors may temporarily render a communication edge between two antennas useless. In the setting of SDS we can model this through a probability space of graphs $\mathcal{Y}$ (i.e., a random graph [107]) whose elements correspond to the various realizations of communication networks. We can now consider, for example, induced SDS over these graphs with induced probabilities. Just as before this leads to Markov chains or probabilistic phase spaces. Through this model we may be able to answer questions on system reliability and expected communication capacities.

We conclude this section by remarking that probabilistic analysis and techniques (ergodic theory and statistical mechanics) have been used to analyze deterministic, infinite, one-dimensional CA [36, 37]. The area of *probabilistic cellular automata* (PCA) deals with cellular automata with random variables as local update functions [38, 124]. PCA over Circ_n have been studied in [39] focusing on conservation laws. The use of Markov chains to study PCA was established in the 1970s; see, e.g., [125, 126]. Examples of applications of PCA (finite and infinite) include hydrodynamics/lattice gases [41] and traffic modeling [6, 7, 39]. In addition, both random Boolean networks (Section 2.2) and interacting particle systems [25] are stochastic systems. These frameworks may provide guidance in the development of a theory of stochastic SDS.

8.2 Gene-Regulatory Networks

8.2.1 Introduction

A *gene-regulatory network* (GRN) is a network composed of interacting genes. Strictly speaking, it does not represent the direct interactions of the involved genes since there are in fact *three* distinct biochemical layers that factor in those interactions: the genes, the ribonucleic acids (RNA), and the proteins. RNA is created from the genes via *transcription*, and proteins are created from RNA via the *translation* process. However, when we speak of a GRN in the following, we identify these three biochemical layers. After completing the sequencing of basically the entire human genome, it has become apparent that more than the knowledge about single, isolated genes is necessary in order to understand their complex regulatory interactions. The purpose of this section is to show that SDS-modeling is a natural modeling concept, as it allows one to capture asynchronous updates of genes.

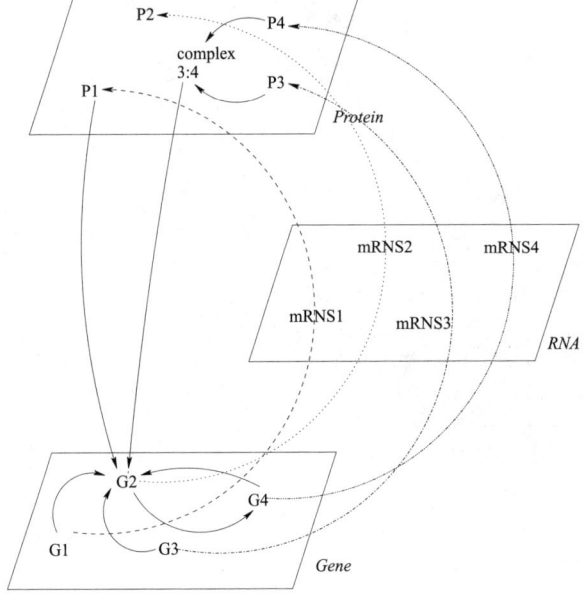

Fig. 8.2. Schematic representation of a GRN.

8.2.2 The Tryptophan-Operon

In this section we discuss a GRN that is typical for the regulation of the tryptophan- (*trp*) operons or asparagin (*asn*) system in *E. coli*: Below we have

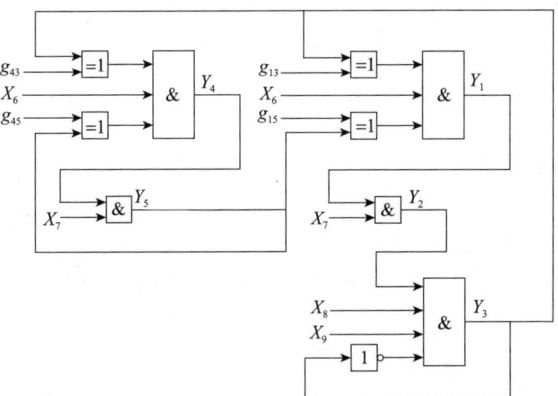

Fig. 8.3. The repressible GRN.

the binary input parameters x_6, x_7, x_8, and x_9, the binary system parameters g_{13}, g_{15}, g_{43}, and g_{45}, and the intracellular components: *effector*-mRNA x_1, *enzyme* x_2, *product* x_3, *regulator*-mRNA x_4, and *regulator protein* x_5 with the following set of equations:

$$x_1(t+1) = (g_{13} + x_3(t)) \cdot x_6(t) \cdot (g_{15} + x_5(t)) \,,$$
$$x_2(t+1) = x_1(t) \cdot x_7(t) \,,$$
$$x_3(t+1) = x_2(t) \cdot (1 + x_3(t)) \cdot x_8(t) \cdot x_9(t) \,,$$
$$x_4(t+1) = (g_{43} + x_3(t)) \cdot x_6(t) \cdot (g_{45} + x_5(t)) \,,$$
$$x_5(t+1) = x_4(t) \cdot x_7(t) \,.$$

Figure 8.4 shows the phase spaces of four specific system realizations.

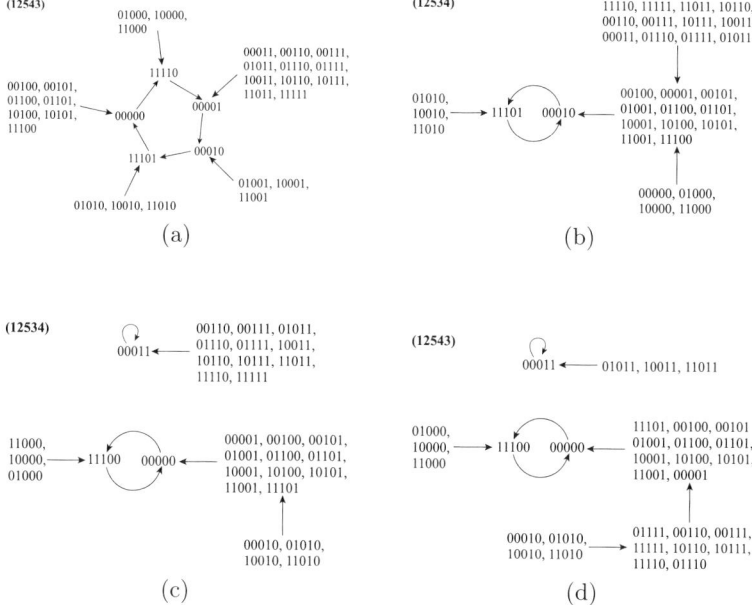

Fig. 8.4. In (a) and (b) all system parameters are 1. In (c) and (d) $g_{45} = 0$ while all other system parameters are 1.

It is interesting to rewrite the above relations in the **SDS**-framework: the graph Y expressing the mutual dependencies of the variables relevant for the time evolution $(x_1, \ldots, x_5)$:

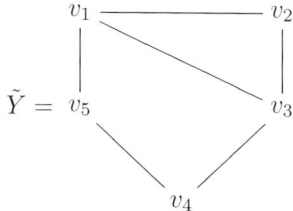

and the associated family of Y-local functions $\mathbf{F}_{\tilde{Y}}$ given by $(F_{v_i}(x_1,\ldots,x_5))_{v_i}$ $= x_i(t+1)$ with $i = 1,\ldots,5$. In the following we will restrict ourselves to permutation-words where the bijection

$$S_5/\sim_{\tilde{Y}} \longrightarrow \mathsf{Acyc}(\tilde{Y})$$

provides an upper bound on the number of different system types (Corollary 7.18). Clearly, we have $|\mathsf{Acyc}(\tilde{Y})| = 42$.

The system size allows for a complete classification of system dynamics obtained through exhaustive enumeration. Explicitly, we proceed as follows: We ignore transient states and consider the induced subgraph of the periodic points. In fact there are exactly 12 different induced subgraphs over the periodic points, each of which is characterized by the quintuple $(z_1,\ldots,z_5)$ where z_i denotes the number of cycles of length i. We detail all system types in Table 8.1.

An ODE modeling ansatz for the network in Figure 8.3 yields exactly one fixed point, which, depending on system parameters, can be either stable or unstable. This finding is reflected in the observation that the majority of the SDS-systems exhibits exactly one fixed point. There are, however, 11 additional classes of system dynamics, which are missed entirely by the ODE-model. These are clearly a result of the asynchronous updates of the involved genes, and there is little argument among biologists about the fact that genes do update their states asynchronously.

8.3 Evolutionary Optimization of SDS-Schedules

8.3.1 Neutral Networks and Phenotypes of RNA and SDS

In theoretical evolutionary biology the evolution of single-stranded RNA molecules has been studied in great detail. The field was pioneered by Schuster et al., who systematically studied the mapping from RNA molecules into their *secondary structures* [127–132]. Waterman et al. [133–135] did seminal work on the combinatorics of secondary structures. A paradigmatic example for evolutionary experiments with RNA is represented by the SELEX method (*systematic evolution of ligands by exponential enrichment*), which allows one to create molecules with optimal binding constants [136]. The SELEX experiments have motivated our approach for studying the evolution of SDS. SELEX is a protocol that isolates high-affinity nucleic acid ligands for a given target such as a protein from a pool of variant sequences. Multiple rounds of replication and selection exponentially enrich the population that exhibits the highest affinity, i.e., fulfills the required task. Paraphrasing the situation, SELEX is a method to perform molecular computations by *white noise*.

One natural choice for an SDS-*genotype* is its word update order w whose structure as a linear string has an apparent similarity to single-stranded RNA molecules. We have seen several instances where a variety of update orders

Table 8.1. System classification.

Systems with Only Fixed Points — Schedule Independent Systems

No. of Systems	$P(S)$ z_1	Fixed Points
185	1	00000
12	1	00010
12	1	10000
4	1	10010
6	1	11000
7	1	00000, 00011
3	1	00000, 11011
3	1	00011, 11000

Systems with Orbits of Length > 1 — Schedule Dependent Systems Class I

Label	$P(S)$ z_1 z_2	No. of Systems
195	0 2	42
199	0 2	42
203	0 2	42
207	0 2	42
211	0 2	42
215	0 2	42
219	0 2	42
223	0 2	42
227	0 2	42
231	0 2	42
235	0 2	42
239	0 2	42

Schedule Dependent Systems Class II

Label	z_1	z_2	z_3	z_4	z_5	No. of Systems
241	1	1	0	0	0	16
	1	0	0	0	0	26
243	0	1	0	0	0	31
	0	2	0	0	0	11
244	1	1	0	0	0	27
	1	0	0	0	0	15
245	1	0	0	0	0	31
	1	1	0	0	0	11
246	2	1	0	0	0	27
	2	0	0	0	0	15
247	0	2	0	0	0	16
	0	1	0	0	0	26
249	1	0	0	0	0	16
	1	0	1	0	0	17
	1	0	0	1	0	1
	1	1	0	0	0	8
251	0	0	1	0	0	23
	0	0	0	1	0	11
	0	0	0	0	1	1
	0	1	0	0	0	7
252	0	0	1	0	0	27
	0	1	0	0	0	15
253	0	0	1	0	0	24
	0	0	0	1	0	11
	0	1	0	0	0	7
254	1	0	1	0	0	27
	1	1	0	0	0	15
255	0	0	0	1	0	15
	0	0	1	0	0	12
	0	0	1	1	0	1
	0	1	0	0	0	8
	0	0	0	0	1	6

produced either an identical or an equivalent dynamical system. Our findings in Section 3.1.4 provide us with a notion of adjacency: Two SDS update orders are adjacent if they differ exactly by a single flip of two consecutive coordinates. We call the set of all update orders producing one particular system its *neutral network*. In the case of RNA we have a similar situation: Two RNA sequences are called adjacent if they differ in exactly one position by a *point mutation*, and a neutral network consists of all molecules that fold into a particular coarse-grained structure. In this section we will investigate the following aspects of SDS evolution: (1) the fitness-neutral [137], stochastic transitions between two SDS-phenotypes [138] and (2) critical mutation rates originally introduced in [139] and generalized to phenotypic level in [131,138,140].

Let us begin by discussing *phenotypes* as they determine our concept of neutrality. RNA exhibits generic phenotypes by forming 2D or 3D structures. One example of RNA phenotypes is their secondary structures [133], which are planar graphs over the RNA nucleotides and whose edges are formed by base pairs subject to specific conditions [141]. Choosing minimum free energy as a criterion, we obtain (fold) a unique secondary structure for a given single-stranded RNA sequence. The existence of phenotypically neutral mutations is of relevance for the success of *white noise* computations as it allows for the preservation of a high average fitness level of the population while simultaneously reproducing errorproneously. In Section 7.1.3 we have indeed encountered one particular form of neutral mutations of SDS-genotypes. In Proposition 7.8 we have shown that for any two $\sim_\varphi$-equivalent words w and w' ($w \rightsquigarrow w'$) we have the identity of SDS-maps $[\mathbf{F}_Y, w] = [\mathbf{F}_Y, w']$. Adopting this combinatorial perspective we observe that SDS over words exhibit phenotypes that are in fact very similar to molecular structures. To capture these observations we consider the dependency graph $G(w, Y)$ as introduced in Section 7.1.1. The phenotype in question will now be the equivalence class of acyclic orientations of $G(w, Y)$ induced by $\sim_w$ (Section 7.1.4). That is, the equivalence is induced by $G(w, Y)$-automorphisms that fix w and two acyclic orientations $\mathcal{O}$ and $\mathcal{O}'$ are w-equivalent ($\mathcal{O} \sim_w \mathcal{O}'$) if and only if

$$\exists\, \rho \in S_k;\ (w_{\rho^{-1}(1)}, \ldots, w_{\rho^{-1}(k)}) = w;\qquad \rho(\mathcal{O}(\{r, s\})) = \mathcal{O}'(\{\rho(r), \rho(s)\})\,.$$

As for neutral networks, Theorem 7.17 of Chapter 7

$$\mathcal{O}'_Y : W_k / \sim_Y \longrightarrow \dot{\bigcup}_{\varphi \in \Phi} [\mathsf{Acyc}(G(\varphi, Y)) / \sim_\varphi]$$

shows that $w \sim_Y w'$ if and only if the words w and w' can be transformed into each other by successive transpositions of consecutive pairs of letters that are Y-independent. In other words $\sim_Y$ is what amounts to the transitive closure of neutral flip-mutations "$\rightsquigarrow$". Accordingly, the $\sim_Y$-equivalence class of the word w, denoted by $[w]$, is the neutral network of $\mathcal{O}'_Y(w)$, which is the equivalence class of acyclic orientations.

8.3.2 Distances

The goal of this section is to introduce a distance measure D for words w and w' that captures the distance of the associated SDS-maps $[\mathbf{F}_Y, w]$ and $[\mathbf{F}_Y, w']$. In our construction the distance measure D is independent of the particular choice of family of Y-local functions $(F_v)_{v \in v[Y]}$.

Let $\sigma \cdot w = (w_{\sigma^{-1}(1)}, \ldots, w_{\sigma^{-1}(k)})$ be the S_k-action on W_k as defined in Section 7.1.1. Its orbits induce the partition $W_k = \bigcup_{\varphi \in \Phi} S_k(\varphi)$ where Φ is a set of representatives. Let $w, w' \in S_k(\varphi)$ and let $\sigma, \sigma' \in S_k$ such that $w = \sigma \cdot \varphi$ and $w' = \sigma' \cdot \varphi$. We consider $\mathcal{O}_Y(\sigma_1)$ and $\mathcal{O}_Y(\sigma_2)$ [as defined in Eq. (7.20)] as acyclic orientations of $G(\varphi, Y)$ and define their distance d as

$$d(\mathcal{O}_Y(\sigma_1), \mathcal{O}_Y(\sigma_2)) = |\{\{i,j\} \mid \mathcal{O}_Y(\sigma_1)(\{i,j\}) \neq \mathcal{O}_Y(\sigma_2)(\{i,j\})\}| . \quad (8.1)$$

According to Theorem 7.17 each word naturally induces an equivalence class $\mathcal{O}_Y(w) = [\mathcal{O}_Y(\sigma)]_\varphi$ of acyclic orientations of $G(\varphi, Y)$, and Lemma 7.12 describes this class completely by $[\mathcal{O}_Y(\sigma)]_\varphi = \{\mathcal{O}_Y(\sigma\rho) \mid \rho \in \mathsf{Fix}(\varphi)\}$. Based on the distance d between acyclic orientations [Eq. (8.1)], we introduce $D \colon S_k(\varphi) \times S_k(\varphi) \longrightarrow \mathbb{Z}$ by

$$D(w, w') = \min_{\rho, \rho' \in \mathsf{Fix}(\varphi)} \{d(\mathcal{O}_Y(\sigma\rho), \mathcal{O}_Y(\sigma'\rho'))\} . \quad (8.2)$$

In the following we will prove that D naturally induces a metric for $\sim_Y$-equivalence classes of words. For RNA secondary structures similar distance measures have been considered in [138, 142].

According to Proposition 7.8, we have the equality

$$[\mathbf{F}_Y, w] = [\mathbf{F}_Y, w']$$

for any two $\sim_Y$-equivalent words $w \sim_Y w'$. In Lemma 8.2 we show that D does indeed capture the distance of SDS since for any two $\sim_Y$-equivalent words w and w' we have $D(w, w') = 0$.

Lemma 8.2. *For $w, w' \in S_k(\varphi)$*

$$w \sim_Y w' \quad \Longleftrightarrow \quad D(w, w') = 0 \quad (8.3)$$

holds.

Proof. Suppose we have $w = \sigma \cdot \varphi \sim_Y \sigma' \cdot \varphi = w'$. By induction on the W_k-distance (Section 7.1.3) between w and w', we may without loss of generality assume that w and w' are adjacent in W_k, that is, we have $\tau \cdot w = w'$ with $\tau = (i, i{+}1)$. Since we have $\sigma'^{-1}\tau\sigma \cdot \varphi = \varphi$, or equivalently $\rho = \sigma'^{-1}\tau\sigma \in \mathsf{Fix}(\varphi)$ and $\sigma' \cdot w = (\sigma'\rho) \cdot w$, we can replace σ' by $\sigma'\rho$. Without loss of generality we may thus assume $\tau\sigma = \sigma'$.

In Lemma 7.14 we have shown that for $\tau\sigma = \sigma'$ we have the equality $\mathcal{O}_Y(\sigma) = \mathcal{O}_Y(\sigma')$. Hence, we have

$$D(w, w') = \min_{\rho, \rho' \in \mathsf{Fix}(\varphi)} \{d(\mathcal{O}_Y(\sigma\rho), \mathcal{O}_Y(\sigma'\rho'))\} = d(\mathcal{O}_Y(\sigma), \mathcal{O}_Y(\sigma')) = 0 \ .$$

Suppose now $D(w, w') = 0$, that is, there exist $\rho, \rho' \in \mathsf{Fix}(\varphi)$ such that $\mathcal{O}_Y(\sigma\rho) = \mathcal{O}_Y(\sigma'\rho')$. In Lemma 7.16 we have proved

$$\mathcal{O}_Y(\sigma\rho) = \mathcal{O}_Y(\sigma'\rho') \implies (\sigma\rho)\cdot\varphi \sim_Y (\sigma'\rho')\cdot\varphi, \qquad (8.4)$$

and since $(\sigma\rho)\cdot\varphi = \sigma\cdot\varphi = w$ and $(\sigma'\rho')\varphi = \sigma'\cdot\varphi = w'$, the lemma follows. $\square$

Proposition 8.3 shows that D satisfies the triangle inequality and will lay the foundations for Proposition 8.4, where we prove that D induces a metric over word equivalence classes or neutral networks. Its proof hinges on the facts that (1) $\mathsf{Fix}(\varphi)$ is a group and (2) the $\mathcal{O}_Y(\sigma)$ orientations have certain compatibility properties (see Lemma 7.12). As for the proof, Eq. (8.5) is key for being able to derive Eq. (8.6) from (8.8).

Proposition 8.3. *Let $w = \sigma\cdot\varphi$ and $w' = \sigma'\cdot\varphi$. Then we have*

$$D(w, w') = \min_{\rho \in \mathsf{Fix}(\varphi)} \{d(\mathcal{O}_Y(\sigma\rho), \mathcal{O}_Y(\sigma'))\} \ . \qquad (8.5)$$

Furthermore for any three words $w, w', w'' \in S_k(\varphi)$

$$D(w, w') \le D(w, w'') + D(w'', w') \qquad (8.6)$$

holds.

Proof. We first prove Eq. (8.5) and claim

$$\min_{\rho, \rho' \in \mathsf{Fix}(\varphi)} \{d(\mathcal{O}_Y(\sigma\rho), \mathcal{O}_Y(\sigma'\rho'))\} = \min_{\rho \in \mathsf{Fix}(\varphi)} \{d(\mathcal{O}_Y(\sigma\rho), \mathcal{O}_Y(\sigma'))\} \ . \qquad (8.7)$$

Suppose that for some $\{i, j\} \in G(w, Y)$: $\mathcal{O}_Y(\sigma\rho)(\{i, j\}) \ne \mathcal{O}_Y(\sigma'\rho')(\{i, j\})$ holds. Since $\rho\colon G(\varphi, Y) \longrightarrow G(\varphi, Y)$ is an automorphism, we may replace $\{i, j\}$ by $\{\rho^{-1}(i), \rho^{-1}(j)\}$ and obtain

$$\mathcal{O}_Y(\sigma\rho)(\{\rho^{-1}(i), \rho^{-1}(j)\}) = \rho^{-1}(\mathcal{O}(\sigma)(\{i, j\})),$$
$$\mathcal{O}_Y(\sigma'\rho')(\{\rho^{-1}(i), \rho^{-1}(j)\}) = \rho^{-1}(\mathcal{O}(\sigma'\rho'\rho^{-1})(\{i, j\})) \ .$$

Hence, we have proved

$$\mathcal{O}_Y(\sigma\rho)(\{\rho^{-1}(i), \rho^{-1}(j)\}) \ne \mathcal{O}_Y(\sigma'\rho')(\{\rho^{-1}(i), \rho^{-1}(j)\})$$
$$\iff \mathcal{O}(\sigma)(\{i, j\}) \ne \mathcal{O}(\sigma'\rho'\rho^{-1})(\{i, j\}) \ ,$$

and, accordingly,

$$d(\mathcal{O}_Y(\sigma\rho), \mathcal{O}_Y(\sigma'\rho')) = d(\mathcal{O}_Y(\sigma), \mathcal{O}_Y(\sigma'\rho'\rho^{-1})) .$$

Equation (8.7) now follows from the fact that $\mathsf{Fix}(\varphi)$ is a group. For arbitrary, fixed $\rho, \rho' \in \mathsf{Fix}(\varphi)$ we have

$$d(\mathcal{O}_Y(\sigma\rho), \mathcal{O}_Y(\sigma'\rho')) \leq d(\mathcal{O}_Y(\sigma\rho), \mathcal{O}_Y(\sigma'')) + d(\mathcal{O}_Y(\sigma''), \mathcal{O}_Y(\sigma'\rho')) . \quad (8.8)$$

We now use $D(w, w') = \min_{\rho \in \mathsf{Fix}(\varphi)} \{d(\mathcal{O}_Y(\sigma\rho), \mathcal{O}_Y(\sigma'))\}$ and Eq. (8.5), and choose ρ and ρ' such that

$$d(\mathcal{O}_Y(\sigma\rho), \mathcal{O}_Y(\sigma'')) = \min_{\rho \in \mathsf{Fix}(\varphi)} \{d(\mathcal{O}_Y(\sigma\rho), \mathcal{O}_Y(\sigma''))\} = D(w, w''),$$

$$d(\mathcal{O}_Y(\sigma''), \mathcal{O}_Y(\sigma'\rho')) = \min_{\rho \in \mathsf{Fix}(\varphi)} \{d(\mathcal{O}_Y(\sigma''), \mathcal{O}_Y(\sigma'\rho'))\} = D(w'', w') .$$

Obviously, we then have

$$D(w, w') = \min_{\rho, \rho' \in \mathsf{Fix}(\varphi)} \{d(\mathcal{O}_Y(\sigma\rho), \mathcal{O}_Y(\sigma'\rho'))\} \leq d(\mathcal{O}_Y(\sigma\rho), \mathcal{O}_Y(\sigma'\rho')) .$$

$\square$

Proposition 8.4. *The map*

$$D': S_k(\varphi)/\sim_Y \times S_k(\varphi)/\sim_Y \longrightarrow \mathbb{Z}, \quad \text{where} \quad D'([w], [w']) = D(w, w'), \quad (8.9)$$

is a metric.

Proof. We first show that D' is well-defined. For this purpose we choose $w_1 \sim_Y w$ and $w_1' \sim_Y w'$ and compute using Eq. (8.6) of Proposition 8.3

$$D(w, w') \leq D(w, w_1) + D(w_1, w'),$$
$$D(w_1, w') \leq D(w_1, w) + D(w, w'),$$

from which, in view of $D(w, w_1) = D(w_1, w) = 0$, we obtain $D(w, w') = D(w_1, w')$. Clearly, $D(w_1, w') = D(w_1, w_1')$ follows in complete analogy and we derive $D(w, w') = D(w_1, w_1')$; thus, D' is well-defined.

D' consequently has the following properties: (a) for any $w, w' \in S_k(\varphi)$ we have $D'([w], [w']) \geq 0$; (b) $D'([w], [w']) = 0$ implies $w \sim_Y w'$ (by Lemma 8.2); (c) for any $w, w' \in S_k(\varphi)$ we have $D'([w], [w']) = D'([w'], [w])$, and finally (d) for any $w, w', w'' \in S_k(\varphi)$

$$D'([w], [w']) \leq D'([w], [w'']) + D'([w''], [w'])$$

holds according to Proposition 8.3, and it follows that D' is a metric. $\square$

8.3.3 A Replication-Deletion Scheme

It remains to specify a replication-deletion process for the SDS genotypes. We will choose a process based on the *Moran model* [143] that describes the time evolution of populations. A population M over a graph X is a mapping from vertices of X into natural numbers, and we call a vertex of X an element "present" in M if its multiplicity $M(x)$ satisfies $M(x) > 0$. We call the quantity $s(M) = \sum_x M(x)$ the size of M and define $\mathsf{M}[m]$ to be the set of populations of size m. A replication-deletion scheme $\mathcal{R}$ is a mapping $\mathcal{R}\colon \mathsf{M}[m] \longrightarrow \mathsf{M}[m]$, and we call the mapping $\mu\colon X \longrightarrow [0,1]$ the *fitness landscape*. The mapping μ assigns fitness values to elements of the population.

The specific replication-deletion scheme $\mathcal{R}_0\colon \mathsf{M}[m] \longrightarrow \mathsf{M}[m]$ that we will use in the following sections basically consists of the removal of an ordered pair (w, w') of elements from M and its replacement by the ordered pair $(w, \tilde{w})$:

- For w we pick an M-element with probability $M(x) \cdot \mu(x)/[\sum_x M(x)\mu(x)]$. The word w is subsequently subject to a replication event that maps w into $\tilde{w}$.
- For w' we select an M-element with probability $M(x)/s(M)$.

The replication-map, which maps w into $\tilde{w}$, is obtained by the following procedure: With probability q we independently select each index-pair of the form

$$\forall\, i;\ 1 \leq i \leq k-1;\quad \tau_i = (i, i+1), \tag{8.10}$$

of $w = (w_1, \ldots, w_k)$ and obtain the sequence $(\tau_{i_1}, \ldots, \tau_{i_m})$ of transpositions where $i_t \leq i_{t+1}$. We then set

$$\tilde{w} = (\tau_{i_1}, \ldots, \tau_{i_m})(w). \tag{8.11}$$

Accordingly, M and $\mathcal{R}_0(M)$ differ exactly in that the element w' of M is replaced by $\tilde{w}$.

So far there is no notion of *time* in our setup. We consider the applications of $\mathcal{R}_0$ to be independent events. The time interval Δt which elapses between two such events is assumed to be exponentially distributed, that is,

$$\mathsf{Prob}(\Delta t > \tau) = \exp\left[-\tau \sum_x M(x)\mu(x)\right]. \tag{8.12}$$

Intuitively, $\sum_x M(x)\mu(x)$ can be interpreted as a mean fitness of the population M at time t, which can only change after application of $\mathcal{R}_0$, since new elements in the population potentially emerge and others are being removed. According to Eq. (8.12), the population undergoes mutational changes in shorter periods of time if its mean fitness is higher.

8.3.4 Evolution of SDS-Schedules

In this section we study the evolution of SDS-schedules. We limit ourselves to presenting a few aspects of SDS-evolution, a detailed analysis can be found in [144]. In the following we consider the base graphs Y to be sampled from the random graph $G_{n,p}$, and we assume that the word w is a fixed word in which each vertex of Y occurs at least once, i.e., w is a fair word. Transitions of word populations between two phenotypes are obviously of critical importance to understand SDS evolution as they constitute the basic evolutionary steps. They are a result of the stochastic drift and can occur even when the two phenotypes in question have identical fitness. In [144] we investigated neutral evolution of SDS schedules. Explicitly, we selected a random fair word w_0 and generated a random mutant w_i in distance class i [i.e., $D(w_0, w_i) = i$]. We set the fitness of all words on the neutral network of w_0 and w_i to be 1 and to 0.1 for those words that are not on the neutral network. The protocol for the computer experiments is presented in Sections 8.3.5 and 8.5.

We monitored the fractions of the population on the neutral networks of w_0 and w_i. It turned out that for a wide spectrum of parameter sets the population is concentrated almost entirely on one of the two neutral networks and then switches between them.

The findings can be categorized as follows:

Case (a): The two neutral networks are "close," that is, the population at almost all times has some large fraction on both neutral networks. This scenario is generic (i.e., typically occurs for extended parameter ranges) in case $D = 1$.

Case (b): The population is almost always on either one of the two neutral networks for extended periods of time (epochs). Then rapid, fitness-neutral transitions between the two neutral networks are observed. This scenario is generic in case $D = 2$.

Case (c): The population is almost entirely on either one of the neutral networks, but transitions between the nets are very rare. This situation is generic for $D > 2$.

Accordingly, the distance measure D captures the closeness of neutral networks of words and appears to be of relevance to describe and analyze the time evolution of populations.

Next let us study the role of the mutation rate q. For this purpose we consider *single-net-landscapes* (of ratio $r > 1$), which is a mapping $\mu_{w_0} : W_k \longrightarrow [0,1]$ such that every word w with $w \sim_Y w_0$ satisfies $\mu_{w_0}(w) = \mu_{w_0}(w_0) = x$, and $\mu_{w_0}(w') = x'$ otherwise where $x/x' = r$. We set $x = 1$ and $x' = 0.1$, that is, $r = 10$.

In the following we show that there is a *critical mutation rate* $q_*(n, k, p, s)$ characterized as follows: In a single-net landscape a word-population replicating with error probability $q > q_*$ is essentially randomly distributed, and a population replicating with $q < q_*$ remains localized on its neutral network. We refer to words that are on the neutral network of w_0 as "masters" and set their fitness to be 1, while any other word has fitness 0.1. We now gradually increase the mutation probability q of the replication event and study the parts of the population in the distance classes $D_i(w_0)$, where $w \in D_i(w_0)$ if and only if $D(w_0, w) = i$ holds. Similar studies for RNA sequences as genomes and RNA secondary structures as phenotypes can be found in [131,138]. That particular analysis was motivated by the seminal work of Eigen et al. on the molecular quasi-species in [139]. Clearly, for $q = 0$ the population consists of m identical copies of w_0, but as q increases, mutations of higher distance classes emerge. It is evident from Figure 8.5 that there exists a critical mutation probability $q_*(n, k, p, s)$ at which the population becomes essentially randomly distributed. The protocol for the computer experiments is given in Sections 8.3.5 and 8.6.

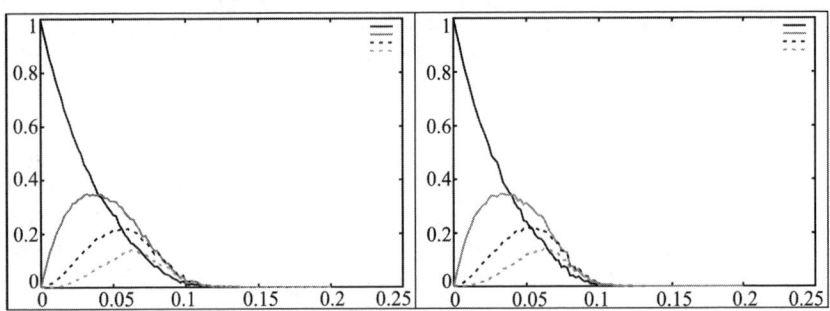

Fig. 8.5. The critical mutation rate for $p = 0.50$. The x-axis gives p and the y-axis denotes the percentage of the population in the respective distance classes. The parameters are $n = 25$, and $k = 51$. In the figure on the left a fixed random fair word was used for all samples, and in the figure on the right a random fair word was used for each sample point.

8.3.5 Pseudo-Codes

Algorithm 8.5. (Parameters: n = 52, k = 103, q = 0.0625, p, D, s)

- Generate a random fair word w_0.
- Generate a random mutant w_i of w_0 with Distance(w_0, w_i) = i.
- Generate an element Y of G_n,p.
- Initialize a pool with s copies of w_i all with fitness 1.
- Repeat
    - Compute average fitness lambda.
    - Sample Delta T from exponential distribution with
        parameter lambda and increment time by Delta T.
    - Pick w_a at random from the pool weighted by fitness.
    - Pick w_b at random from pool minus w_1.
    - Replace w_b by a copy of w_a.
    - Mutate the copy of w_a with probability q.

- Update fitness of mutated copy.
- At every 100th iteration step output fractions of pool with
    (i) distance 0 to w_0, and (ii) distance 0 to w_i.

Algorithm 8.6. (Parameters: n = 25, k = 51, p = 0.50)
The line preceeded by [fix] is only used for the runs with a fixed word, the line preceeded by [vary] is only used for the run with a varying word.

- [fix] Generate a fair word w of length k over Y
- for q = 0.0 to 0.2 using stepsize 0.02 do
{
    - repeat 100
        - generate a random graph Y in G(n,p)
        - [vary] Generate a fair word w of length k over Y
        - initialize pool with 250 copies of w
        - perform 10000 basic replication/mutation steps
        - accumulate distance distribution relative to w
    - output average fractions of distance class 0, 1, 2, and 3.
}

8.4 Discrete Derivatives

The concept of derivatives is of central importance for the theory for classical dynamical systems. This motivates the question of whether there are analogue operators in the context of sequential dynamical systems and finite discrete dynamical systems in general. In fact, various definitions of discrete derivatives have been developed. In case of binary states the notion of *Boolean derivatives* [145, 146] has been introduced.

Definition 8.7. Let $f \colon \mathbb{F}_2^n \longrightarrow \mathbb{F}_2$, and let $x = (x_1, x_2, \ldots, x_n) \in \mathbb{F}_2^n$. The partial Boolean derivative of f with respect to x_i at the point x is

$$D_i f(x) = \frac{\partial f}{\partial x_i}(x) = f(\bar{x}^i) + f(x) , \qquad (8.13)$$

where $\bar{x}^i = (x_1, \ldots, 1 + x_i, \ldots, x_n)$.

Thus, $D_i f(x)$ can be viewed as to measure the sensitivity of f with respect to the variable x_i at the point x.

Example 8.8. Consider $f = \text{parity}_3 \colon \mathbb{F}_2^3 \longrightarrow \mathbb{F}_2$ given by $f(x_1, x_2, x_3) = x_1 + x_2 + x_3$. In this case we see

$$D_2 f(x) = f(\bar{x}^2) + f(x)$$
$$= (x_1 + (1 + x_2) + x_3) + (x_1 + x_2 + x_3) = 1 .$$

◇

Basic Properties

Note first that $D_i f(x)$ does not depend on x_i in the sense that

$$D_i^2 f(x) \equiv 0 .$$

This is straightforward to verify by applying the preceding definition. The Boolean derivative has similarities with the "classical" derivative. For example, it is easy to see that it is a linear operator in the sense that for $c_1, c_2 \in \{0, 1\}$ we have

$$D_i(c_1 f_1 + c_2 f_2) = c_1 D_1 f_1 + c_2 D_2 f_2 ,$$

and that partial derivatives commute:

$$\frac{\partial^2 F}{\partial x_i \partial x_j} = \frac{\partial^2 F}{\partial x_j \partial x_i} .$$

In addition, if $f \colon \mathbb{F}_2^{n-1} \longrightarrow \mathbb{F}_2$ and $g \colon \mathbb{F}_2^n \longrightarrow \mathbb{F}_2$ where $g(x_1, \ldots, x_n) = x_n f(x_1, \ldots, \ldots, x_{n-1})$, then as a special case we have

$$\frac{\partial g}{\partial x_n}(x) = f(x_1, \ldots, x_{n-1}) .$$

The product rule or Leibniz's rule differs from the classical form since

$$\frac{\partial(fg)}{\partial x_i} = \frac{\partial f}{\partial x_i} g + f \frac{\partial g}{\partial x_i} + \frac{\partial f}{\partial x_i} \frac{\partial g}{\partial x_i} .$$

We give a detailed derivation of this formula since it nicely illustrates what it implies not to have the option of taking limits:

$$\begin{aligned}
\frac{\partial(fg)}{\partial x_i}(x) &= f(\bar{x}^i)g(\bar{x}^i) + f(x)g(x) \\
&= f(\bar{x}^i)g(\bar{x}^i) + f(x)g(\bar{x}^i) + f(x)g(\bar{x}^i) + f(x)g(x) \\
&= \frac{\partial f}{\partial x_i}(x)g(\bar{x}^i) + f(x)\frac{\partial g}{\partial x_i}(x) \\
&= \frac{\partial f}{\partial x_i}(x)g(\bar{x}^i) + \frac{\partial f}{\partial x_i}(x)g(x) + \frac{\partial f}{\partial x_i}(x)g(x) + +f(x)g(x) \\
&= \frac{\partial f}{\partial x_i}(x)g(x) + f(x)\frac{\partial g}{\partial x_i}(x) + \frac{\partial f}{\partial x_i}(x)\frac{\partial g}{\partial x_i}(x).
\end{aligned}$$

The last term is the "$\mathcal{O}((\Delta h)^2)$" term that would vanish when taking the limit in the continuous case. For the generalized chain rule and Boolean derivatives the number of such additional terms becomes excessive. To illustrate this let $F, G, f_i \colon \mathbb{F}_2^n \longrightarrow \mathbb{F}_2$ for $1 \leq i \leq n$ with $G(x) = F(f_1(x), \ldots, f_n(x))$, and let $P = \{k_1, \ldots, k_l\} \subset \mathbb{N}_n = \{1, 2, \ldots, n\}$. Using multi-index-style notation,

$$D_P H = \Big[\prod_{i \in P} D_i\Big] H,$$

we have the *chain rule*

$$\frac{\partial G}{\partial x_i}(x) = \sum_{\emptyset \neq P \subset \mathbb{N}_n} \left[\frac{\partial^{|P|} F}{\partial^P x_i}(f_1(x), \dots, f_n(x)) \prod_{k \in P} \frac{\partial f_k}{\partial x_i}(x) \right]. \qquad (8.14)$$

Note that the sum over singleton subsets $P \subset \mathbb{N}_n$ gives

$$\sum_{k=1}^n \frac{\partial F}{\partial x_k}(f_1(x), \dots, f_n(x)) \frac{\partial f_k}{\partial x_i}(x) \ ,$$

which has the same structure as the classical chain rule.

8.2. Write explicit expressions for the chain rule when $F, G, f_i \colon \mathbb{F}_2^3 \longrightarrow \mathbb{F}_2$.
[1]

For more details on the derivation of the chain rule, see, for example, [147]. In [148] you can also find results on Boolean Lie algebras.

Computing the Boolean partial derivatives even for a small discrete finite dynamical system is nontrivial. For SDS matters get even more involved: Because of the compositional structure of SDS, the chain rule will typically have to be applied multiple times in order to compute the partial derivative $[\mathbf{F}_Y, w]_j / x_i$. Even computing a relatively simple partial derivatives such as $D_1([\mathbf{Nor}_{\text{Wheel}_4}, (1, 0, 2, 4, 3)])$ is a lengthy process. The notion of a Boolean derivative in its current form may be conceptually useful, but it is challenging to put it to effective use for, e.g., SDS. The identification of operators aiding the analysis of finite dynamical system would be very desirable.

8.5 Real-Valued and Continuous SDS

Real-valued SDS allow for the use of conventional calculus. Some versions of real-valued SDS have been studied in the context of *coupled map lattices* (CML) in [149]. As a specific example let $Y = \text{Circ}_n$, and take vertex states $x_i \in \mathbb{R}$. We set

$$f_i(x_{i-1}, x_i, x_{i+1}) = \epsilon x_{i-1} + f(x_i) + \epsilon x_{i+1} \ ,$$

where $f \colon \mathbb{R} \longrightarrow \mathbb{R}$ is some suitable function and $\epsilon \geq 0$ is the coupling parameter. For $\epsilon = 0$ the dynamics of each vertex evolves on its own as determined by the function f. As ϵ increases the stronger the dynamics of the vertices are coupled. For CML the vertex functions are applied synchronously and not in a given order as for SDS. This particular form of a CML may be viewed as an elementary cellular automaton with states in $\mathbb{R}$ rather than $\{0, 1\}$. The work on CML over circle graphs have been extended to arbitrary directed graphs in, e.g., [24] — for an ODE analogue see [150]. By considering real-valued differentiable vertex functions, it seems likely that the structure of the Y-local maps should allow for interesting analysis and insight.

8.3. Show that without loss of generality a real-valued permutation SDS over $Y = \mathsf{Line}_2$ can be written as

$$x_1 \mapsto f_1(x_1, x_2),$$
$$x_2 \mapsto f_2(f_1(x_1, x_2), x_2) \,.$$

What can you say about this system? You may assume that f_1 and f_2 are continuously differentiable or smooth. Can you identify interesting cases for special classes of maps f_1 and f_2? What if f_1 and f_2 are polynomials of degree at most 2? What is the structure of the Jacobian of the composed SDS?   [**3**]

We close this section with an example of a real-valued SDS. It is an SDS version of the Hénon map [74] arising in the context of chaotic classical discrete dynamical systems.

Example 8.9 (A real-valued SDS). In this example we consider the SDS over Circ_3 with states $x_i \in \mathbb{R}$.

$$F_1(x_1, x_2, x_3) = (1 + x_2 - a\,x_1^2, x_2, x_3),$$
$$F_2(x_1, x_2, x_3) = (x_1, b\,x_3, x_3), \qquad\qquad (8.15)$$
$$F_3(x_1, x_2, x_3) = (x_1, x_2, x_1),$$

where $a, b > 0$ are real parameters. We use the update order $\pi = (3, 1, 2)$ and set $a = 1.4$ and $b = 0.3$ with initial value $(0.0, 0.0, 0.0)$. The projection onto the first two coordinates of the orbit we obtain is shown in Figure 8.6.   ◇

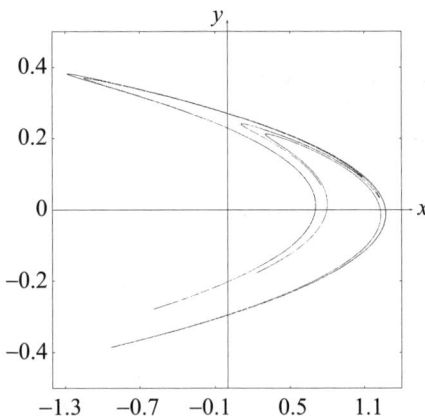

Fig. 8.6. The projection of the orbit of Example 8.9.

This example also illustrates the fact that any system using a parallel update order with maps F_i can be embedded in a sequential system as illustrated in Figure 1.5.

8.6 L-Local SDS

In the cases of sequentially updated random Boolean networks, asynchronous cellular automata, and SDS, exactly one vertex state is potentially altered per vertex update, and this is done based on the states of the vertices in the associated ball of radius 1. It is clear, for instance, in the case of communication networks where discrete data packets are exchanged, that *simultaneous* state changes occur. That is, two or more vertex states are altered at one update step. Parallel systems represent an extreme case in which *all* vertex states may change at a single update step. The framework described in the following is a natural generalization of SDS and it allows one to consider hybrid systems, which may be viewed to certain degrees as sequential *and* parallel at the same time. In Section 8.7 we will show in detail how to model routing protocols via L-local SDS .

Let

$$L : Y \longrightarrow \{X \mid X \text{ is a subgraph of } Y \}, \quad v_i \mapsto L(v_i), \qquad (8.16)$$

be a mapping assigning to each vertex of Y a subgraph of Y, and let $\lambda(v_i)$ denote the cardinality of the vertex set of the subgraph $L(v_i)$. Furthermore, we define the vertex functions as

$$f_{v_i} : K^{\lambda(v_i)} \longrightarrow K^{\lambda(v_i)} . \qquad (8.17)$$

For each vertex $v_i \in Y$ we consider the sequence

$$(x_{v_{j_1}}, \dots, x_{v_{j_s}}, x_{v_{j_{s+1}}} = x_{v_i}, x_{v_{j_{s+2}}}, \dots, x_{v_{j_r}}), \qquad (8.18)$$

where $j_t < j_{t+1}$ and $v_{j_h} \in L(v_i)$. We next introduce the map

$$n_L[v_i] : \{1, \dots, \lambda(v_i)\} \longrightarrow v[Y], \quad t \mapsto v_{j_t} , \qquad (8.19)$$

and define the L-local map of v_i, $F_{v_i}^L : K^n \longrightarrow K^n$:

$$F_{v_i}^L(x) = (y_{v_1}, \dots, y_{v_n}), \quad y_{v_h} = \begin{cases} (f_{v_i}(n_L[v_i]))_{v_h} & \text{for } v_h \in L(v_i), \\ x_{v_h} & \text{otherwise.} \end{cases} \qquad (8.20)$$

We are now prepared to define an L-local SDS over a word:

Definition 8.10. Let w be a word and $L : Y \longrightarrow \{X < Y\}$ be a map assigning Y-vertices to Y-subgraph. The triple $(Y, (F_{v_i})_{v_i \in Y}, w)$ is the L-local SDS. The composition of the L-local maps $F_{v_i}^L$ according to w,

$$[(F_{v_i})_{v_i \in v[Y]}, w] = \prod_{i=k}^{1} F_{w_i} : K^n \longrightarrow K^n, \qquad (8.21)$$

is the L-local SDS-map.

8.7 Routing

The scope of this section is to cast packet-switching problems arising in the context of ad hoc *networks* in the framework of SDS. We believe that such a formulation has the potential to shed new light on networking in general, and routing at the networking layer in particular. We restrict ourselves to presenting only some of the core ideas of how to define these protocols as L-local maps. The interested reader can find a detailed analysis of these protocols in [151–153].

In the following we adopt an end-to-end perspective: The object of study is the flow of data packets "hopping" from node to node from a source to a given destination. In contrast to the common approach where perceived congestion is a function of the states of all nodes along a path, we introduce SDS-based protocols, which are *locally load-sensing*. We assume that all links are static and perfectly reliable (or error-free) with zero delay. Furthermore, packets can only be transmitted from one vertex v to another vertex v' if v and v' are adjacent in Y.

Our main objective is to describe the dynamical evolution of the data-queue sizes of the entire system. We consider unlabeled packets, that is, packets do not contain explicit routing information in their headers and cannot be addressed individually. We assume that some large number of packets is injected into the network via the source vertex and that the destination has enough capacity to receive all data packets. The source successively loads the network and after some finite number of steps the system reaches an orbit in phase space. Particular observables we are interested in are the total number of packets located at the vertices (total load) and the throughput, which is the rate at which packets arrive at the destination.

8.7.1 Weights

We will refer to vertex $v_i \in v[Y]$ by its index i. Let Q_k denote the number of packets located at vertex k, let m_k be the queue capacity for packets located at vertex k, and let $m_{\{k,i\}}$ be the edge capacity of the edge $\{k, i\}$. We assume uniform edge capacities, i.e., $m_{\{k,i\}} = \mu$. Clearly, we have $Q_k \in \mathbb{Z}/m_k\mathbb{Z}$ since Q_k cannot exceed the queue-capacity m_k, and we take Q_k as the state of vertex k. Suppose we want to transmit packets from vertex k to its neighbors. In the following we introduce a procedure by which we assign weights to the neighbors of k. Our procedure is generic, parameterized, and in its parameterization location-invariant. Its base parameters are (1) the distance to the destination, δ, (2) the relative load, and (3) the absolute queue size.

Let k be a vertex of degree $d(k)$ and $B_1'(k) = \{i_1, \ldots, i_{d(k)}\}$ be the set of neighbors of k. We set

$$c_h = \{i_j \in B_Y'(k) \mid d(i_j, \delta) = h\} \ .$$

Let $(h_1, \ldots, h_s)$ be the tuple of indices such that $c_{h_j} \neq \varnothing$ and $h_j < h_{j+1}$. In view of $B'_Y(k) = \bigcup_h c_h$, we can now define the rank of a k-neighbor:

$$\text{rnk} \colon B'_Y(k) \longrightarrow \mathbb{N}, \qquad \text{rnk}(i_j) = r, \text{ where } i_j \in c_{h_r}, \ h_r = (h_1, \ldots, h_s)_r \, . \tag{8.22}$$

The weight w_{i_j} of vertex i_j is given by

$$w(i_j) = e^{-a \, \text{rnk}(i_j)} \left[1 - \frac{Q_{i_j}}{m_{i_j}} \right]^b \left[\frac{m_{i_j}}{m_{\max}} \right]^c, \quad \text{where } a, b, c > 0 \tag{8.23}$$

and $w^*(i_j) = w(i_s)/(\sum_{j \in B'_1(k)} w(j))$.

8.7.2 Protocols as Local Maps

Suppose we have Q_k packets at vertex k and the objective is to route them via neighbors of k. For this purpose we first compute for $B'_Y(k) = \{i_1, \ldots, i_{d(k)}\}$ the family

$$W^*(k) = (w^*(i_1), \ldots, w^*(i_{d(k)}))$$

of their relative weights. Without loss of generality we may assume that $w^*_{i_1}$ is maximal and set

$$y_{i_r} = \begin{cases} \lfloor Q_k \cdot w^*(i_r) \rfloor & \text{for } r \neq 1, \\ Q_k - \sum_{r=2}^{d(k)} \lfloor Q_k \cdot w^*(i_r) \rfloor & \text{for } r = 1 \, . \end{cases} \tag{8.24}$$

The y_{i_r} can be viewed as idealized flow rates, that is, where edge capacities and buffer sizes of the neighbors are virtually infinite. However, taking into account the edge-capacity μ, the queue-capacity and actual queue-size of vertex i_r, we observe that

$$s_{i_r} = \min\{y_{i_r}, \mu, (m_{i_r} - Q_{i_r})\} \tag{8.25}$$

is the maximal number of packets that can be forwarded to vertex i_r. This is based on the system state and $W^*(k)$. We are now prepared to update the states of the vertices contained in $B_Y(k)$ *in parallel* as follows:

$$\tilde{Q}_a = \begin{cases} Q_k - \sum_{r=1}^{d(k)} s_{i_r} & \text{for } a = k, \\ Q_a + s_a & \text{for } a \in B'_{1,Y}(k) \, . \end{cases} \tag{8.26}$$

That is, vertex k sends the quantities s_{i_r} [Eq. (8.25)] in parallel to its neighbors and consequently $\sum_{r=1}^{d(k)} s_{i_r}$ packets are subtracted from its queue. Instantly, the queue size of each neighbor i_r increases by exactly s_{i_r}. It is obvious that this protocol cannot lose data packets.

In view of Eq. (8.26), we can now define the L-local map (Section 8.6) F_k^{L} as follows:

$$F_k^{\mathsf{L}} \colon \prod_{k \in Y} (\mathbb{Z}/m_k\mathbb{Z}) \longrightarrow \prod_{k \in Y} (\mathbb{Z}/m_k\mathbb{Z}), \qquad F_k^{\mathsf{L}}((Q_h)_h) = (\tilde{Q}_h)_h \, , \tag{8.27}$$

where

$$
\tilde{Q}_a = \begin{cases} Q_k - \sum_{r=1}^{d(k)} s_{i_r} & \text{for } a = k, \\ Q_a + s_a & \text{for } a \in B'_{1,Y}(k), \\ Q_a & \text{for } a \notin B_Y(k) . \end{cases}
$$

Indeed, F_k^{L} is a L-local map as defined in Eq. (8.20) of Section 8.6: it (1) potentially alters the states of *all* vertices contained in $B_Y(k)$ in parallel and (2) it does so based exclusively on states associated to vertices in $B_Y(k)$.

References

1. David R. Jefferson. Virtual time. *ACM Transactions on Programming Languages and Systems*, 7(3):404–425, July 1985.
2. J. Misra. Distributed discrete-event simulation. *ACM Computing Surveys*, 18(1):39–65, March 1986.
3. Shawn Pautz. An algorithm for parallel s_n sweeps on unstructured meshes. *Nuclear Science and Engineering*, 140:111–136, 2002.
4. K. Nagel, M. Rickert, and C. L. Barrett. Large-scale traffic simulation. *Lecture Notes in Computer Science*, 1215:380–402, 1997.
5. M. Rickert, K. Nagel, M. Schreckenberg, and A. Latour. Two lane traffic simulations using cellular automata. *Physica A*, 231:534–550, October 1996.
6. K. Nagel, M. Schreckenberg, A. Schadschneider, and N. Ito. Discrete stochastic models for traffic flow. *Physical Review E*, 51:2939–2949, April 1995.
7. K. Nagel and M. Schreckenberg. A cellular automaton model for freeway traffic. *Journal de Physique I*, 2:2221–2229, 1992.
8. Kai Nagel and Peter Wagner. *Traffic Flow: Approaches to Modelling and Control*. John Wiley & Sons, New York, 2006.
9. Randall J. LeVeque. *Numerical Methods for Conservation Laws*, 2nd ed. Birkhauser, Boston, 1994.
10. Tommaso Toffoli. Cellular automata as an alternative to (rather than an approximation of) differential equations in modeling physics. *Physica D*, 10:117–127, 1984.
11. Justin L. Tripp, Anders Å. Hansson, Maya Gokhale, and Henning S. Mortveit. Partitioning hardware and software for reconfigurable supercomputing applications: A case study. In *Proceedings of the 2005 ACM/IEEE Conference on Supercomputing (SC|05)*, September 2005. Accepted for inclusion in proceedings.
12. Eric Weisstein. Mathworld. http://mathworld.wolfram.com, 2005.
13. Anthony Ralston and Philip Rabinowitz. *A First Course in Numerical Analysis*, 2nd ed. Dover Publications, 2001.
14. C. L. Barrett, H. B. Hunt III, M. V. Marathe, S. S. Ravi, D. J. Rosenkrantz, and R. E. Stearns. On some special classes of sequential dynamical systems. *Annals of Combinatorics*, 7:381–408, 2003.
15. M. R. Garey and D. S. Johnson. *Computers and Intractability: A Guide to the Theory of NP-Completeness*. W.H. Freeman, San Francisco, 1979.

16. C. L. Barrett, H. H. Hunt, M. V. Marathe, S. S. Ravi, D. Rosenkrantz, and R. Stearns. Predecessor and permutation existence problems for sequential dynamical systems. In *Proc. of the Conference on Discrete Mathematics and Theoretical Computer Science*, pages 69–80, 2003.

17. K. Sutner. On the computational complexity of finite cellular automata. *Journal of Computer and System Sciences*, 50(1):87–97, 1995.

18. Jarkko Kari. Theory of cellular automata: A survey. *Theoretical Computer Science*, 334:3–33, 2005.

19. Jarkko Kari. Reversibility of 2D CA. *Physica D*, 45–46:379–385, 1990.

20. C. L. Barrett, H. H. Hunt, M. V. Marathe, S. S. Ravi, D. Rosenkrantz, R. Stearns, and P. Tosic. Gardens of Eden and fixed point in sequential dynamical systems. In *Discrete Models: Combinatorics, Computation and Geometry*, pages 95–110, 2001. Available via LORIA, Nancy, France. http://www.dmtcs.org/dmtcs-ojs/index.php/proceedings/article/view/dmAA0106/839.

21. Richard P. Stanley. *Enumerative Combinatorics: Volume 1.* Cambridge University Press, New York, 2000.

22. Kunihiko Kaneko. Pattern dynamics in spatiotemporal chaos. *Physica D*, 34:1–41, 1989.

23. York Dobyns and Harald Atmanspacher. Characterizing spontaneous irregular behavior in coupled map lattices. *Chaos, Solitions and Fractals*, 24:313–327, 2005.

24. Chai Wah Wu. Synchronization in networks of nonlinear dynamical systems coupled via a directed graph. *Nonlinearity*, 18:1057–1064, 2005.

25. Thomas M. Liggett. *Interacting Particle Systems.* Classics in Mathematics. Springer, New York, 2004.

26. Wolfgang Reisig and Grzegorz Rozenberg. *Lectures on Petri Nets I: Basic Models: Advances in Petri Nets.* Number 1491 in Lecture Notes in Computer Science. Springer-Verlag, New York, 1998.

27. John von Neumann. *Theory of Self-Reproducing Automata.* University of Illinois Press, Chicago, 1966. Edited and completed by Arthur W. Burks.

28. E. F. Codd. *Cellular Automata.* Academic Press, New York, 1968.

29. G. A. Hedlund. Endomorphisms and automorphisms of the shift dynamical system. *Math. Syst. Theory*, 3:320–375, 1969.

30. Erica Jen. Aperiodicity in one-dimensional cellular automata. *Physica D*, 45:3–18, 1990.

31. Burton H. Voorhees. *Computational Analysis of One-Dimensional Cellular Automata*, volume 15 of *A*. World Scientific, Singapore, 1996.

32. O. Martin, A. Odlyzko, and S. Wolfram. Algebraic properties of cellular automata. *Commun. Math. Phys.*, 93:219–258, 1984.

33. René A. Hernández Toledo. Linear finite dynamical systems. *Communcations in Algebra*, 33:2977–2989, 2005.

34. Mats G. Nordahl. *Discrete Dynamical Systems.* PhD thesis, Institute of Theoretical Physics, Göteborg, Sweden, 1988.

35. Kristian Lindgren, Christopher Moore, and Mats Nordahl. Complexity of two-dimensional patterns. *Journal of Statistical Physics*, 91(5–6):909–951, 1998.

36. Stephen J. Willson. On the ergodic theory of cellular automata. *Mathematical Systems Theory*, 9(2):132–141, 1975.

37. D. A. Lind. Applications of ergodic theory and sofic systems to cellular automata. *Physica D*, 10D:36–44, 1984.

38. P. A. Ferrari. Ergodicity for a class of probabilistic cellular automata. *Rev. Mat. Apl.*, 12:93–102, 1991.

39. Henryk Fukś. Probabilistic cellular automata with conserved quantities. *Nonlinearity*, 17:159–173, 2004.

40. Michele Bezzi, Franco Celada, Stefano Ruffo, and Philip E. Seiden. The transition between immune and disease states in a cellular automaton model of clonal immune response. *Physica A*, 245:145–163, 1997.

41. U. Frish, B. Hasslacher, and Y. Pomeau. Lattice-gas automata for the Navier-Stokes equations. *Physical Review Letters*, 56:1505–1508, 1986.

42. Dieter A. Wolf-Gladrow. *Lattice-Gas Cellular Automata and Lattice Bolzmann Models: An Introduction*, volume 1725 of *Lecture Notes in Mathematics*. Springer-Verlag, New York, 2000.

43. J.-P. Rivet and J. P. Boon. *Lattice Gas Hydrodynamics*, volume 11 of *Cambridge Nonlinear Science Series*. Cambridge University Press, New York, 2001.

44. Parimal Pal Chaudhuri. *Additive Cellular Automata. Theory and Applications*, volume 1. IEEE Computer Society Press, 1997.

45. Palash Sarkar. A brief history of cellular automata. *ACM Computing Surveys*, 32(1):80–107, 2000.

46. Andrew Ilichinsky. *Cellular Automata: A Discrete Universe*. World Scientific, Singapore, 2001.

47. Stephen Wolfram. *Theory and Applications of Cellular Automata*, volume 1 of *Advanced Series on Complex Systems*. World Scientific, Singapore, 1986.

48. B. Schönfisch and A. de Roos. Synchronous and asynchronous updating in cellular automata. *BioSystems*, 51:123–143, 1999.

49. Stephen Wolfram. Statistical mechanics of cellular automata. *Rev. Mod. Phys.*, 55:601–644, 1983.

50. Bernard Elspas. The theory of autonomous linear sequential networks. *IRE Trans. on Circuit Theory*, 6:45–60, March 1959.

51. William Y. C. Chen, Xueliang Li, and Jie Zheng. Matrix method for linear sequential dynamical systems on digraphs. *Appl. Math. Comput.*, 160:197–212, 2005.

52. Ezra Brown and Theresa P. Vaughan. Cycles of directed graphs defined by matrix multiplication (mod n). *Discrete Mathematics*, 239:109–120, 2001.

53. Wentian Li. *Complex Patterns Generated by Next Nearest Neighbors Cellular Automata*, pages 177–183. Elsevier, Burlington, MA, 1998. (Reprinted from *Comput. & Graphics Vol. 13, No 4, 531–537, 1989.*)

54. S. A. Kauffman. Metabolic stability and epigenesis in randomly constructed genetic nets. *Journal of Theoretical Biology*, 22:437–467, 1969.

55. I. Shmulevich and S. A. Kauffman. Activities and sensitivities in Boolean network models. *Physical Review Letters*, 93(4):048701:1–4, 2004.

56. E. R. Dougherty and I. Shmulevich. Mappings between probabilistic Boolean networks. *Signal Processing*, 83(4):799–809, 2003.

57. I. Shmulevich, E. R. Dougherty, and W. Zhang. From Boolean to probabilistic Boolean networks as models of genetic regulatory networks. *Proceedings of the IEEE*, 90(11):1778–1792, 2002.

58. I. Shmulevich, E. R. Dougherty, S. Kim, and W. Zhang. Probabilistic Boolean networks: A rule-based uncertainty model for gene regulatory networks. *Bioinformatics*, 18(2):261–274, 2002.

59. Carlos Gershenson. Introduction to random Boolean networks. arXiv:nlin.AO/040806v3-12Aug2004, 2004. (Accessed August 2005.)

60. Mihaela T. Matache and Jack Heidel. Asynchronous random Boolean network model based on elementary cellular automata rule 126. *Physical Review E*, 71:026231:1–13, 2005.

61. Michael Sipser. *Introduction to the Theory of Computation*. PWS Publishing Company, Boston, 1997.

62. John E. Hopcroft and Jeffrey D. Ullman. *Introduction to Automata Theory, Languages, and Computation*. Addison-Wesley, Reading, MA, 1979.

63. Mohamed G. Gouda. *Elements of Network Protocol Design*. Wiley-Interscience, New York, 1998.

64. J. K. Park, K. Steiglitz, and W. P. Thruston. Soliton-like behavior in automata. *Physica D*, 19D:423–432, 1986.

65. N. Bourbaki. *Groupes et Algebres de Lie*. Hermann, Paris, 1968.

66. J. P. Serre. *Trees*. Springer-Verlag, New York, 1980.

67. Sheldon Axler. *Linear Algebra Done Right*, 2nd ed. Springer-Verlag, New York, 1997.

68. P. Cartier and D. Foata. *Problemes combinatoires de commutation et reárrangements*, volume 85 of *Lecture Notes in Mathematics*. Springer-Verlag, New York, 1969.

69. Volker Diekert. *Combinatorics on Traces*, volume 454 of *Lecture Notes in Computer Science*. Springer-Verlag, New York, 1990.

70. Richard P. Stanley. Acyclic orientations of graphs. *Discrete Math.*, 5:171–178, 1973.

71. Morris W. Hirsch and Stephen Smale. *Differential Equations, Dynamical Systems, and Linear Algebra*. Academic Press, New York, 1974.

72. Lawrence Perko. *Differential Equations and Dynamical Systems*. Springer-Verlag, New York, 1991.

73. Erwin Kreyszig. *Introductory Functional Analysis with Applications*. John Wiley and Sons, New York, 1989.

74. Michael Benedicks and Lennart Carleson. The dynamics of the Hénon map. *Annals of Mathematics*, 133:73–169, 1991.

75. John B. Fraleigh. *A First Course in Abstract Algebra*, 7th ed. Addison-Wesley, Reading, MA, 2002.

76. P. B. Bhattacharya, S. K. Jain, and S. R. Nagpaul. *Basic Abstract Algebra*, 2nd ed. Cambridge University Press, New York, 1994.

77. Nathan Jacobson. *Basic Algebra I*, 2nd ed. W.H. Freeman and Company, San Francisco, 1995.

78. Thomas W. Hungerford. *Algebra*, volume 73 of *GTM*. Springer-Verlag, New York, 1974.

79. B. L. van der Waerden. *Algebra Volume I*. Springer-Verlag, New York, 1971.

80. B. L. van der Waerden. *Algebra Volume II*. Springer-Verlag, New York, 1971.

81. Warren Dicks. *Groups Trees and Projective Modules*. Springer-Verlag, New York, 1980.

82. Reinhard Diestel. *Graph Theory*, 2nd ed. Springer-Verlag, New York, 2000.

83. Chris Godsil and Gordon Royle. *Algebraic Graph Theory*. Number 207 in *GTM*. Springer-Verlag, New York, 2001.

84. John Riordan. *Introduction to Combinatorial Analysis*. Dover Publications, Mineola, NY, 2002.

85. J. H. van Lint and R. M. Wilson. *A Course in Combinatorics*. Cambridge University Press, New York, 1992.

86. John Guckenheimer and Philip Holmes. *Nonlinear Oscillations, Dynamical Systems, and Bifurcations of Vector Fields*. Springer-Verlag, New York, 1983.

87. Earl A. Coddington and Norman Levinson. *Theory of Ordinary Differential Equations*. McGraw-Hill, New York, 1984.

88. Robert L. Devaney. *An Introduction to Chaotic Dynamical Systems*, 2nd ed. Reading, MA, Addison-Wesley, 1989.

89. Welington de Melo and Sebastian van Strien. *One-Dimensional Dynamics*. Springer-Verlag, Berlin, 1993.

90. Reinhard Laubenbacher and Bodo Paraigis. Equivalence relations on finite dynamical systems. *Adv. Appl. Math.*, 26:237–251, 2001.

91. J. S. Milne. *Étale Cohomology*. Princeton University Press, Princeton, NJ, 1980.

92. Reinhard Laubenbacher and Bodo Pareigis. Update schedules of sequential dynamical systems. *Discrete Applied Mathematics*, 154(6):980–994, 2006.

93. C. M. Reidys. The phase space of sequential dynamical systems. *Annals of Combinatorics*. Submitted in 2006.

94. C. L. Barrett, H. S. Mortveit, and C. M. Reidys. Elements of a theory of simulation II: Sequential dynamical systems. *Appl. Math. Comput.*, 107(2–3):121–136, 2000.

95. Saunders Mac Lane. *Category Theory for the Working Mathematician*, 2nd ed. Number 5 in *GTM*. Springer-Verlag, 1998.

96. N. Kahale and L. J. Schulman. Bounds on the chromatic polynomial and the number of acyclic orientations of a graph. *Combinatorica*, 16:383–397, 1996.

97. N. Linial. Legal colorings of graphs. *Proc. 24th Symp. on Foundations of Computer Science*, 24:470–472, 1983.

98. U. Manber and M. Tompa. The effect of number of Hamiltonian paths on the complexity of a vertex-coloring problem. *SIAM J. Comp.*, 13:109–115, 1984.

99. R. Graham, F. Yao, and A. Yao. Information bounds are weak in the shortest distance problem. *J. ACM*, 27:428–444, 1980.

100. C. L. Barrett, H. S. Mortveit, and C. M. Reidys. Elements of a theory of simulation IV: Fixed points, invertibility and equivalence. *Appl. Math. Comput.*, 134:153–172, 2003.

101. C. M. Reidys. On certain morphisms of sequential dynamical systems. *Discrete Mathematics*, 296(2–3):245–257, 2005.

102. Reinhard Laubenbacher and Bodo Pareigis. Decomposition and simulation of sequential dynamical systems. *Adv. Appl. Math.*, 30:655–678, 2003.

103. William S. Massey. *Algebraic Topology: An Introduction*, volume 56 of *GTM*. Springer-Verlag, New York, 1990.

104. William S. Massey. *A Basic Course in Algebraic Topology*, volume 127 of *GTM*. Springer-Verlag, New York, 1997.

105. Warren Dicks and M. J. Dunwoody. *Groups Acting on Graphs*. Cambridge University Press, New York, 1989.

106. F. T. Leighton. Finite common coverings of graphs. *Journal of Combinatorial Theory*, 33:231–238, 1982.

107. Béla Bollobás. *Graph Theory. An Introductory Course*, volume 63 of *GTM*. Springer-Verlag, New York, 1979.

108. J. H. van Lint. *Introduction to Coding Theory*, 3rd ed. Number 86 in *GTM*. Springer-Verlag, New York, 1998.

109. C. L. Barrett, H. S. Mortveit, and C. M. Reidys. Elements of a theory of simulation III, equivalence of SDS. *Appl. Math. Comput.*, 122:325–340, 2001.

110. Erica Jen. Cylindrical cellular automata. *Comm. Math. Phys.*, 118:569–590, 1988.

111. V.S. Anil Kumar, Matthew Macauley, and Henning S. Mortveit. Update order instability in graph dynamical systems. Preprint, 2006.

112. C. M. Reidys. On acyclic orientations and sequential dynamical systems. *Adv. Appl. Math.*, 27:790–804, 2001.

113. A. Å. Hansson, H. S. Mortveit, and C. M. Reidys. On asynchronous cellular automata. *Advances in Complex Systems*, 8(4):521–538, December 2005.

114. The GAP Group. Gap — groups, algorithms, programming — a system for computational discrete algebra. http://www.gap-system.org, 2005.

115. G. A. Miller. Determination of all the groups of order 96. *Ann. of Math.*, 31:163–168, 1930.

116. Reinhard Laue. Zur konstruktion und klassifikation endlicher auflösbarer gruppen. *Bayreuth. Math. Schr.*, 9, 1982.

117. H. S. Mortveit. *Sequential Dynamical Systems*. PhD thesis, NTNU, 2000.

118. C. M. Reidys. Sequential dynamical systems over words. *Annals of Combinatorics*, 10, 2006.

119. C. M. Reidys. Combinatorics of sequential dynamical systems. *Discrete Mathematics*. In press.

120. Luis David Garcia, Abdul Salam Jarrah, and Reinhard Laubenbacher. Sequential dynamical systems over words. *Appl. Math. Comput.*, 174(1):500–510, 2006.

121. A. M. Law and W. D. Kelton. *Simulation Modeling and Analysis*. McGraw-Hill, Singapore, 1991.

122. Christian P. Robert and George Casella. *Monte Carlo Statistical Methods*, 2nd ed. Springer Texts in Statistics. Springer-Verlag, New York, 2005.

123. G. Korniss, M. A. Novotny, H. Guclu, Z. Toroczkai, and P. A. Rikvold. Suppressing roughness of virtual times in parallel discrete-event simulations. *Science*, 299:677–679, January 2003.

124. P.-Y. Louis. Increasing coupling of probabilistic cellular automata. *Statist. Probab. Lett.*, 74(1):1–13, 2005.

125. D. A. Dawson. Synchronous and asynchronous reversible markov systems. *Canad. Math. Bull.*, 17(5):633–649, 1974.

126. L. N. Vasershtein. Markov processes over denumerable products of spaces describing large system of automata. *Problemy Peredachi Informatsii*, 5(3):64–72, 1969.

127. Walter Fontana, Peter F. Stadler, Erich G. Bornberg-Bauer, Thomas Griesmacher, Ivo L. Hofacker, Manfred Tacker, Pedro Tarazona, Edward D. Weinberger, and Peter K. Schuster. RNA folding and combinatory landscapes. *Phys. Rev. E*, 47:2083–2099, 1993.

128. W. Fontana and P. K. Schuster. Continuity in evolution: On the nature of transitions. *Science*, 280:1451–1455, 1998.

129. W. Fontana and P. K. Schuster. Shaping space: The possible and the attainable in RNA genotype-phenotype mapping. *J. Theor. Biol.*, 1998.

130. Christoph Flamm, Ivo L. Hofacker, and Peter F. Stadler. RNA *in silico:* The computational biology of RNA secondary structures. *Advances in Complex Systems*, 2(1):65–90, 1999.

131. C. M. Reidys, C. V. Forst, and P. Schuster. Replication and mutation on neutral networks. *Bulletin of Mathematical Biology*, 63(1):57–94, 2001.

132. C. M. Reidys, P. F. Stadler, and P. Schuster. Generic properties of combinatory maps: Neutral networks of RNA secondary structures. *Bull. Math. Biol.*, 59:339–397, 1997.

133. W. R. Schmitt and M. S. Waterman. Plane trees and RNA secondary structure. *Discr. Appl. Math.*, 51:317–323, 1994.

134. J. A. Howell, T. F. Smith, and M. S. Waterman. Computation of generating functions for biological molecules. *SIAM J. Appl. Math.*, 39:119–133, 1980.

135. M. S. Waterman. Combinatorics of RNA hairpins and cloverleaves. *Studies in Appl. Math.*, 60:91–96, 1978.

136. C. Tuerk and L. Gold. Systematic evolution of ligands by exponential enrichment: RNA ligands to bacteriophage T4 DNA polymerase. *Science*, 249:505–510, 1990.

137. M. Kimura. *The Neutral Theory of Molecular Evolution.* Cambridge University Press, Cambridge, 1983.

138. C. V. Forst, C. M. Reidys, and J. Weber. *Lecture Notes in Artificial Intelligence V 929*, pages 128–147. Springer-Verlag, New York, 1995. Evolutionary Dynamics and Optimization: Neutral Networks as Model Landscapes for RNA Secondary Structure Landscapes.

139. M. Eigen, J. S. McCaskill, and P. K. Schuster. The molecular quasi-species. *Adv. Chem. Phys.*, 75:149–263, 1989.

140. M. Huynen, P. F. Stadler, and W. Fontana. Smoothness within ruggedness: The role of neutrality in adaptation. *PNAS*, 93:397–401, 1996.

141. I. L. Hofacker, P. K. Schuster, and P. F. Stadler. Combinatorics of RNA secondary structures. *Discrete Applied Mathematics*, 88:207–237, 1998.

142. C. M. Reidys and P. F. Stadler. Bio-molecular shapes and algebraic structures. *Computers and Chemistry*, 20(1):85–94, 1996.

143. U. Göbel, C. V. Forst, and P. K. Schuster. Structural constraints and neutrality in RNA. In R. Hofestädt, editor, *LNCS/LNAI Proceedings of GCB96*, Lecture Notes in Computer Science, Springer-Verlag, Berlin, 1997.

144. H. S. Mortveit and C. M. Reidys. Neutral evolution and mutation rates of sequential dynamical systems over words. *Advances in Complex Systems*, 7(3–4):395–418, 2004.

145. André Thayse. *Boolean Calculus of Differences*, volume 101 of *Lecture Notes in Computer Science*. Springer-Verlag, New York, 1981.

146. Gérard Y. Vichniac. Boolean derivatives on cellular automata. *Physica D*, 45:63–74, 1990.

147. Fülöp Bazsó. Derivation of vector-valued Boolean functions. *Acta Mathematica Hungarica*, 87(3):197–203, 2000.

148. Fülöp Bazsó and Elemér Lábos. Boolean-Lie algebras and the Leibniz rule. *Journal of Physics A: Mathematical and General*, 39:6871–6876, 2006.

149. Kunihiko Kaneko. Spatiotemporal intermittency in couple map lattices. *Progress of Theoretical Physics*, 74(5):1033–1044, November 1985.

150. M. Golubitsky, M. Pivato, and I. Stewart. Interior symmetry and local bifurcations in coupled cell networks. *Dynamical Systems*, 19(4):389–407, 2004.

151. S. Eidenbenz, A. Å. Hansson, V. Ramaswamy, and C. M. Reidys. On a new class of load balancing network protocols. *Advances in Complex Systems*, 10(3), 2007.

152. A. Å. Hansson and C. M. Reidys. A discrete dynamical systems framework for packet-flow on networks. *FMJS*, 22(1):43–67, 2006.

153. A. Å. Hansson and C. M. Reidys. Adaptive routing and sequential dynamical systems. Private communication.

Index

Universitext

Printed in the United States of America.